AF324874

Second Edition

Environmental Pollution
HEALTH AND TOXICOLOGY

Second Edition

Environmental Pollution

HEALTH AND TOXICOLOGY

S V S Rana

Alpha Science International Ltd.
Oxford, U.K.

Environmental Pollution: *Health and Toxicology*
Second Edition
350 pgs. | 104 figs. (57 in colour) | 52 tbls.

S V S Rana
Department of Zoology
Ch. Charan Singh University
Merrut, India

ALPHA SCIENCE INTERNATIONAL LTD.
7200 The Quorum, Oxford Business Park North
Garsington Road, Oxford OX4 2JZ, U.K.

www.alphasci.com

Printed from the camera-ready copy provided by the Author.

ISBN 978-1-84265-484-2

Printed in India

PREFACE TO THE SECOND EDITION

Science of environment and toxicology though recognized as separate disciplines unite to form a new subject known as Environmental Health Sciences. The scope of this science embraces several topics viz: pollution and its health effects, ecotoxicology, occupational health, environment and community health, environmental carcinogenesis etc. that were amptly described in the first edition of this book. During course of time both the disciplines were included in the syllabi of courses i.e. Biological Sciences or Life Sciences, Pharmacy, Public Health, Community medicine etc. Therefore, carefully designed contents of this book helped a large section of global student community. Encouraged by the response I decided to add a few chapters in the second edition of the book.

Even though this edition resembles earlier edition in some essential ways, I have added information on chemical warfare agents, biological warfare threats and animal and plant toxins. The updated subject matter is supported by suitable case studies and examples. Information on their clinical management opens the scope of the book to students of Medical Toxicology, as well. With this additional information, I hope the book will acquire a new niche amongst teachers and students.

I am thankful to the Shri N. K. Mehra of Narosa Publishing House for his keen interest in the publication of this book.

S V S RANA

PREFACE TO THE FIRST EDITION

Every author has a reason to write a textbook. It should address the course content of the subject. It should meet the need of its reader. Moreover, it should be up-to-date and easy to understand. Environmental pollution specially after the UN conference on Human Environment held at Stockholm in 1972, was paid attention by several foreign and Indian authors. Subject matter kept on changing with the changing status of pollution from country to country. Various laws and policies were framed and the subject slowly-slowly acquired multidisciplinary status as today. It is realized by every student of ecology and environmental science that pollution accompanies various health problems and disease. Most often, key to the nature, occurrence, prevention and control of a disease lies in the environment. It was Hippocrates who first related disease to environment, e.g. climate, air, water etc. Centuries later, Pettenkofer in Germany revived the concept of relationship between environment and health. Health according to ecological concepts is visualized as a state of dynamic equilibrium between man and his environment. Environmental hazards like pollution disturb this equilibrium and manifest into disease. Therefore, it becomes necessary to understand environmental pollution and health together. Many Universities teach environmental pollution in courses dealing with ecology, environmental science, atmospheric sciences, soil sciences, hydrology, marine biology, agriculture and medicine. However, problem related with public health, community health, occupational health etc. are not considered as an environmental problem. Subject matter on these topics is available to students.

It simultaneously deals with three interrelated disciplines viz.: environmental pollution, environmental health and toxicology, chapters 1 to 8 deal extensively with the problems of air pollution, water pollution, marine pollution, noise pollution, radioactive pollution, solid waste pollution, soil and land pollution and their effects on man, animals and plants. Whereas chapter 9 deals with community health. It offers detailed information on environmental borne diseases like malaria, cholera, tuberculosis, filariasis, lishmaniasis and AIDS. Chapter 10 deals with ecological consequences of dangerous chemicals viz.: DDT, acid rain and cholrofluorocarbons. Chapter 11 describes occupational health problems in man e.g. pneumoconiosis, silicosis, asbestosis, metal fume fever etc.

Science of toxicology that has been defined as the study of the effects of poisons has been dealt in chapter 12 – 20. Adverse effects of chemicals viz.: heavy metals, pesticides, organic solvents, pigments and dyes have been described in separate chapters. Environmental carcinogenesis has been discussed in chapter 17. The book ends with human toxicology and looks forward to new developments in toxicology that are expected to occur in new millennium.

pesticides, organic solvents, pigments and dyes have been described in separate chapters. Environmental carcinogenesis has been discussed in chapter 17. The book ends with human toxicology and looks forward to new developments in toxicology that are expected to occur in new millennium.

Thus a comprehensive information on three interrelated disciplines viz.: environmental pollution, health and toxicology is available in this book. I hope it will meet the demands of a large section of student community undertaking environmental education at undergraduate and graduate level in Indian Universities.

No book can be completed without the support and assistance of many individuals. I offer my sincere appreciation to all those who so freely offered their advice and encouragement in this endeavour. I am particularly thankful to my students Dr. Yeshvendra Verma and Km. Sohni for organizing the chapters. Nitin Sharma deserves appreciation for typing the entire manuscript. Completion of this work required lot of solitude. But solitude has a steep price, the most costly component of which is the time away from family and friends. To my wife Usha, and our children Sushmit, Shilpa, Shitij and Master Avayay, and to all others whom I've neglected, unbounded appreciation for your understanding and support is my only coin. Finally I thank Mr. N.K. Mehra of Narosa Publishing House for his timely assistance and keen interest in bringing out this book.

S V S RANA

CONTENTS

Chapter 1

POLLUTION

The *American Heritage Dictionary of the English Language* (1992 edition) defines pollution as, "the act or process of polluting or the state of being polluted, especially the contamination of soil, water or the atmosphere by the discharge of harmful substances". An introductory college textbook may define pollution in its glossary as a process, to make foul, unclean, dirty; any physical, chemical, or biological change that adversely affects the health, survival, or activities of living organisms or that alters the environment in undesirable ways (Saigo and Cunningham, 1992).

As we enter the 21st century our definition of pollution becomes wider and more complex as to include natural and social systems affecting the quality of our environment. An expanded definition of pollution shall include terms such as ecological restoration, ecosystem rehabilitation and revitalization as well as aesthetic pollution, cultural eutrophication and anthropogenic impacts.

Since the Neolithic revolution, the shift from food collecting to food producing, man has modified natural systems to fit his needs of whims. Thus age old relationship within the biosphere are being altered by the activities of human mind. Vernadsky coined a word "Nooshere" (Greek noos = mind) for the entire man dominated or affected part of the biosphere. The main characteristic feature of noosphere is pollution. Human activities increased dramatically, frequently and chronically resulting into unforeseen problems of diverse nature. He "aesthetically" polluted natural landscapes, "culturally" eutrophied lakes and deliberately allowed poisonous gases to enter the atmosphere.

Beginning with the industrial revolution in the 18th century, pollution became a more noticeable phenomenon in the middle of 19th century, particularly following the American Civil War. The word smog was probably introduced by a physician in 1905.

People have long recognized the existence of noxious or bad air. Perhaps one of the first laws to control air pollution was enacted in 1273 when the King of England passed an act prohibiting the burning of soft coal. Leonardo da Vinci in 1550 noted a blue haze in the atmosphere. The phenomenon of acid rain was first described in the seventeenth century. Two major events, one in the Meuse Valley in Belgium in 1930 and the other in Donora, Pennsylvania, in 1948 were responsible for raising alarm against air pollution in scientific community. London smog crisis in 1952 reinforced the formulations needed to control air quality. Bhopal Gas Tragedy (1984) affected 3.5 lakh people of Bhopal (India).

Ever since 1962, when Rachel Carson drew our attention to the hazards of DDT in *Silent Spring,* we have learned to be much more careful about how we use pesticides. But accidents and oversights still occur as a result of which people are seriously harmed by pesticides once thought to be safe.

1.1 Types of Pollution

The Earth has always experienced natural sources of pollution. For example, volcanic activity has been found to contribute in long-term alterations in local climates; algal blooms, low dissolved oxygen in water causes fish kills. Periodic yet dramatic changes in ocean currents known as *El Nino* can cause disturbances in ocean water temperatures which affect nutrient cycles resulting in population variations of several species. These natural pollution phenomena significantly affect the environment activities.

1.2 Sources of Pollution

Virtually all human activities generate potential pollutants. They involve the creation or concentration of substances and energy. These processes can lead to accidents and the generation of wastes. The scope of these activities is so large that we study them as different subjects (*Agricultural Impact on Environment, Hazardous Materials Transportation and Accidents, Incineration of Waste Products, Mines, Mining Hazards, Mine Drainage, Mine Waste, Petroleum Production and Its Environmental Impacts etc.)* Research on industrial sites has shown that virtually all of them are polluted to some extent. Even relatively simple activities can, when concentrated together by urbanization create pollution.

Pollutants may be released accidentally or deliberately into the environment in a single slug, such as an oil spill. or continuously as an effluent, a leak, or by leaching of wastes. Whether such releases constitute pollution will depend on the critical load present in the ecosystem, or on the acceptable standards for such pollutants.

Environmental pollutants can be divided into three types viz: *biological, physical* or *chemical.* Disease-causing organisms, such as the cholera bacterium, are spread by sewage pollution of waters and cause the greatest number of human illnesses and deaths. Despite the UN Water Supply and Sanitation Decade (the 1980s) and continuing efforts to provide safe drinking water in developing countries, water-borne diseases kill thousands of people every year.

Physical pollutants include energy forms, such as heat, noise and radioactivity and particulate matter such as smoke and dust For example, the waste heat from power plants is often disposed off by discharging hot water into an open water body. The rise in temperature will alter the local ecosystem and so is a form of pollution. Radionuclides cause harm to living organisms by the energy released by their disintegration deep within the organism. Radioactive substances, such as plutonium, are also chemically harmful.

Most naturally occurring substances can be pollutants if their concentrations in the environment increase to the level at which they cause health problems or environmental (eco-system) damage. For example, chloride is one of the most common ions dissolved in water. The high concentration seen in seawater is in harmony with marine ecosystems, so chloride is not treated as a marine pollutant. However, a much weaker solution leached out of a road salt depot into a stream shall damage the fresh water ecosystem. Similarly carbon dioxide is an abundant gas causing no harm at natural concentrations. However, a slight rise in concentration in the atmosphere due to fossil fuel consumption harm us and change the global ecosystem through the greenhouse effect.

However, chemical pollutants are intrinsically harmful. They include heavy metals and a wide range of man-made organic chemicals like pesticides. Over 10 million organic compounds have been synthesized and perhaps 10,000 are in regular use. Many have been created specifically for their biological effects, while others have useful properties, for example resistance to degradation, and their polluting effects are considered acceptable by users. Toxicological data are not available for many compounds, although modern testing and licensing arrangements are fairly strict, and the information shortages are more for older compounds. A good source of information on the effects of particular compounds is Lewis (1992).

1.3 Priority Pollutants

A number of countries have complied lists of pollutants. Although there are no complete listings of problem compounds, they provide useful starting points for further studies, particularly related to industry, waste disposal and agriculture. The US Environmental Protection Agency has listed 129 priority pollutants. They are substances which are known to have dangerous health or environmental effects, and which are found frequently in the environment.

1.4 Persistent Organic Pollutants (POPs)

The modern history of POPs began in 1945 with the large-scale manufacture of DDT as an insecticide to eradicate disease carrying insects from fields, and to control mosquitoes. It took another 20 years for the other side of this wonder chemical to emerge: e.g. evidence of dying wildlife. Birds of prey were particularly vulnerable. Birds with sub-lethal concentrations of DDT in their bodies produced eggs with unusually thin shells, which were consequently prone to break during incubation.

Growing scientific concern in the North peaked with the 1962 publication of Rachel Carson's *Silent Spring*. Soon DDT was found to be widespread in the environment. By 1970, it had been detected in the blubber of seals in the Arctic, thousands of kilo-metres away from its spraying grounds. Since then, concern about POPs has been growing.

During the 1980s, the Canadian and the US governments made efforts to reduce the pollution in the Great lakes of the US. Though their efforts met with success, they found that pristine areas of the Great lakes system, not greatly affected by pollutants like fertilisers and industrial discharges, were suffering from a decline in predator species such as eagle, trout and mink. In 1991, the Great lakes Science Advisory Board reported to a joint US-Canadian intergovernmental panel, the International Joint Commission, that the populations of these species were declining, and that they showed reproductive defects, thinning of eggshells, and tumours. The report noted that the main cause of these disorders were POPs, which magni-fied along the species' food chains. The report also observed that the health effects do not appear only in the exposed adult population, but in their offspring.

Table 1.1 Environmental Protection Agency list of priority pollutants (organic compounds are subdivided into four categories according to the method of analysis).

Base-neural extractables	Hexachlorocyclopentadiene	cis-1,3-Dichloropropene	4,4' -DDE
Acenaphthene	Hexachloroethane	trans-1,3-Dichloropropene	4,4' -DDT
Acenaphthylene	Indeno[1,2,3-cd]pyrene	Ethylbenzene	Dieldrin
Anthracene	Isophorone	Methylene chloride	α-Endosulfan
Benzidine	Naphthalene	1,1,2,2- Tetrachloroethane	β-Endosulfan
Benzo[a]anthracene	Nitrobenzene	1,1,2,2- Tetrachloroethane	Endosulfan-sulfate
Benzo[b]fluoranthene	N-Nitrosodimethylamine	Toluene	Endrin
Benzo[k]fluoranthene	N-Nitrosodiphenylamine	1,1,1- Trichloroethane	Endrin aldehyde
Benzo[*gbi*]perylene	N-Nitrosodi-n-propylamine	1,1,2- Trichloroethane	Heptachlor
Benzo[a]pyrene	Phenanthrene	Trichloroethylene epoxide	Heptachlor -
Bis(2-chloroethoxy)methane	Pyrena	Trichlorofluoromethane	PCB – 1016[a]
Bis(2-chloroetbyl)ether	2,3,7,8-Tetrachlorodibenzo-p-dioxin	Vinyl chloride	PCB-1221[a]
Bis(2-chloroisopropyl)ether			PCB-I232[a]
Bis(2-ethylhexyl)phthalate	1,2,4- Trichlorobenzene	*Acid extractables*	PCB-1242[a]
4-Bromophenyl phenyl ether	p-Chloro-m-cresol	PCB-1248[a]	
Butyl benzyl phthalate	*Volatines*	2-Chlorophenol	PCB-1254[a]
2-Chloronaphthalene	Acrolein	2,4-Dichlorophenol	PCB-1260[a]
4-Chlorophenyl phenyl ether	Acrylonitrile	2,4-Dimethylphenol	Toxaphene
Chrysene	Benzene	4,6-Dinitro-o-cresol	
Dibenzo[a,b]anthracene	Bis(chloromethyl)ether	2,4-Dinitrophenol	*Inorganics*
Di-n-butyl phthalate	Bromodichloromethane	2-Nitrophenol	Antimony
1,2-Dichlorobenzene	Bromoform	4-Nitrophenol	Arsenic
1,3-Dichlorobenzene	Bromomethane	Pentachlorophenol	Asbestos
1,4-Dichlorobenzene	Carbon tetrachloride	Phenol	Beryllium
3,3' -Dichlorobenzene	Chlorobenzene	2,4,6-Trichlorophenol	Cadmium
Diethyl phthalate	Chloroethane	Total phenols	Chromium
Dimethyl phthalate	2-Chloroethyl vinyl ether		
2,4-Dinitrotoluene	Chloroform	*Pesticides*	Cyanide
2,6-Dinitrotoluene	Chloromethane	Aldrin	Lead
Di-n-octyl phthalate	Dibromochloromethane	α-BHC	Mercury
1,2-Diphenylhydrozine	Dichlorodifluoromethane	β-BHC	Nickel
Fluoroanthene	1,1-Dichloroethane	γ-BHC	Selenium
Fluorene	1,2-Dichloroethane	δ-BHC	Silver
Hexachlorobenzene	1,1-Dichloroethylene	Chlordane	Thallium
Hexachlorobutadiene	*trans*-1,2-Dichloroethylene	4,4'-DDD	Zinc
	1,2-Dichloropropane		

Growing toxicological data on the harm done by POPs and on their ability to accumulate in the local environment caused industrialized countries to impose increasing controls and outright bans on their domestic use. The use of DDT, for example, was banned in the USA in 1971. The use of aldrin and dieldrin was also banned three years later because of their carcinogenic properties.

While some developing countries also adopted those controls, many have been increasingly manufacturing POPs themselves. Eventually, Northern countries grew concerned, and they have been largely responsible for the initiatives towards a POPs treaty.

Global concern over these pollutants has been growing for decades, largely because of emerging scientific information which confirms the serious threats POPs pose to wildlife and human health. Countries have now come together to negotiate a new global treaty, which will aim to control and eventually remove them from the environment. The proposed treaty, which is currently being negotiated under UNEP, is known as *An International Legally Binding Instrument* for Implementing International Action on Certain Persistent Organic Pollutants (POPs), and seeks to 'reduce and/or eliminate' these pollutants worldwide.

The proposed global treaty has prepared an agreement on POPs which was signed in June 1998. Its signatories are the 42 members of the United Nations. Economic Commission for Europe (UNECE), which covers both eastern and western Europe, besides Canada and the US. It is known as the **Aarhus Protocol** after the Danish town where it was signed. The agreement sets bans on, or tightly controls the production, use and disposal of, 16 toxic organic chemicals. This follows an emerging trend in global environmental negotia-tions, wherein Northern nations negotiate a treaty amongst themselves and invite Southern countries to use it as a model for a global treaty.

Three formal meetings of an intergovernmental negotiating committee (INC) have taken place in 1998 and 1999. At the first meeting, in Montreal from June 29 to July 3, 1998, it was agreed by the 92 nations present, that an initial list of twelve POPs, dubbed 'the dirty dozen', would be the treaty's main targets for now. A criteria expert group was established during the meeting to formulate scientific criteria and procedure for identifying POPs, and to help more candidates join the list. The second INC meeting was held in Nairobi from January 25-29, 1999, and a third session in Geneva from September 6-11, 1999. Two more INC sessions took place, followed by the signing of a global treaty in late 2000.

1.5 Dirty Dozens

The ongoing negotiations for a POPs treaty are initially targeting 12 of the most notorious POPs contributing to persistent global pollution.

1. **Dichlorodiphenyl Trichloroethane (DDT):** Small amounts of DDT are today used on crops; however, DDT is still widely employed as the cheapest means of controlling mosquitoes that carry malaria and other diseases. It is banned in 34 countries because of toxicity to mammals, including humans, but it is imported and used in many, including Mexico, India, the Philippines and Thailand. Global cumulative production of DDT is estimated to be at 1 .36 million tones.

2. **Toxaphene**: It took over from DDT in 1970s as the world's most popular insecticide, particularly used in cotton-growing countries. It is now banned in 37 countries including India, Mexico and Kenya. A report by the World Wildlife Fund (WWF) claims that toxaphene is still manufactured in China, Nicaragua and Pakistan.

3. **Polychlorinated Biphenyls (PCBs)**: PCBs are the leading industrial POPs. Being chemically stable and heat-resistant, they are widely used in electrical equipment oils such as transformers and capacitors: as a heat-exchange fluid, and as a lubricant and plasticiser. Acutely toxic in high concentrations, they damage immune and reproductive systems and cause birth defects. Though only small quantities are now produced, they are still being released into the environment due to the breakup of old electrical equipment. Russia still produces and uses PCBs because the country's electric power grid depends on transformers manufactured with PCBs. During the negotiations leading up to a regional agreement on POPs, members of the United Nations Economic Commission for Europe: (UNECE) granted permission to Russia to produce PCBs till 2005. Existing stocks have to be destroyed by 2020.

4-5. **Dioxins and Furans**: These chlorinebased compounds are produced in high-temperature incinerators used for burning organic materials including plastics, as an unwanted byproduct in industrial processes ranging from metallurgy to the bleaching of paper and as a trace contaminant in chlorophenol compounds (such as wood preservatives) and herbicides (such as agent orange). They are more potent than PCBs but the effects are similar.

6. **Hexachlorobenzene (HCB):** Once widely used as a fungicide to protect seeds, HCB is also released into the environment as a byproduct during the manufacture of chlorine gas and some chlorinated pesticides, and during waste incineration. It causes *porphyria cutanea tarda,* a metabolic disorder and damages reproductive and immune systems.

7. **Chlordane**: It is a carcinogenic pesticide widely used to control termites. Though it has been banned .in many countries including the Philippines and India, its restricted use is permitted in Mexico, China, UK, Belgium, Canada, Belzium and Cyprus.

8. **Dieldrin**: It is an insecticide with probably the strongest carcinogenic effect amongst all organochlorine pesticides.

9. **Endrin**: Used mainly on field crops such as cotton and grains. It is also used as a rodenticide and to keep away birds.

10. **Aldrin**: Used on crops like corn and cotton for termite control.

11. **Mirex**: Used as an insecticide and fire retardant in the US till 1978. Known to cause cancer in laboratory animals.

12. **Heptachlor**: Used in Mexico, Bulgaria and the US to kill Insects in seed grains and on crops and to fight termites.

1.6 Environmental Mobility of Pollutants

Environmental Compartments: Our environment can be divided into several physical and living compartments. The main compartments are the atmosphere, water, soil sediment and biota, although all can be subdivided further. Pollutants enter a compartment through which they can be transported and dispersed; they may be transformed into other compounds or transferred to another compartment. These processes, which are discussed below, lead to complex and changing distributions of pollutants.

Transport Through Compartments: The transport of pollutants through environmental compartments is often described as the sum of three processes: advection, dispersion and diffusion. Advection (sometimes called convection) is movement with the bulk movement of the mobile fluid of the compartment, as a slug of dye is carried along by a river. Dispersion (mixing) spreads the pollutant out as a result of the fluid movement as the smoke plume spreads while being blown downwird; if there is no advection there is no dispersion. It always acts to reduce concentrations. In faster moving fluids (e.g., air or rivers), dispersion is a result of turbulent eddies created by the fluid's movement. Groundwater is an unusual fluid in this context. In that the fluid movement is laminar rather than turbulent, and dispersion arises from differences in water velocities through the complex pore networks combined with diffusion (Fetter, 1993).

Transfers: Pollution will enter one compartment, but may reach an equilibrium on being transferred to another. For example, over the Earth's present lifetime, most helium has been lost to space, most sodium chloride has ended up in the oceans, and most iron is in geological materials. The equilibrium distribution of a chemical between compartments is controlled by partitioning laws, which are all very similar in style and which reflect the properties of the chemical and the compartment. For example, the partitioning between air and water of a volatile compound like trichloroethylene, a solvent widely used for degreasing metal, is described by Henry's law:

$$C_{air} = HC_{water}$$

where C is concentration in a particular phase, and H is Henry's constant, defined by properties of the compound as

$$H = V/S$$

where V is vapor pressure and S is its solubility.

Similar laws can predict the sorption of a pollutant to soil or to body tissue. The actual distribution at any time will also depend on transport, mixing and transfer rates, and is much more complex to calculate. Equilibrium distributions do have a value because they allow rapid assessment of the likely impact of a pollutant. For example, the pesticide DDT has a very low solubility in water, low vapor pressure, and high affinity for other organic matter. The consequences are that it will partition to soils and sediments, rather than air or water, and to living matter. Therefore, it is not mobile in the environment, except in association with moving sediment. When it gets into the food chain, it remains, accumulating in higher organisms which consume lower species.

Transformations: Part of the philosophical justification for the routine release of pollutants into the environment, for example in a sewage effluent, is that they will be transformed and become less harmful in time. This attenuation may be dilution, or it may be transformation into other chemicals, hopefully less polluting in character. For example, one of the major problems with sewage is that it contains degradable organic compounds. If discharged to a river, these will be consumed by bacteria, using up oxygen dissolved in the river and making the water unfit for many higher species. Thus the acceptable standards for discharge of treated municipal sewage are often phrased in terms of their biological oxygen demand. Further downstream, where degradation is complete, the river will become re-oxygenated and return to its original ecological character. Of course, there may be other, more persistent and troublesome, chemicals in the effluent which continue to damage the ecosystem. Understanding how pollutants are transformed in the environment is important in predicting their fate and impact on health and ecosystems. We find that, some of the reactions are abiotic, entirely chemical and do not involve any organisms, such as the interactions between chlorofluorocarbons and ozone in the high atmosphere. Others are biologically mediated, such as the sewage degradation discussed above. Either type of reaction requires certain environmental conditions to occur, and these vary between pollutants. For example, the most dangerous components of modern petroleum are the BTEX group (benzene, toluene, ethylbenzene, and xylenes). Research shows that all are biodegraded in the presence of oxygen, but at different rates, with the xylenes being the most refractory. In anaerobic conditions (without oxygen) benzene does not appear to degrade at all, but the others still do. The end products of complete biodegradation of such hydrocarbons are carbon dioxide and water in aerobic, or methane in anaerobic conditions.

The transformation of organic pollutants destroys the original molecules, but degradation does not always proceed to completion, and may create more dangerous molecules in the process. For example, the anaerobic degradation of tetrachloroethane $[(CCl_2)_2]$ proceeds by successively removing chlorine atoms. One of the intermediates, chloroethane or vinyl chloride (CH_2CCl) is a potent carcinogen, and is frequently found in landfill leachates and groundwaters.

Many inorganic pollutants undergo reversible transformations in the environment, and are not completely destroyed. Chromium can exist in natural waters in several ionic forms, based on two oxidation states, tri and hexavalent chromium. In general, trivalent chromium is insoluble and immobile, while the hexavalent form is mobile. Hexavalent chromium is widely used in the metal plating industry, and so discharges of this more dangerous form do occur. Transformation to the trivalent forms requires a reduction of pH to below 6. Nitrogen is harmless, but two of the ions that from it have different pollution effects in water. Nitrate (NO_3^-) is harmful to babies under six months, and can stimulate algal blooms in open waters. Reduction of nitrate can create nitrogen gas. or go further to create ammonium ions (NH^{4+}) which are toxic to many aquatic species.

Environmental pollution defies abstract territorial boundaries. Pesticides sprayed on one farmer's crops can be carried in the air or water to long distances causing serious problems. Thus understanding the interdependence of the environment is essential to understanding the impact of pollution activities on us. As Lewis Thomas has pointed out, understanding interdependence may be the only way to perceive correctly our own role on this planet.

SUGGESTED FURTHER READINGS

Saigo, B.W. and Cunningham, W.P. 1992. Environmental Science, a Global Concern (2nd edn.). W. C. Brown. Dubuque. Ia.
Fetter, C.W. 1993. Contaminant Hydrogeology, Macmillon New York.
Lewis, RJ. Sr. 1992. Sax's dangerous properties of Industrial Materials (8th edn.) Nostrand Reinhold, New York.

STUDY QUESTIONS

1. Define pollution. Discuss major events of pollution that occurred in recent past.
2. Write in details on environmental mobility of pollutants.
3. Write briefly on the following:
   - i. Priority pollutants
   - ii. Henry's law
   - iii. Transport and transfer of pollutants
   - iv. POPs
   - v. Dirty dozens
   - vi. Aarhus protocol

Chapter 2

AIR POLLUTION

Air pollution may be defined as any atmospheric condition in which certain substances are present in such concentrations that they can produce undesirable effects on man and his environment. These substances include gases (sulphur oxides, nitrogen oxides, carbonmonoxide, hydrocarbons, etc.), particulate matter (smoke, dust, fumes, aerosols), radioactive materials and many others. Most of these substances are naturally present in the atmosphere in low (background) concentrations and are usually considered to be harmless. The background concentrations of various components of dry air near sea level and their estimated residence times are given in Table 2.1. Thus, a particular substance can be considered as air pollutant only when its concentration is relatively high compared with the background value and causes adverse effects. For example, sulphur dioxide, if present in the atmosphere in concentrations greater than the background value of 2×10^{-4} ppm and causes measurable effects on humans, animals, plants, or property, then only it is classified as an air pollutant.

The concentration of a pollutant in the atmosphere can be expressed in a number of ways involving units of weights or volume per unit weight or volume of air. Four concentration scales are generally used to describe the concentrations of either gaseous or particulate pollutants.

The first is the mass concentration, ω_p, defined as the ratio of the mass of pollutant to the mass of air plus mass of pollutant.

$$\omega_p = \frac{m_p}{m_a + m_p} \qquad 2.1$$

where m_p is the mass of the pollutant and m_a is the mass of pure air in a given volume of air-pollutant mixture. The second concentration scale is the volume consideration, v_p, defined as the ratio of the volume of pollutant to the volume of air plus volume of pollutant:

$$y_p = \frac{v_p}{v_a + v_p} \qquad 2.2$$

Table 2.1 Composition of clean, dry atmospheric air

Component	Concentration*	Estimated residence time
Nitrogen	78.09×10^4	Continuous
Oxygen	20.94×10^4	Continuous
Argon	93×10^2	Continuous
Carbon dioxide	3.2×10^2	2-4 years
Neon	18	Continuous
Helium	5.2	~2 million years
Krypton	0.1	Continuous
Xenon	8×10^{-2}	Continuous
Carbon monoxide	1×10^{-1}	0.5 years
Methane	1.2	4-7 years
Nitrous oxide	6×10^{-4}	5 days
Ammonia	6×10^{-3}	7 days
Hydrogen sulphide	2×10^{-4}	2 days
Sulphus dioxide	2×10^{-4}	4 days
Hydrogen	5×10^{-1}	?**
Ozone	2×10^{-2}	~60 days

* Single values for concentrations, instead of ranges of concentrations, are given to indicate order of magnitude, not specific and universally accepted concentrations.
** Little is known about the residence time.

The third concentration scale is the volume concentration in parts per million (ppm), y_{ppm}:

$$y_{ppm} = v_p \times 10^6 \hspace{3cm} 2.3$$

2.1 Classification and Properties of Air Pollutants

2.1.1 Classification

The variety of matter emitted into the atmosphere by natural and anthropogenic sources is so diverse that it is difficult to classify air pollutants neatly. However, usually they are divided into two categories of *primary pollutants* and *secondary pollutants*. The primary pollutants are those that are emitted directly from the sources. Typical pollutants included under this category are particulate matter such as ash, smoke, dust, fumes, mist and spray; inorganic gases such as sulphur dioxide, hydrogen sulphide, nitric oxide, ammonia, carbon monoxide, carbon dioxide, and hydrogen fluoride; olefinic and aromatic hydrocarbons; and radioactive compounds. The secondary pollutants are those that are formed in the atmosphere by chemical interactions among primary pollutants and normal atmospheric constituents. Pollutants such as sulphur trioxide, nitrogen dioxide, PAN (peroxyacetyl nitrate), ozone, aldehydes, ketones, and various sulphate and nitrate salts are included in this category.

Of the large number of primary pollutants emitted into the atmosphere, only a few are present in sufficient concentrations to be of immediate concern. These are the five major types - particulate matter, sulphur oxides, oxides of nitrogen, carbon monoxide and hydrocarbons. Carbon dioxide is generally not considered an air pollutant but, because of its increased global background concentration, its influence on global climatic patterns is of great concern. The radioactive pollutants are of specialized nature and they are discussed separately in this book.

Secondary pollutants are formed from chemical and photochemical reactions in the atmosphere. The reaction mechanisms and various steps involved in the process are influenced by many factors such as concentration of reactants, the amount of moisture present in the atmosphere, degree of photo activation, meteorological forces, and local topography.

2.1.2 Properties of Air Polltants

2.1.2.1 Particulate Matter

In general the term "particulate" refers to all atmospheric substances that are not gases. They can be suspended droplets or solid particles or mixtures of the two. Particulates can be composed of inert or extremely reactive materials ranging in size from 100 ìm down to 0.1 ìm and less. The inert materials do not react readily with the environment nor do they exhibit any morphological changes as a result of combustion or any other process, whereas the reactive materials could be further oxidized or may react chemically with the environment.

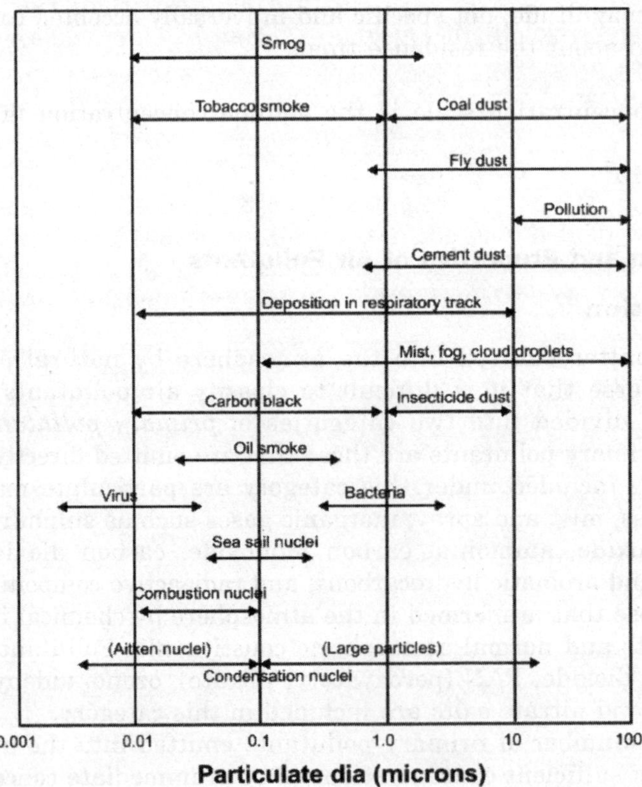

Fig. 2.1: Sizes of atmospheric particulate matter.

The classification of various particulates may be made as follows:

Dust - It contains particles of the size ranging from 1 to 200 ì m. These are formed by natural disintegration of rock and soil or by the mechanical processes of grinding and spraying. They have large settling velocities and are removed from the air by gravity and other inertial processes. Fine dust particles act as centers of catalysis for many of the chemical reactions taking place in the atmosphere.

Smoke - It contains fine particles of the size ranging from 0.01 to 1 ì m which can be liquid or solid, and are formed by combustion or other chemical processes. Smoke may have different colours depending on the nature of material burnt

Fumes – These are solid particles of the size ranging from 0.1 to 1 ì m and are normally released from chemical or metallurgical processes.

Mist - It is made up of liquid droplets generally smaller than 10 ì m which are formed by condensation in the atmosphere or are released from industrial operations.

Fog - It is the mist in which the liquid is water and is sufficiently dense to obscure vision.

Aerosol - Under this category are included all air-borne suspensions either solid or liquid; these are generally smaller than 1mm.

Particles in the size range 1 - 10 mm have measurable setting velocities but are readily stirred by air movements, whereas particles of size 0.1 - 1 ì m have small settling velocities. Those below 0.1 mm. a submicroscopic size found in urban air, undergo random Brownian motion resulting from collisions among individual molecules (Fig 2.1). Of all the different types of particulates in the atmosphere, the presence of trace elements such as cadmium, lead, nickel and mercury may constitute the greatest health hazard. Many of the trace metals are toxic and are concentrated in the finest of particulate matter in a variety of combined forms such as oxides, hydroxides, sulphates, and nitrates.

2.1.2.2 Oxides of Sulphur

The most important oxide emitted by pollution sources is sulphur dioxide (SO_2). SO_2 is a colourless gas with a characteristic, sharp, pungent odour. It is moderately soluble in water (11.3 g/100 cm) forming weakly acidic sulphurous acid (H_2SO_3). It is oxidized slowly in clean air to sulphur trioxide. In a polluted atmosphere, SO_2 reacts photochemically or catalytically with other pollutants or normal atmospheric constituents to form sulphur trioxide, sulphuric acid and salts of sulphuric acid.

Sulphur trioxide (SO_3) is generally emitted along with SO_2, at about 1-5 percent of the SO_2 concentration. SO_3 rapidly combines with moisture in the atmosphere to form sulphuric acid which has a low dew point. Both SO_2 and SO_3 are relatively quickly washed out of the atmosphere by rain or settle out as aerosols. This is the reason why SO_2 and SO_3 are relatively quickly washed out of the atmosphere by rain or settle out as aerosols. This is the reason why SO_2 mass in clean dry air is so small compared to nnual emissions from anthropogenic sources (Table 2.2).

Table 2.2 Comparison of the amounts of anthropogenic pollutants with the amounts naturally present in dry, clean air.

Pollutant	Amount (millions of tones)	
	Anthropogenic (per year)	Dry, Clean air
Particulates	269	-
SO_2	132	2
NO_2	48	8
CO	400	500

2.1.2.3 Nitrogen Oxide

Of the six or seven oxides of nitrogen, only three-nitrous oxide (N_2O), nitric oxide (NO), and nitrogen dioxide (NO_2)⁻ are formed in any appreciable quantities in the atmosphere. Often NO and NO_2 are analyzed together in air and are referred to as NO_x.

Nitrous oxide is a colourless, odourless nontoxic gas present in the natural atmosphere in relatively large concentrations (0.25 ppm). The major source of N_2O in the atmosphere is the biological activity of the soil and there are no significant anthropogenic sources. It has a low reactivity in the lower atmosphere and is generally not considered an air pollutant.

Nitric oxide is a colourless, odourless gas produced largely by fuel combustion. It is oxidized to NH_2 in a polluted atmosphere through photochemical secondary reactions. Nitrogen dioxide is a brown pungent gas with a irritating odour which can be detected at concentrations of about 0.12 ppm. It absorbs sunlight and initiates a series of photochemical reactions. Small concentrations of NO_2 have been detected in the lower stratosphere; NO_2 is probably produced by the oxidation of NO by ozone. Nitrogen dioxide is of major concern as a pollutant; it is emitted by fuel combustion and nitric acid plants.

2.1.2.4 Carbon Monoxide

It constitutes the single largest pollutant in the urban atmosphere. CO is colourless, odourless, and tasteless, and has a boiling point of 192°C. It has a strong affinity towards the hemoglobin of the blood and is a dangerous asphyxiant. The rate of oxidation of carbon monoxide to carbon dioxide in the atmosphere seems to be very slow; mixtures of CO and O_2 exposed to sunlight for several years have been found to remain almost unchanged. Carbon monoxide is present in small concentrations (0.1 ppm). The main sources of CO in the urban air are smoke and exhaust fumes of burning coal, gas or oil.

2.1.2.5. Hydrocarbons

The gaseous and volatile liquid hydrocarbons are of particular interest as air pollutants. Hydrocarbons can be saturated or unsaturated, branched or straight-chain, or can have a ring structure as in the case of aromatics and other cyclic compounds. In the saturated class, methane is by far the most abundant hydrocarbon constituting about 40 to 80 percent of the total hydrocarbons present in an urban atmosphere. The unsaturated class includes alkenes (olefins) and acetylenes. Among the alkenes the prominent pollutants are ethylene and propene. The first member of the aromatic class is benzene, but some of its substituted derivatives such as toluene and *m*-xylene are usually present in larger concentrations in the urban atmosphere. Terpenes are a particular class of volatile hydrocarbons emitted largely by natural sources. These are cyclic non-aromatic hydrocarbons found in pine tar and in other wood sources.

The hydrocarbons in air by themselves alone cause no harmful effects. They are of concern because the hydrocarbons undergo chemical reactions in the presence of sunlight and nitrogen oxides forming photochemical oxidants of which the predominant one is ozone. Methane has very low photochemical activity as compared to that of other hydrocarbons. For this reason, it is the non-methane hydrocarbon concentration that is of interest while considering air pollution.

2.2 Behaviour and Fate of Air Pollutants

Although large amounts of pollutants are discharged annually into the atmosphere, the very fact that their ambient levels have remained very much the same throughout the world suggests that there are certain pathways of exchange from the atmosphere to the earth, whereby the pollutants are continually removed. These pathways or the scavenging processes, as they are called, may be grouped as follows for both particulates and gases:

Particulates:
(a) Wet removal by precipitation
(b) Dry removal by sedimentation, impaction and diffusion.
Gases:
(a) Wet removal by precipitation.
(b) Chemical reaction in the atmosphere to produce aerosols and/or absorption on aerosols with subsequent removal.
(c) Absorption or reaction at land and ocean surfaces.

2.2.1 Wet Precipitation

Wet precipitation has two distinct mechanisms- "rainout" and "washout". The first includes various processes taking place inside clouds, where the contaminants serve as condensation nuclei on which droplets condense. The second mechanism refers to the removal of pollutants below the cloud level by falling rain. Wet precipitation is one of the most effective scavenging processes for both particulate and gaseous pollutants in a global sense. The rainout mechanism is particularly effective for Aitken particles whose size is less than 0.1 μm. These particles are captured by cloud droplets by Brownian diffusion, and the cloud droplets (typically 30 μm) in turn grow in size by coalescence and precipitate. Washout is most effective in removing particles larger than 2 μm. Its scavenging efficiency, however, is influenced by the rain-drop cross-sectional area and the intensity of rain fall. Particles smaller than 2 mm are not usually collected by rain drops (typically 500 μm) because they are brushed aside by the diverging air ahead of the drop.

The scavenging of gaseous pollutants by wet precipitation is more complex and is less well understood. Soluble gas molecules such as SO_2 can migrate to the rain drop by Brownian motion or to the surface of the cloud droplets by diffusion due to the concentration gradient of the gas across the liquid-air interface. However, wet precipitation is very effective in removing the acid droplets and sulphate particles formed after chemical reactions in the atmosphere.

2.2.2 Dry Deposition

Particulate matter smaller than 0.1 µm often coagulates through mutual collisions and forms larger aggregates which are effectively removed by gravitational settling. Brownian motion is the major mechanism of coagulation, although atmospheric turbulence also enhances the diffusive motion of particles. Atmospheric turbulance is particularly effective for coagulating larger particles whose Brownian motion is less pronounced. The rate of settling of the particles depends on their setting velocities according to Stokes' law:

$$v_t = \frac{gd_p^2}{18\mu_a}\ (\rho_p - \rho_a)\quad 1 + \frac{2C}{d_p p} \qquad\qquad 2.4$$

where
v_t = terminal settling velocity;
d_p = particle diameter;
ρ_p and ρ_a = density of particle and air, respectively;
μ_a = viscosity of air
p = air pressure
and
C = constant [where p is given in millibars and d_p in centimeters C == 0.0084

From the above equation, it is seen that the rate of sedimentation is strongly influenced by the particle size. Particles larger than 10µm have high settling rates and, hence have short residence time in the atmosphere. The terminal velocity is also related to the particle's density, but only by a less sensitive linear dependence. Deviations from the Stoke's law occur due to irregular particle shapes, turbulance in the wake of large particles, and atmospheric vertical velocities, but it is clear that small particles must aggregate to form larger ones if they are to be removed effectively from the atmosphere.

In addition to sedimentation, the mechanisms of inertial impaction and diffusion also contribute to the removal of particulate matter. In inertial impaction, windborne particles strike an obstacle and are deposited; whereas in diffusion small particles migrate to land and ocean surfaces. Dry deposition accounts for about 20 percent of the total particulate removal from the atmosphere.

2.2.3 Interaction at the Earth's Surface

Gaseous pollutants can be transported to the Earth's surface by atmospheric turbulence where they interact with the ocean surface, vegetation and upper layers of the soil, and are removed by absorption or chemical reaction. Assimilation at these surfaces depends on many factors about which little is known for many pollutants. Two major pollutants, about which some information is available, are sulphur dioxide and carbon monoxide.

At the ocean surface, SO_2 first diffuses through the gas phase, crosses the gas-liquid interface, and finally diffuses into the bulk of the ocean where it is absorbed. Vegetation and upper layers of the soil also act as sinks for SO_2. Here the mode of transfer involved is adsorption. SO_2 first diffuses to the external surface of the solid, penetrates into the pores of the solid, and is subsequently adsorbed on the pore site. An estimated removal of 4×10^{-7} tonnes of SO_2 per year has been calculated for oceanic absorption and the solid surfaces account for 5.6×10^{-7} tonnes per year. The importance of these sinks can be seen when their performance is compared with the estimated emission of 14×10^{-7} tonnes per year SO_2 from the anthropogenic sources.

For carbon monoxide, biological action in soils seems to be an important sink; the role played by soil in the removal of CO has been firmly established. It is theorized that the soil contains certain bacteria which can make use of CO in their metabolism, producing either CO_2 or CH_4:

$$CO + \tfrac{1}{2} O_2 \rightarrow CO_2 \qquad\qquad 2.5$$
$$CO + 3H_2 \rightarrow CH_4 + H_2O \qquad\qquad 2.6$$

Nitrogen dioxide also seems to be absorbed by the ocean and other surface waters, but the extent of the contribution of this sink to the overall removal of nitrogen oxides is not clearly established.

2.2.4 Chemical Reactions in the Atmosphere

Many of the gaseous pollutants undergo chemical reactions within the atmosphere and form either new compounds or aerosols. This mode of removal is of great importance for sulphur dioxide. A large part of SO_2 in the atmosphere is oxidized to sulphur trioxide which quickly combines with moisture to form sulphuric acid mist. The reaction is represented as

$$2SO_2 + 2H_2O + O_2 \rightarrow 2H_2SO_4 \qquad\qquad 2.7$$

This process has been shown to be catalyzed by metal salts such as iron and manganese, commonly found in the flyash. These particles serve as nucleation sites for droplet formation and the sulphuric acid droplet may in turn react with metal salts (such as NaCl from sea salt particles), metal oxides such as MgO, Fe_2O_3, ZnO and Mn_2O_3 or ammonia to produce sulphates:

$$2NaCl + H_2SO_4 \rightarrow Na_2SO_4 + 2HCl \qquad\qquad 2.8a$$
$$MgO + H_2SO_4 \rightarrow MgSO_4 + H_2O \qquad\qquad 2.8b$$
$$2NH_3 + H_2SO_4 \rightarrow (NH_4)_2SO_4 \qquad\qquad 2.8c$$

With these reactions, the sulphuric acid droplet is neutralized and the solubility of SO_2 in the droplets is further increased thereby enchancing the oxidation process. Both the acid droplets and the sulphate particles are rapidly removed from the atmosphere primarily by wet precipitation.

Similarly, a major process of nitrogen oxides appears to be through their conversion to form nitric acid; however, the mechanism of such a conversion has not been clearly determined. The direct conversion of NO_2 to HNO_3 in the presence of moisture appears to be too slow to account for the observed rate of removal:

$$2NO_2 + H_2O \Leftrightarrow HNO_3 + HNO_2 \qquad \text{2.9a}$$
$$3HNO_2 \rightarrow HNO_3 + 2NO + H_2O \qquad \text{2.9b}$$
OR
$$3NO_2 + H_2O \rightarrow 2HNO_3 + NO \qquad \text{2.9c}$$

Since the direct conversion of sulphur dioxide to sulphur trioxide and nitrogen dioxide to nitric acid is slow, other mechanisms have been postulated, based mainly on theoretical conjecture and to a lesser extent on experimental observations. These mechanisms involve the oxidation by reactive species such as radicals, atomic oxygen, ozone, and hydroxyl radicals.

The first step in the oxidation of SO_2 is the photoexcitation of the SO_2 molecule through its absorption of solar radiation:

$$SO_2 + h_v \rightarrow SO_2^* \qquad \text{2.10}$$

where SO_2^* represents an excited SO_2 molecule. These molecules in their excited state react more readily with molecular oxygen. Then several reactions follow to complete the oxidation of SO_2 to SO_3.

$$SO_2^* + O_2 \rightarrow SO_4^{2-} \qquad \text{2.11a}$$
$$SO_4^{2-} + O_2 \rightarrow SO_3 + O_3 \qquad \text{2.11b}$$
$$SO_2 + O_3 + h_v \rightarrow SO_3 + O_2 \qquad \text{2.11c}$$

A three-body reaction with atomic oxygen has also been suggested as a possible mechanism:

$$O_2 + h_v \rightarrow 2O \qquad \text{2.12a}$$
$$SO_4 + O + M \rightarrow SO_3 + M \qquad \text{2.12b}$$

The third body, M, is required in order to carry off excess energy of reaction.

A proposed rapid mechanism for the conversion of NO_2 to HNO_3 is the reaction of NO_2 with atmospheric ozone to give nitrogen trioxide:

$$NO_2 + O_3 \rightarrow NO_3 + O_2 \qquad \text{2.13}$$

In addition, NO_2 can be formed by reaction with atomic oxygen

$$NO_2 + O + M \rightarrow NO_3 + M \qquad \text{2.14}$$

The NO_2 radical is removed by reaction with NO_2 forming N_2O_5, which in the presence of moisture forms nitric acid:

$$NO_3 + NO_2 \rightarrow N_2O_5 \qquad\qquad 2.15a$$
$$N_2O_5 + H_2O \rightarrow HNO_3 \qquad\qquad 2.15b$$

The HNO_3 so formed is then washed out of the atmosphere in the form of nitrate salts by precipitation.

For carbon monoxide, its reaction with atmospheric oxygen in the presence of sunlight is found to be very slow and accounts for the removal of only 0.1 percent of available CO for each hour of sunlight. Of major interest as an atmospheric sink for CO is the relatively fast reaction of CO with hydroxyl radicals present in the atmosphere.

$$CO + OH^{\bullet} \rightarrow CO_2 + H^{\bullet} \qquad\qquad 2.16$$

The above mechanism may account for the removal of a substantial portion of CO from the troposphere depending upon the concentration of $OH^{\bullet}$ radicals. Another possible mechanism is the migration of CO into the stratosphere, where the oxidation to CO_2 may subsequently take place by the hydroxyl radicals. In fact, all these gaseous pollutants, including the hydrocarbons, inter-react by photochemical processes in the phenomenon mown as the photochemical smog.

2.3 Photochemical Smog

Photochemical smog was first observed in Los Angeles, U.S.A in the mid-1940's and since then the phenomenon has been detected in most major metropolitan cities of the world The conditions for the formation of photochemical smog are air stagnation, abundant sunlight, and high concentrations of hydrocarbon and nitrogen oxides in the atmosphere. In India, Bombay and Calcutta are ideal candidates for the formation of photochemical smog, but it may be masked by smoke and sulphur dioxide.

Smog arises from photochemical reactions in the lower atmosphere by the interaction of hydrocarbons and nitrogen oxide released by exhausts of automobiles and some stationary sources. This interaction results in a series of complex reactions producing secondary pollutants such as ozone, aldehydes, ketones, and peroxyacyl nitrates. The reaction mechanisms are complex and are not fully understood.

A broad outline of the principal reactions that occur in a photochemical process are illustrated in Fig. 2.2. The starting mechanism is the absorption of ultraviolet light from the Sun by NO_2. This causes the nitrogen dioxide to decompose into nitric oxide and highly reactive atomic oxygen

$$NO_2 + h_v \rightarrow NO + O \qquad\qquad 2.17$$

The atomic oxygen initiates oxidizing processes or quickly combines with molecular oxygen to form ozone, which itself is reactive and acts as an oxidant:

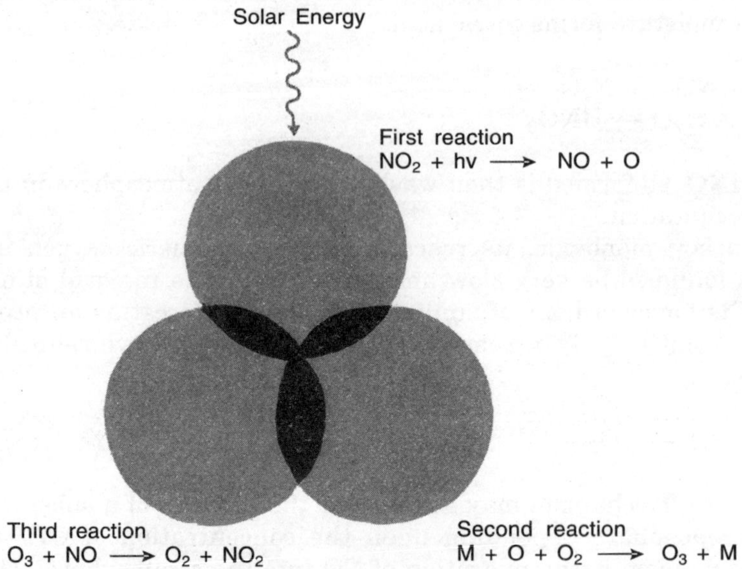

Fig. 2.2 The cycle of chemical reactions initiated by NO_2. In the absence of hydrocarbons, the cycle initiated by NO_2 consists of three reactions. First, NO_2 absorbs solar energy, dissociating into a molecule of NO and a free oxygen atom, which is highly reactive. Second, the free oxygen atom then reacts with an ordinary oxygen molecule (O_2) to form ozone (O_3). Since the ozone formed in this way is unstable, having too much energy to exist for very long, it dissipates this excess energy by colliding with another molecule, transferring the excess energy by colliding with another molecule, transferring the excess as kinetic energy to that other molecule. If another molecule is not handy at the proper time, the ozone molecule will dissociate, leaving the free oxygen atom to try again. When a stable ozone molecule is formed, it can participate in the third reaction, reacting with the NO molecule produced in the first reaction, which forms nitrogen dioxide again, in this way all the molecular species exist in an equilibrium, and the primary result is that the energy absorbed by the NO_2 is transferred to the air as kinetic energy, warming the air.

$$O + O_2 + M \rightarrow O_3 + M \qquad\qquad 2.18$$
$$O_3 + NO \rightarrow NO_2 + O_2 \qquad\qquad 2.19$$

In eq. 2.18 an energy-absorbing molecule or particle M is required to stabilize O_3 or else it will rapidly decompose. Under normal conditions, the ozone formed will be quickly removed by reaction with NO to provide NO_2 and O_2 according to eq. 2.19; however, when hydrocarbons are present in the atmosphere this mechanism is partially eliminated as NO reacts with the hydrocarbon radical peroxyacyl (RCO_3) according eq. 2.23 and as a result ozone concentration builds up to dangerous levels.

Hydrocarbons, indicated by symbol HC, -compete for free oxygen released by NO_2 decomposition to form oxygen-bearing free radicals such as the acyl radical.

$$HC + O \rightarrow RCO^{\bullet} \text{ (acyl radical)} \qquad\qquad 2.20$$

This radical takes part in a series of reactions involving the formation of still more reactive species, which in turn react with O_2, hydrocarbons and nitric oxide.

$$RCO^{\bullet} + O_2 \rightarrow RCO_3^{\bullet} \text{ (peroxyacyl radical)} \qquad\qquad 2.21$$

$$RCO_3^{\bullet} + HC \rightarrow RCHO^{\bullet} \text{ (aldehydes), } R_2CO \text{ (ketones)} \qquad 2.22$$

$$RCO_3^{\bullet} + NO \rightarrow RCO_2^{\bullet} + NO_2 \qquad 2.23$$

$$RCO_3^{\bullet} + O_2 \rightarrow RCO_2 + O_3 \qquad 2.24$$

Reactions represented by eq. 2.22 are termination reactions forming aldehydes and ketones; however, in eqs. 2.23 and 2.24 the peroxyacyl radical reacts with NO and O_2 to produce another oxidized hydrocarbon radical (RCO_2) as well as more NO_2 and O_3. Further, the acylate radical (RCO_2) can react with NO to generate even more NO_2.

$$RCO_2^{\bullet} + NO \rightarrow RCO^{\bullet} + NO_2 \qquad 2.25$$

The NO level in the atmosphere eventually drops off with the accmnulation of NO_2 and O_3. When reactions such as these increase the NO_2 level sufficiently, another reaction begins to compete for the peroxyacyl radical.

$$RCO_3^{\bullet} + NO_2 \rightarrow RCO_3NO_2 \text{ (PANS)} \qquad 2.26$$

The end products are known as peroxyacyl nitrates or PANS. Numerous PANS could be formed, corresponding to the different possible R groups. Three of the common members of PAN family are:

$$\overset{\displaystyle O}{\overset{\displaystyle \|}{HCOO}} NO_2 : \text{Peroxyformyl nitrate (PFN)}$$

$$\overset{\displaystyle O}{\overset{\displaystyle \|}{CH_8COO}} NO_2 : \text{Peroxyacetyl nitrate (PAN)}$$

$$\overset{\displaystyle O}{\overset{\displaystyle \|}{C_6H_5 - COONO_2}} : \text{Peroxybenzoyl nitrate (PBzN)}$$

The ozone formed according to eq. 2.19 and 2.24 reacts with the hydrocarbons to generate more aldehydes and ketones.

$$HC + O_3 \rightarrow RCO_2^{\bullet} + RCHO, R_2CO \qquad 2.27$$

The above equations represent in a broad sense the nature of the overall photochemical reactions leading to formation of smog and they are by no means the only important mechanisms. It has been observed that carbon monoxide and sulphur dioxide also play a significant part in the process of formation of smog by strongly interacting with many species present in the smog and accelerate the oxidation processes. For example, carbon monoxide does this through a series of reactions whose net effect is to convert CO, NO and O_2 into CO_2 and NO_2 thus accelerating the oxidation of NO. First, CO is oxidized to CO_2 by the OH radical. In the smoggy atmosphere the OH radical may be produced when aldehydes are attacked by atomic oxygen.

$$O + CH_3 CHO \rightarrow CH_3 CO + OH^{\bullet} \qquad 2.28$$

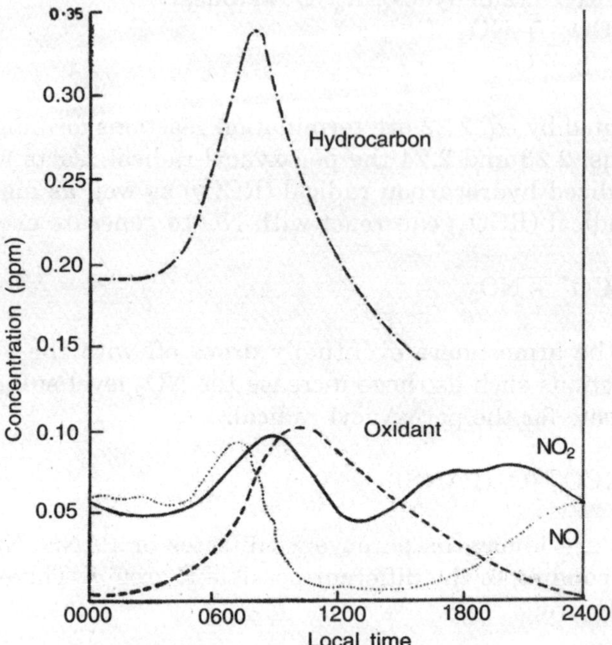

Fig. 2.3: Dynamic behaviour of photochemical smog.

$$CO + OH^\bullet + CO_2 + H^\bullet \qquad\qquad 2.16$$
$$H^\bullet + O_2 + M \rightarrow HO_2^\bullet \qquad\qquad 2.29$$

The H radical reacts with O_2 to form the hydroperoxyl radical HO_2 which is a principal agent for the rapid conversion of NO into NO_2.

$$HO_2 + NO \rightarrow OH^\bullet + NO_2 \qquad\qquad 2.30$$

The overall reaction is

$$CO + O_2 + NO \rightarrow CO_2 + NO_2 \qquad\qquad 2.31$$

This sequence of reactions provides another route for the oxidation of NO without the participation of O_3 (eq. 2.19).

Similarly, the reaction of SO_2 with the HO_2 radical may be an important step in the mechanism of the oxidation of SO_2, which in turn is converted to H_2SO_4 droplets resulting in the formation of haze.

$$HO_2^\bullet + SO_2 \rightarrow OH^\bullet + SO_3 \qquad\qquad 2.32$$

In addition, the hydrocarbon radicals may give off an oxygen atom to SO_3 to form SO_3, which in turn is converted to H_2SO_4 droplets resulting in the formation of haze.

2.3.1 Smog Behaviour

The formation of photochemical smog is a dynamic process whose nature is illustrated in Fig.2.3. In the morning the NO and hydrocarbon levels increase followed quickly by increase in NO_2.NO_2 reacts with the sunlight leading to various chain reactions and ultimately to the production of ozone and other oxidations. Ozone concentration now increases until, sometime in 1he afternoon, it reaches a maximum, and then decreases gradually. NO_2 concentration diminishes from its peak as ozone concentration builds up and is usually low by late afternoon. The typical smog episode occurs in hot, sunny weather under low humidity conditions. The characteristic symptoms of the smog are the brown haze in the atmosphere, reduced visibility, eye irritation, respiratory distress and plant damage.

The control of photochemical smog may require a substantial reduction in No_x produced in urban areas. At the same time it is necessary to control the release of hydrocarbons from numerous mobile and stationary sources.

2.4 Nature and Development of Acid Rain

Acid rain is now one of the most serious environmental problems in developed countries. Interest in the problem has increased significantly since the 1990s, especially since reports of extensive forest dieback in the then West Germany started to appear in 1975. Concern quickly spread to Scandinavia, where large scale fish deaths were believed to be caused by acid rain, and where more recently concern has focused on forest changes and heavy metal mobilization in stream, lakes and groundwater.

Acid rain is widely believed to result from the washout from the atmosphere of oxides of sulphur and nitrogen. Although these oxides exist naturally in environmental cycles and ultimately have natural sources. The main sources today are coal fired

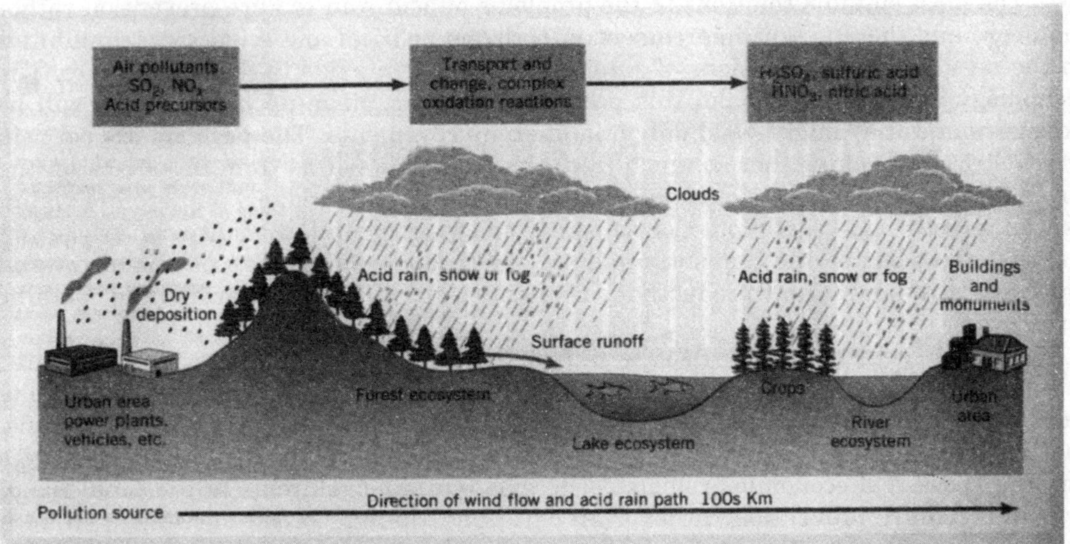

Fig. 2.4: Idealized diagram showing selected aspects of acid rain formation and paths.

power stations and smelter and motor vehicle exhausts. These gaseous pollutants are light and immiscible and they can be carried hundreds- if not thousands of kilometers by prevailing winds. The oxides may be mixed with other chemicals in the atmosphere to produce the poisonous and corrosive substances that either settle as dry fallout or are washed out by rain as acid deposition (Fig. 2.4). Emissions and fallout were previously extremely localized, but since the introduction of 'tall stacks' policies in both Britain (since 1958) and the USA (since 1971) paradoxically to disperse particulate pollutants and hence reduce local damage, emissions are now lifted into the upper air currents and carried long distances downwind. The tall stacks policies have thus turned a smoke problem into an acid rain problem. Neither winds nor acids respect political boundaries, so the pollutants are often country across state and national frontiers. Acid rain thus represents a hidden export from one country to another- the exporter gains through not having to install costly pollution control activity to limit emissions and the importer suffers from the adverse impacts on its environment Thus acid rain is a critical problem for international relations and it rightly claims a high priority on the agenda of political debate both within and between countries.

The acid rain debate now embraces many Western countries, including Canada, the United States, England, Scotland, Wales, Sweden, Norway, Denmark, Germany, the Netherlands. Austria, Switzerland and a growing number of eastern countries including Russia, Poland and Czechoslovakia.

Those countries that are hardest hit by the problem (such as Canada and Scandinavian countries) are convinced that something must be done, without further delay, to reduce emissions and thus save their forests and lakes. With the belief that 'charity begins at home' some 21 governments have now resolved to cut their own emissions of SO_2 by at least 30% (over 1980 levels) by 1993 and hope to persuade the main producers of the culprit oxides to join them in a truly international attack on the problem.

The most suitable solutions to the problems of acid rain require prevention rather than cure and there is broad agreement in both the political and scientific communities on the need to reduce emissions of sulphur and nitrogen oxide to the atmosphere. The technology now exists to make this possible. However, clean-up programmes will be expensive and they might yield only qualified improvements. The problem of acid rain arises strictly speaking, not so much from the rainfall itself as from its effects on the environment. Runoff affects surface water (such as rivers and lakes) and ground water, as well as soils and vegetation. Consequently changes in rainfall acidity can trigger off a range of impacts on the chemistry and ecology of lakes and rivers, soil chemistry and processes, the health and productivity of plants and building materials and metallic structures.

Human health might also be affected via the intake of food and water contaminated with toxic heavy metals mobilized by fallout acid rain. The significance of such impacts is more than purely scientific- they affect the quality of life for humans, they threaten environmental stability and the sustainability of food and timber reserves and they pose real economic problems. The most suitable solutions to the problems of acid rain require prevention rather than cure, and there is broad agreement in both the political and scientific communities on the need to reduce emissions of SO_2 & NO_x to the atmosphere.

2.4.1 Acid Rain and Geology

The impact of acid rain on the environment depends not only on the level of acidity in the rain, but also on the nature of the environment itself. Areas underlain by granitic or quartatic bedrock for example, are particularly susceptible to damage, since soils and water already acidic and lack the ability to buffer or neutralize additional acidity from the precipitation. Acid levels, therefore, rise, the environmental balance is disturbed and serious ecological damage is the inevitable result In contrast, areas which are geologically basic- underlain by limestone or chalk for example - are much less sensitive and may even benefit from the additional acidity. The highly alkaline soils and water of these areas ensure that the acid added to the environment by the rain is very effectively neutralized. In areas covered by glacial drift or some other unconsolidated deposit the susceptibility of the environment to damage by acid rain will be determined by the nature of the superficial material rather than by the composition of the bedrock. In theory, it is important to establish background levels of acidity or alkalinity, so that the vulnerability of the environment to acidification can be estimated.

2.4.2 Acid Rain and Aquatic Environment

The earliest concerns over the impact of acid rain on the environment were expressed by Robert Smith in England, as long as 1852, but modern interest in the problem dates only from the 1960s. Initial attention concentrated on the impact of acid rain on the aquatic environment, which can be particularly sensitive to even moderate increase in acidity and it was in the lakes and streams on both sides of the Atlantic that the effects were first apparent.

In general, there is a tendency for all lakes to become more acidic with time, as a result of natural ageing processes, but studies of acid sensitive lakes suggest that observed rates of change in pH value since the middle of the nineteenth century have

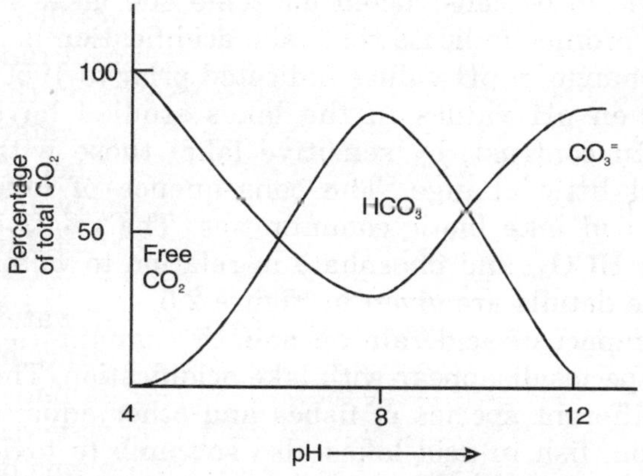

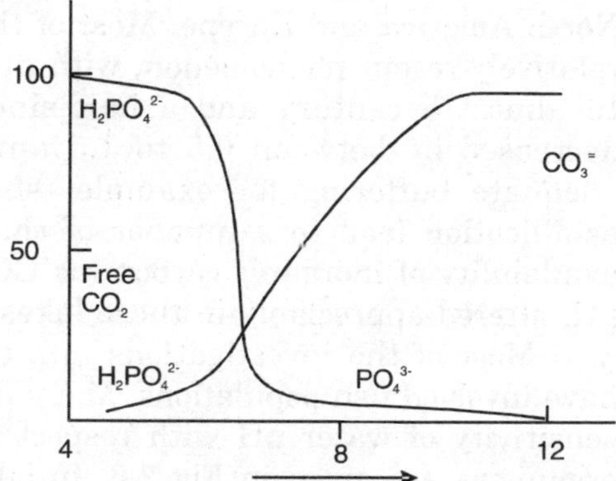

Fig. 2.5a The relative propartions of different forms of inorganic carbon in relation to the pH of water under normal conditions.

Fig. 2.5b The equilibrium of different forms of phosphate in relation to the pH of pure fresh water.

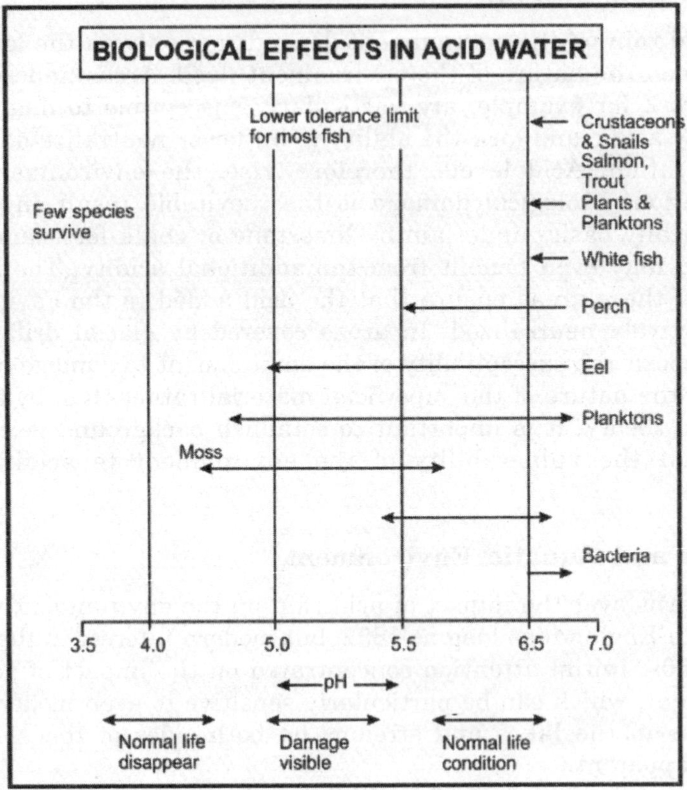

Fig 2.6: Changes in tolerance of fresh water.

exceeded the expected natural rates. The analyses of acid sensitive diatom species in lake sediments has allowed pH- age profile to be constructed for some 800 lakes in North America and Europe. Most of these profiles indicate that lake acidification is a relatively recent phenomenon, with little change in pH values indicated prior to 1950. In almost a century and a half since then pH values in the lakes studied have decreased by between 0.5 to 1.5 units. In contrast, by sensitive lake- those with adequate buffering for example- showed little change. The consequence of lake acidification lead to a number of changes in lake biotic communities. The relative availability of inorganic carbon (as CO_3 or HCO_3) and phosphate in relation to water pH, altered appreciably in these lakes. The details are given in Figure 2.5.

Most of the investigations into the impact of acid rain on aquatic communities have involved fish populations. Many fish species disappear with lake acidification. The sensitivity of water pH with respect to different species of fishes and other aquatic organisms are given in Fig 2.6. In addition, fish in acid lakes also succumb to toxic concentrations of metals, such as aluminium, mercury, manganese, zinc and lead, leached from the surrounding rocks by the acids.

2.4.3 Acid Rain and Terrestrial Environment

Terrestrial ecosystem takes much longer period to show the effects of acid rain than aquatic ecosystem. As a result, the nature and magnitudes of the impact of acid precipitation on the terrestrial environment has been recognized only in recent decades. Acid precipitation damages the forests and other vegetation cover in a number of ways. The details are given in Fig 2.7. The changes include crown dieback, necrosis of foliages, defoliation, soil acidification and various other related processes. As a result,

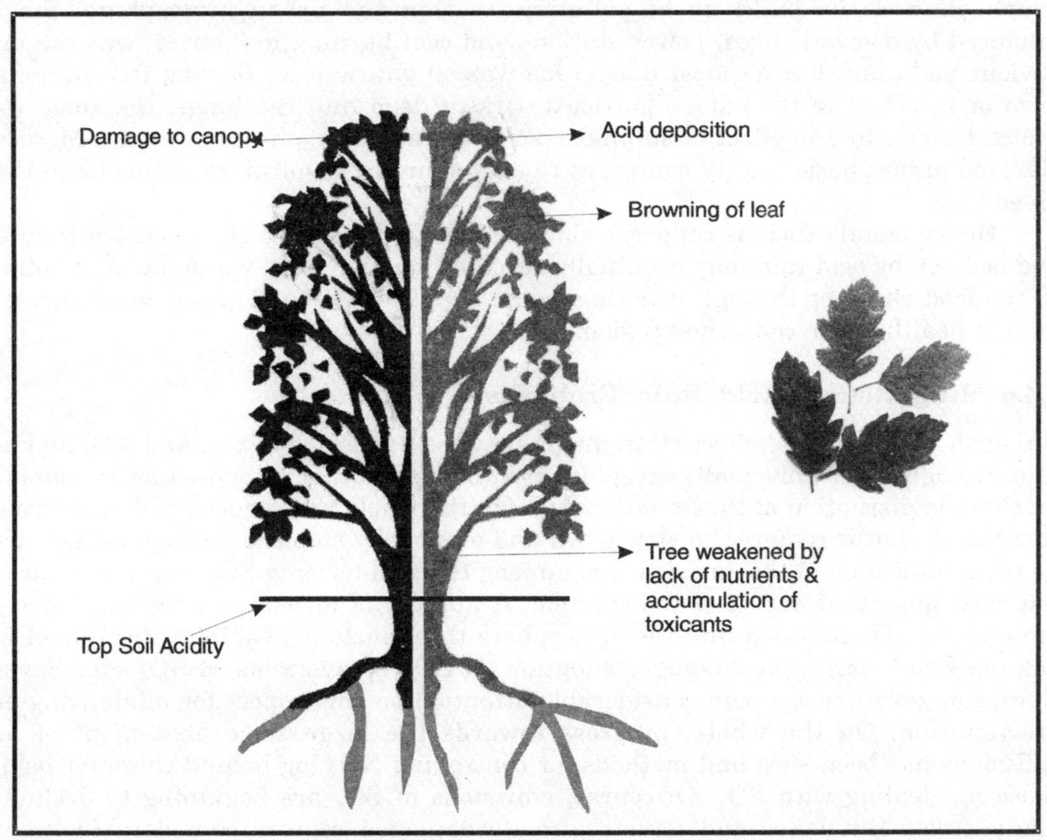

Fig. 2.7: Damage to vegetation by acid rain.

vegetation cover or forest declines on long term basis. Mobilization of toxic metals, such as aluminium, cadmium, zinc, mercury, lead and cooper and iron is another feature which accompanies soil acidification. Thus detrimental effects of these toxic metals were seen with times in affected areas.

2.4.4 Acid Rain and the Build Environment

Acid rain also contributes to deterioration of the build environment; particularly in limestone and marble rock buildings. Several monuments thus get affected with time by continued acid precipitation in this area. Crystals of calcium and magnesium sulphate begin to form on or beneath the surface of the stone, as corrosion effect induced by acid rain. By attacking the fabric of buildings acid rain causes physical and economic damage, but it does more than that, it also threatens the world's cultural heritage.

2.4.5 Acid Rain and Human Health

The famous London Smog of 1952 developed as a result of meteorological conditions which allowed the build up of pollutants within the urban atmosphere. Smoke, produced by domestic fires, power stations and coal burning industries, was the most obvious pollutant, but he most dangerous was sulphuric acid, flowing free in aerosol form or attached to the smoke particles. Drawn deep into the lungs, the smog were brought about by the effect of sulphuric acid on human respiratory system. Moreover elevated atmospheric acidity continues to cause chronic respiratory problems in these areas.

Heavy metals such as copper, cadmium, zinc and mercury etc. liberated from soil and bedrock by acid rain may eventually reach the human body via plants and animals in the food chain or through drinking water supplies. These changes poses threat to human health of the concerned region.

2.4.6 Mitigation of Acid Rain Problems

Although the cause and effect relationship between emissions of SO_2 and NO_x and acid rain damage is not universally accepted, most of the solutions proposed for the problem involve the disruption of that relationship. On the whole the reduction of acid-forming gases is all that is required to slow down and eventually stop the damage being caused by the acidification of the environment. Among the acid-forming SO_2 emission control is the most important aspect of the solution. A number of measures were suggested for minimizing SO_2 emission into the atmosphere these include desulphurification of fuel gas emission control technology, adoption & SO_2 recovery as H_2SO_4 etc. Several European governments paid considerable attention on this aspect for minimizing acid precipitation. On the whole, progress towards the large scale abatement of acid emissions has been slow and methods for controlling NO_x lag behind those for behind those for dealing with SO_x. Of course, emissions of SO_x are beginning to decline in many areas. But lakes and forests already damaged require considerable time to recover back to the past conditions. A number of ameliorative measures were also undertaken for speedy recovery of these affected areas too.

2.5 Effects of Air Pollution

Substantial evidence has accumulated that air pollution affects the health of human beings and animals, damages vegetation, soils and deteriorates materials, affects climate, reduces visibility and solar radiation, impairs production processes, contributes to safety hazards, and generally interferes with the enjoyment of life and property. Although some of these effects are specific and measurable, such as damages to vegetation and material and reduced visibility: most are difficult to measure, such as health effects on human beings and animals and interference with comfortable living. Each of the effects listed above has been the subject of considerable attention, and a number of comprehensive reviews of the effects of air pollution have been written. In this section we shall present a brief summary of some of the most important established effects of air pollution.

2.5.1 Effects of Air Pollution on Atmospheric Properties

Air pollutants affect atmospheric properties in the following ways:

1. Visibility reduction
2. Fog formation and precipitation
3. Solar radiation reduction
4. Temperature and wind distribution alteration

These effects are primarily associated with the urban atmosphere. In addition, there is much current interest in possible effects of air pollutants, mainly carbon dioxide and particles, on the atmosphere as a whole.

In addition to reducing visibility, air pollution affects urban climates with respect to increased fog formation and reduced solar radiation. The frequency of fog formation has been observed to be higher in cities than in the country in spite of the fact that air temperatures tend to be higher and relative humidities tend to be lower in cities as opposed to the country. The explanation for this observation lies in the mechanism of fog formation.

Scattering and absorption of both solar and infrared radiation, as well as emission of radiation, occur within the polluted layer. The net effect of these radiative processes during the night is a marked cooling of the pointed layer.

Studies in London, for example, have shown that the average duration of bright sunshine in central London (in hours per day) is discernibly less than in the surrounding countryside. In general, the decrease in direct solar radiation due to a polluted layer amounts to 10 to 20 percent.

2.5.2 Effects of Air Pollution of Materials

Air pollutants can affect materials by soiling or chemical deterioration. High smoke and particulate levels are associated with soiling of clothing and structures, and acid or alkaline particles, especially those containing sulfur, corrode materials, such as paint, masonry, electrical contacts, and textiles. Ozone is particularly effective in deteriorating rubber.

2.5.3 Effects of Air Pollution on Vegetation

Pollutants which are known phytotoxicants (substances harmful to vegetation) are sulfur dioxide, peroxyacetyl nitrate (an oxidation product in photochemical smog) and ethylene. Of somewhat lesser severity are chlorine, hydrogen chloride, ammonia, and mercury. In general, the gaseous pollutants enter the plant with air through the stomata in the course of the stomatal respiration of the plant. Once in the leaf of the plant, pollutants destroy chlorophyll and disrupt photosynthesis. Damage can range from a reduction in growth rate to complete death of the plant. Symptoms of damage are usually manifested in the leaf, and the particular symptoms often provide the evidence for the responsible pollutant. Table 2.3 summarizes the symptoms characteristic of plant damage by several pollutants.

Table 2.3 Summary of Symptoms and injury thresholds for air pollution damage to vegetation

Pollution	Symptom	Sustained time	Exposure (ppm)
Ozone (O_3)	Fleck, bleaching, bleached spotting, growth suppression. Tips of conifer needles become brown and necrotic.	0.03	4 hr
SO_2	Bleached spots, bleached areas between veins, chlorosis, growth suppression, reduction in yield.	0.03	8 hr
Peroxyacetyl nitrate (PAN)	Glazing, silvering or bronzing on lower surface of leaves.	0.01	6 hr
HF	Tip and margin burn, chlorosis, dwarfing leaf abscission, lower yield.	0.0001	5 weeks
Cl_2	Bleaching between veins, tip and leaf abscission.	0.01	2 hr
Ethylene (C_2H_4)	Withering, leaf abnormalities, flower dropping, and failure of flower to open.	0.05	6 hr

2.5.4 Effects of Air Pollutants on Human Health

We now come to the most controversial and probably the most important effect of air pollution, that is on human health. First we consider the mechanisms by which pollutants can effect the human body. We then discuss the type of evidence available on the effects of long-term exposure to pollutant levels characteristic of urban areas.

Pollutants enter the body through the respiratory system, which can be divided into the upper respiratory system, consisting of the nasal cavity and the trachea, and the lower respiratory system, consisting of the bronchial tubes and the lungs. At the entrance to the lungs, the trachea divides into two bronchial trees which consist of a series of branches of successively smaller diameter. The entire bronchial tree consists of over 20 generations of bifurcations, ending in bronchioles of diameters of about 0.05 cm. At the end of thebronchioles are large collections of tiny sacs called alveoli. It is across the alveolar membranes that oxygen diffuses in the opposite direction. Although an individual alveolus has a diameter of only about 0.02 cm. there are several hundred million alveoli in the entire lung, providing a total surface area for gas transport of roughly 50 m^2.

The respiratory system has several levels of defense against invasion by foreign material. Large particles are filtered from the airstream by hairs in the nasal passage and are trapped by the mucus layer lining the nasal cavity and the trachea. These large particles are unable to negotiate the sharp bends in the nasal passage, and because of their inertia. impinge on the wall of the cavity as the air rushes down toward the lung. In addition, particles may also be scavenged by fine hairlike cilia

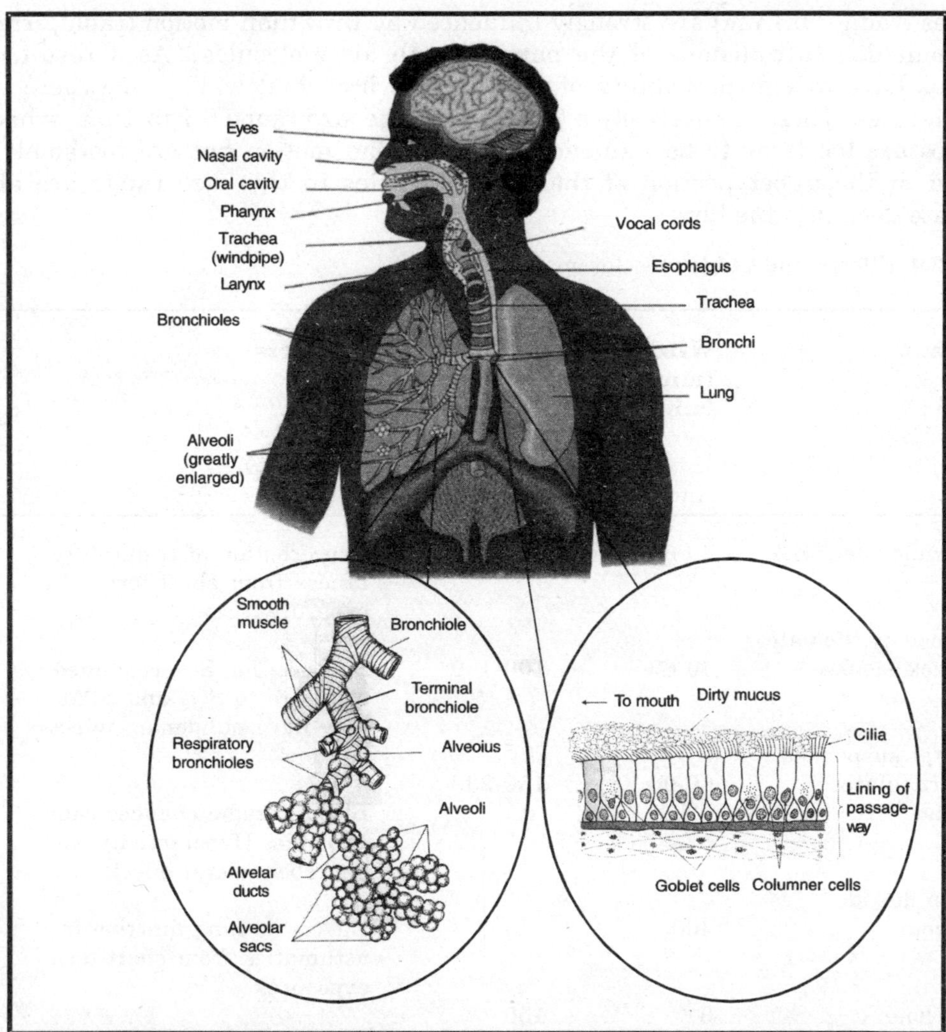

Fig. 2.8: Idealized diagram showing some of the parts (brain and cardiovascular-lung systems) of the human body that may be damaged by common air pollutants. The most severe health risks to normal exposures are related to particulates. Others of concern include carbon monoxide, photochemical oxidants, sulfur dioxide, and nitrogen oxides. Toxic chemicals and tobacco smoke also can cause chronic or acute health problems.

which line the walls of the entire respiratory system These cilia continually move mucus and trapped material to the throat where they are removed by swallowing. Most particles of sizes exceeding 0.5 to 5 μm are effectively removed in the upper respiratory system.

Particles of radii less than a few micrometers generally pass through the upper respiratory system, escaping entrapment. Some of the larger of these particles (about 1 μm in size) are deposited on the bronchial walls immediately behind bifurcations in the bronchial tree. The mechanism for this deposition is believed to be inertial impaction which results from the swirling air motions caused by the bifurcation. Very small particles (radii < 0.1 µm) are strongly influenced by Brownian motion (rapid, irregular movement due to collisions of the particle with air molecules). As a result, these particles have a high probability of striking the bronchial walls somewhere in the bronchial tree. There is effectively a "window" in the size range 0.1 to 1 µm, where the particles are too large to be influenced by Brownian motion but are too small to be trapped in the upper portion of the lung. Particles in this size range are able to penetrate deep into the lung.

Table 2.4 Effects and guidelines for major pollutants.

Pollutant	WHO Guidelines (micrograms per cubic meter of air) (ug/m^3)		Effects
	Annual mean	Percentile	
Sulphur dioxide (SO_2)	40-60	100-150	Exacerbation of respiratory illness from short-term exposures.
Suspended particulate			
Black smoke	40-60	100-150	Same as for SO_2 combined exposure to SO_2 and SPM may have pulmonary effects
Total suspended			
Particulates	60-90	150-230	
Lead	0.5-1	0	Blood enzyme changes cause anaemia, Hyperactivity and neurobehavioral effects
Nitrogen dioxide			
1 hour	400	-	Effects on lung function in asthmatics from short-term exposures
24 hour	0	150	
Carbon monoxide (mg/M^3)			
15 minutes	100	-	Reduces oxygen-carrying capacity of the blood
1 hour	30	-	
30 minutes	-	60	
8 hour	-	10	
Carboxhamoglobin	-	23.5-3%	

WHO: World Health Organization

For gases, 1he solubility governs what proportion is absorbed in the upper airway and what proportion reaches the terminal air sacs of the lungs. For example, SO_2 is quite soluble and consequently, is absorbed early in the airway, leading to airway resistance (swelling) and stimulated mucus secretion. On the other hand, CO, NO_2, and O_3 are relatively insoluble and are able to penetrate deep into the lung to the air sacs. Nitrogen dioxide and ozone cause pulmonary edema (swelling) which inhibits gas transfer to the blood. Carbon monoxide is transported from the air sacs to the blood and combines -with hemoglobin as oxygen does.

It is important to note that more than one pollutant may induce the same effect. For example, sulfur dioxide and formaldehyde both produce irritation and increased airway resistance in the upper respiratory tract, and both CO and NO_2 interfere with oxygen transport by hemoglobin. Several pollutants usually are present at the same time, and as a result, observed effects may actually be attributable to the combined action of more than one pollutant. A good example of this is the case of SO_2 and particulate matter. Health effects become far more serious when both are present then if either occurs separately. A possible explanation for this effect is that SO_2 becomes absorbed on the surface of very small particles and is carried by the particles deep into the lung.

Table 2.5 summarizes some observed relations between pollutant levels and physiological responses. In many cases, we can account for the responses by considering the mechanisms of irritation of individual pollutants.

2.5.5 Carbon Monoxide

The effects of carbon monoxide exposure are reflected in the oxygen-carrying capacity of the blood. In normal functioning, hemoglobin molecules in the red blood cells carry oxygen, which is exchanged for carbon dioxide in the capillaries connecting arteries and veins. Carbon monoxide is relatively insoluble and easily reaches the alveoli along with oxygen. The carbon monoxide diffuses through the alveolar walls and competes with oxygen for one of the four iron sites in the hemoglobin molecule. The affinity of the iron site for CO is about 210 times greater than for O_2, so that this competition is extremely effective. When a hemoglobin molecule acquires a CO molecule it is called carboxyhemoglobin (abbreviated COHb). The presence of carboxyhemoglobin decreases the overall capacity of the blood to carry oxygen to the cells. In addition, the presence of CO on one of the iron sites of a hemoglobin molecule not only removes that site as a potential carrier of an O_2 molecule but also causes the other iron sites of the molecule to hold more tightly onto the O_2 molecules they are carrying. The formation of COHb is a reversible process, with a half-life for dissociation after exposure of about 2 to 4 hr for low concentrations.

The major source of CO in urban areas is automobile exhaust. Levels typical of urban areas range from 5 to 100 ppm. It appears that the most serious danger associated with CO is the exposure of drivers on heavily congested highly congested highways to CO levels of the order of 100 ppm. It has been found experimentally that relatively low COHb levels can affect the ability to estimate time intervals, can delay reaction times, and reduce visual sensitivity in the dark. The supposition that CO leads to increased incidences of traffic accidents, by virtue of effects such as these, is indeed a compelling one.

2.5.6 Oxides or Sulfur

Sulfur dioxide is highly soluble and consequently is absorbed in the moist passages of the upper respiratory system Exposure to SO_2 levels of the order of 1 ppm leads to constriction of the airways in the respiratory tract.

Table 2.5 Observed relationship between pollutants and their health effects.

Pollutant	Concentration level producing adverse health effects	Adverse health effects
Particulate matter and sulfur oxides	1. 80-11 µg/m^3 particulates (annual geometric mean) 2. 130µg/m^3 (0.046 ppm) of SO_2 (annual mean) accompanined by particulate concentrations of 130 µg/m^3. 3. 190 µg/m^3 (0.068 ppm) of SO_2 (annual mean) accompanied by particulate concentrations of about 177µg/m^3. 4. 105-265 µg/m^3 (0.037-0.092 ppm) accompanied by particulate concentrations of about 185 µg/m^3. 5. 140-260 µg/m^3 (0.05-0.09 ppm) of SO_2 (24 hr average). 6. 300-500 µg/m^3 (0.11-0.19 ppm) of SO_2 (24 hr mean) with low particulate levels. 7. 300 µg/m^3 particulates for 24 hr accompanied by SO_2 concentrations of 630 µg/m^3 (0.22 ppm)	1. Increased death rates for persons over 50 years of age 2. Increased frequency and severity of respiratory diseases in schoolchildren. 3. Increased frequency and severity of respiratory diseases in schoolchildren. 4. Increased frequency of respiratory symptoms and lung disease 5. Increased illness rate of older persons with severe bronchitis 6. Increased hospital admissions for respiratory disease and absenteeism from work of older persons 7. Chronic bronchitis patients suffering acute worsening of symptoms
Carbon	1. 58 µg/m^3 (50 ppm) for 90 min (similar effects) upon exposure to 10 to 17 µg/m^3 (10-17 ppm) for 8 or more hr. 2. Effects upon equivalent exposure to 35 µg/m^3 (30 ppm) for 8 or more hr. 3. Effects upon equivalent exposure to 35µg/m^3 (30 ppm) for 8 or more hr.	1. Impaired time-interval discrimination 2. Impaired performance in psychomotor tests 3. Increase in visual threshold
Photochemical oxidants (O_3 and peroxy-organic nitrates)	1. In excess of 130 µg/m^3 (0.07 ppm) 2. 490 µg/m^3 (0.25 ppm) maximum daily value. (This value would be expected to be associated with a maximum hourly average concentration as low as 300µg/m^3 (0.15 ppm) 3. 300µg/m^3 (0.1 ppm) maximum	1. Impairment of performance by student athletes 2. Aggravation of asthma attacks 3. Eye irritation

As we have already noted, high SO_2 levels are often associated with high particulate concentrations. The fact that a three- to fourfold increase in the irritant response to SO_2 is observed in the presence of particulate matter is presumably attributable to the ability of the aerosol particles to transport SO_2 deep into the lung Table 2.6. Summarizes responses to various dosages of SO_2.

Table 2.6 Effects of sulphur dioxide on humans.

Concentrations (ppm)	Effects
0.2	Lowest concentration causing a human response
0.3	Threshold for taste
0.5	Threshold for odour recognition
1.6	Threshold for inducing reversible bronchoconstriction in healthy individuals
8-12	Immediate throat irritation
10	Eye irritation
20	Immediate coughing

2.5.7 Oxides of Nitrogen

There is no available evidence supporting the proposition that nitric oxide (NO) is a health hazard at levels found in urban air. Nitrogen dioxide (NO_2), on the other hand, is transformed in the lungs to nitrosamines, some of which may be carcinogenic. In addition, NO_2 may be transferred to the blood to form a compound called methemoglobin. Nitrogen dioxide is known to irritate the alveoli, leading to symptoms resembling emphysema upon long-term exposure to concentrations of the order of 1 ppm (Project Clean Air. 1970).

2.5.8 Photochemical Oxidants

The term *photochemical oxidants* refers to the secondary pollutants formed in photochemical smog from reactions involving hydrocarbons and oxides of nitrogen. The principal ingredient in this category is ozone, with smaller amounts of oxygen-containing hydrocarbon compounds. The effect of ozone on pulmonary function is still not thoroughly understood. In general, ozone of about 1 ppm produces a narrowing of the airways deep in the lung, resulting in increased airway resistance. The effects of long-term exposure to ozone at levels typical of urban air (about 0.1 to 0.2 ppm) have not been established. Experiments with animals have exhibited irreversible changes in pulmonary function after long-term exposure to levels of 1 ppm. A topic of current speculation is that exposure to low levels of ozone accelerates the aging of lung tissue by the oxidation of certain compounds in proteins.

A widespread effect of photochemical smog is eye irritation. The precise mechanism by which certain compounds cause irritation of eyes is not known; in fact, it appears that the compounds responsible for eye irritation in smog may not all have been identified. Those that are known to be irritants and that have been detected in photochemical smog are formaldehyde (HCHO), acrolein (CH_2CHCHO), and members of the family of peroxyacyl nitrates, two of which are

$$CH_3\overset{\displaystyle O}{\overset{\|}{C}}OONO_2 \qquad \text{Peroxyacetyl nitrate (PAN)}$$

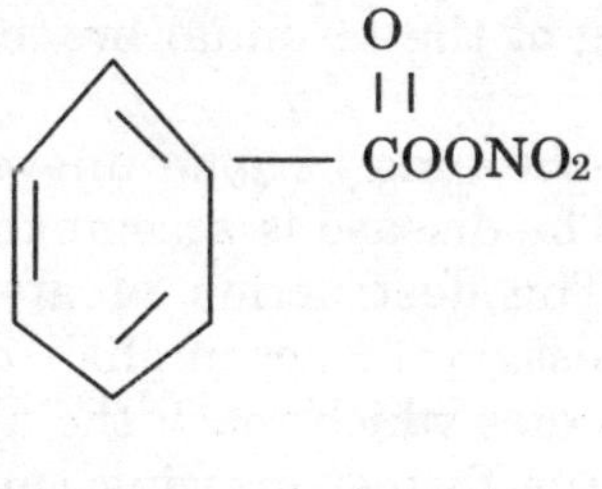

2.5.9 Lead

The mechanisms of lead poisoning are complex. In short, lead inhibits several steps in the formation of hemoglobin. Depending on the mode of entry into the body, up to 60 percent of the total lead ingested can be permanently retained by the body.

Over the years, concentrations of air pollutants in a particular area have, on occasion, reached excessively high levels for periods of several hours to several days. The result has been a number of so-called air pollution episodes, the most serious of which are listed in Table 2.7. During an episode, the person most likely to be seriously injured is one who is either elderly or is in questionable health, perhaps already suffering from a respiratory disease.

Table 2.7 Air Pollution Episodes.

Location	Date	Pollutants	Symptoms and effects
Meuse Valley,	Dec. 1-5, 1930	SO_2	63, excess deaths, chest pain, Belgium (9.6-38.4 ppm) cough, eye and nasal
Donora, Pa.	Oct. 26-31, 1948	SO_2, particles	20 excess deaths, chest pain, (0.5-2 ppm) cough, eye and nasal irritation, older people mainly affected.
Poza Rica,	Nov. 24, 1950	H_2S	22 excess deaths, 320 hospitalized, all ages affected.
London	Dec. 5-9, 1952	SO_2, particles	4000 excess deaths
New York	Nov. 24-30, 1966	SO_2, particles	168 excess deaths

Source : Goldsmith (1968)

Disease of the respiratory system are generally correlated with air pollution. There are two types of reactions to air pollutants by the respiratory system The first is acute reaction, such as irritative bronchitis, and the second is chronic reaction, such as chronic bronchitis and pulmonary emphysema.

Bronchitis refers to a condition of inflammation of the bronchial tree. The inflammation is accompanied by increased mucus production and a cough. Airway resistance is increased because of the presence of the thickened mucus layer. Acute bronchitis is generally a short-lasting disease, caused by a virus or foreign material in the lung. Chronic bronchitis, on the other hand, is a sustained inflammation of the bronchial system, leading to an increase in the volume of mucoid bronchial secretion sufficient to cause expectoration. It is frequently accompanied by a cough and shortness of breath. The persistent inflammation leads to swelling of the terminal bronchi and increased airway resistance.

Emphysema is a condition in which the alveoli in the lung become uneven and over distended due to destruction of the alveolar walls. The disease is accompanied by shortness of breath, particularly following exercise. The destruction of alveoli is progressive, resulting in an increased blood flow necessary to accomplish oxygen transfer and to a decreased ability to eliminate foreign bodies which reach the alveolar region. Emphysema has no known cure and is one of the fastest growing causes of death in the United States.

2.6 Indoor Air Pollution

Human kind faces the most prolonged and worst exposure to pollution within the four walls of his home. Indoor air pollution can have an astonishing variety of sources including cooking gas ranges, space heaters, fire places, wood-or coal burning stores, cigarette smoking, household products like detergents, waxes, polishes, air freshners, pesticides, glues, paints, hair sprays, oven cleaners, permanent press fabrics, synthetic fibres and leaning activities like dusting and vaccum cleaning. And of course whatever pollutes the outside air eventually becomes indoor pollution as well, specially in older leaky buildings.

An open flame produces carbon monoxide and nitrogen oxides both discussed in details above. Biomass fuels (wood, crop residues and animal dung), burnt indoors for daily cooking and space heating in millions of households across the developing world, are slowly turning these homes into death traps. These sources are responsible for almost 60 percent of the total worldwide exposure to particulates. Biofuels generate 10-100 times more respirable particulates per meal owing to their low thermal and heat transfer efficiencies. Biofuel combustion is responsible for the emission of pollutants such as sulphur dioxide, nitrogen dioxide, carbon monoxide, total suspended particulates and polyaromatic hydrocarbons. What makes biofuel smoke particularly deadly is its composition which is similar to tobacco smoke. Women suffer more from these biofuels. Infants carried on their mother's backs in Zambia have been found six times more vulnerable to acute respiratory infections (ARI) opposed those who were not. Still births and chronic obstructive lung disease or cold are the other health impacts of indoor air pollution.

Six major categories of ill health can be attributed to indoor exposures.

i.   Acute respiratory infections (ARI) in children.
ii. Adverse pregnancy outcomes for women exposed during pregnancy.
iii. Lung cancer.
iv. Chronic lung ailments (bronchitis and asthama) and associated heart maladies.
v.   Diseases of the eyes.
vi. Increase in the severity of coronary artery disease.

Not many people are aware that the nail-polish remover they use could be the cause for their diminished mental capacity, or that the batteries from their personal stereos are slowly discharging poisonous mercury into the air. Air inside a building can be laden with a witch's brew of chemicals, gases, smokes, odours, fumes, micro-organisms and other pollutants sometimes in concentrations high enough to pose serious threats to health. Some of the worst offenders are photocopying machines and other electrical equipment (emitting ozone), solvent clearners like paints or paint thinners (sources of benzene), carpets and plastic products (sources of benzene), insect sprays, mothballs, floor waxes, aerosol (sprays cleaners), oven and window cleaners, air freshners and insulation foams and disinfectants (formaldehyde sources). Even the innocuous bricks and walls of some houses shelter and emit gases like randon; paints containing asbestos and asbestos-insulated walls also lead to polluted air.

These pollutants, besides causing irritation of eyes, lungs and throats and retarding mental capacity, can act as catalysts for asthma attacks, and scientists believe that they may even lead to accelerated growth of tumours. Indian consumers are not always told how dangerous some of the products they use are. "With the recent emergence of a multitude of electronic gadgetry and growing consumer market for house hold cleaners, insecticides and cosmetics, it seems getting a breath of fresh air has become a difficult task, whether you are outdoors of indoors.

2.7 Prevention and Control

The control of air pollution is ultimately an engineering problem. The WHO (1968) in its publication (12). "Research into Environmental pollution" recommended the following procedures for the prevention and control of air pollution:

(i) **Containment:** That is prevention of escape of toxic substances into the ambient air. Containment can be achieved by a variety of engineering methods such as enclosure, ventilation and air cleaning. A major contribution in its field is the development of "arrestors" for the removal of contaminants.

(ii) **Replacement:** That is, replacing a technological process causing air pollution by a new process that does not. Increased use of electricity and natural gas in place of coal is an example of replacement.

(iii) **Dilution:** Dilution is valid so long as it is within the self-cleaning capacity of the environment. For example, some air pollutants are readily removed by vegetation. The establishment of "green belts" between industrial areas is an attempt at dilution. The capacity for dilution is, however, limited and trouble occurs when the atmosphere is overburdened with pollutants.

(iv)**Legislation:** Many countries have adopted legislation for control of air pollution. In India, there is a Smoke Nuisance Act effective in a few cities (i.e. Calcutta, Bombay, Ahmedabad and Kanpur). The Government of India have already drafted suitable legislation for control of air pollution.

(v) **International Action:** To deal with air pollution on a world-wide scale, the WHO has established an international network of laboratories for the monitoring and study of air pollution. The network consists of two international centers at London and Washington, three centers at Moscow, Nagpur, and Tokyo and 20 laboratories in various parts of the world (13). These centers will issue warnings of air pollution where and when necessary.

2.7.1 Air (Prevention and Control of Pollution) Act, 1981- A Critical Study

India is a signatory to the UN. Conference on Human Environment held in Stockholm (Sweden) in June 1972, where the participants decided to take appropriate steps for the prevention and control of pollution. Accordingly, the Parliament of India enacted the *Air (prevention and Control of Pollution) Act,* 1981, in the Thirty-second Year of the Republic of India for prevention, control and abatement of air pollution. The Act received assent of the President on March 29, 1981.

The *Central Pollution Control Board* constituted under Section 3 of the Water (Prevention and Control of Pollution) Act, 1974 shall exercise the powers and functions of the Central Pollution Control Board for the prevention of air pollution also under this Act. Similarly, State Pollution Control Boards constituted under Section 4 of the water Act, 1974 for the prevention and Control of Water Pollution deemed to be the State

Boards for the prevention and control of air pollution under this Act. In states, in which water pollution control bounds have not been established, a separate Board for prevention and control of air pollution on similar lines is to be setup. For the Union territories, the Central Pollution Control Board shall exercise the powers and perform functions of a State Board.

2.7.2 Functions of the Boards

The Act empowers the Central and State Boards to perform the following functions:

i.   To advise, plan and execute nationwide programmes, provide technical assistance and guidance, carry out investigations and research, plan and organize the training of persons, organize through mass media comprehensive programmes, collect, compile, and publish technical and statistical data, prepare manuals, codes or guides regarding prevention, control and abatement of pollution;

ii. Lay down standards for quality of air; and

iii. To establish or recognize laboratories for the above purposes. Penalties

Whoever fails to comply with the provisions of the Act under Section 21 (5) or Section 22 or with directions issued under this Act, shall be punishable with imprisonment for a period which many extend to three months or with fine up to Rs. 10,000/- or with both. In case of continuing offences, it provides with an additional fine up to Rs. 100/- for every day during which such failure continues after the conviction for the first such failure.

SUGGESTED FURTHER READINGS

L. Hodges (1973). Environmental pollution. (Holt, Rinehart and Winston, Inc., New York).

S.H. Stoker and S.L. Seager (1976). Environmental chemistry: air and water pollution (Scott Foresman & Co., New York).

P.A. Leighton (1961). Photochemistry of air pollution. (Academic Press, New York).

H M. Dix (1968). Environmental pollution (John Wiley & Sons, New York).

M.K. Das (1977). Air pollution in India and how to tackle it. Chemical age of India. Vol. 28, pp. 433-435.

R.M. Harrison (1995). Pollution causes, effects and control. The royal society of chemistry.

A. Sharma and A. Roychaudhary (1996). Slow Murder. Centre for science and environment. New Delhi.

STUDY QUESTIONS

1. Define air pollution. Classify the air pollutions. Discus their properties.
2. Write in details on the behaviour and fate of different air pollutions.
3. Write briefly on the following-
   i.   Wet precipitation
   ii.   Dry precipitation
   iii. Photochemical smog
   iv.   Air (Prevention & control of pollution act 1981)
4. Discuss in details different chemical reactors that occur in the atmosphere.
5. Write an essay on acid rain.
6. Give a detailed account of the effects of air pollution on vegetation and human health.
7. Discuss the issue of prevention and control of air pollution.

Chapter 3

WATER POLLUTION

Water is a vital natural resource which is essential for a multiplicity of purposes. Its many uses include drinking and other domestic uses, industrial cooling, power generation, agriculture (irrigation), transportation and waste disposal. In the chemical process industry, water is used as a reaction medium, a solvent, a scrubbing medium, and a heat transfer agent. As a source of life for man, plants and other forms of life it cannot be replaced.

Dozens of times a day those of us who live in the industrialized nations of the world enjoy a blessing denied to 75 percent of the world population: abundant supplies of clean water. But water is essential for life on earth. No known organisms can live without it. For centuries water has been used as a dumping ground for human sewage and industrial wastes. Added to them are the materials leached out and transported from land by water percolating through the soil and running off its surface to aquatic ecosystems. Thus the term water pollution refers to *"Water contamination by a variety of chemical substances or eutrophication caused by several nutrients and fertilizers (Southwick, 1976)"*. U.S. Department of Health Education and Welfare defines water pollution as ***"The adding to water of any substance, or the changing of water's physical and chemical characteristics in any way which interferes with its use of legitimate purposes"***.

3.1 Origin of Wastewater

Wastewaters can be classified by their origin as domestic wastewater and industrial wastewater. Any combination of wastewater that is collected in municipal sewers is termed as municipal sewage. Domestic wastewater is that which is discharged from residential and commercial establishments, whereas industrial wastewater is that which is discharged from manufacturing plants. The pollutants in domestic wastewater arise from residential and commercial cleaning operations, laundry, food preparation, body cleaning functions, and body excretions. The composition of domestic wastewater is relatively constant.

Industrial wastewater is formed at industrial plants where water is used for various processes, and also for washing and rinsing of equipment, rooms, etc. These operations result in the pollution of the products and byproducts that are discharged, either deliberately or unintentionally into them.

Normally, wastewaters are conducted to treatment plants for removing undesirable components which include both organic and inorganic matter as well as soluble and insoluble material. These pollutants, if discharged directly or with improper treatment, can interfere with the self-cleaning mechanisms of water bodies. The capacity for self-cleaning is due to the presence of relatively small numbers of different types of micro-organisms in the water bodies. These microorganisms use as food much of the organic pollutants and break them down into simple compounds such as CO_2 or methane, and the micro-organisms produce new cells also. But often either a pollutant does not degrade naturally or the sheer volume of the pollutant discharged is sufficient to overwhelm the self-cleaning process. Also, the microbial population can be destroyed by toxic wastes build up and reach high enough levels that will prevent re-establishment of a microbial population. The water quality thus becomes permanently degraded.

Various constituents of wastewater are potentially harmful to the environment and to human health. In the environment, the pollutants may cause destruction of animal and plant life, and aesthetic nuisance. Drinking water sources are often threatened by increasing concentration of pathogenic organisms as well as by many of the new toxic chemicals disposed of by industry and agriculture. Thus, the treatment of these wastes is of paramount importance.

The major sources of water contamination have been domestic, industrial, agricultural waste products, solid wastes, heat and radioactive materials.

3.2 Domestic Water Pollution

Domestic water requirements differ from season to season and from rural to urban areas. We use more water in summer than winter. Water consumption in cities is greater than rural areas. Factors like standard of living and habits of the people determine the per capita use of water. In India the per capita consumption of water in both rural and urban areas is generally not known because citizens in addition to municipal supplies do use private wells, rivers and lakes for their water demands. Per capita consumption of water has been assumed to increase with the standard of living however, water services are available only for 2 to 3 hours per day. As a result per capita water use has declined in many cities due to tremendous growth in their population.

The starting point of domestic water pollution is urbanization. It is rapidly progressing throughout the world. Urban waters often have elevated nutrient levels, specially phosphorus. Many urban water surfaces are covered with a thin film of oil and grease. In a planned city the run off is channeled into two ways. In the first, storm and domestic sewers are separate systems. In the second, storm and domestic sewers are combined. The leakage of sewers is very common in big cities. The consumption of this water by poor people specially those living in slums close to sewers leads to diseases that often take the shape of an epidemic.

Domestic water pollution is mainly caused by sewage. Sewage is defined as the water-borne waste derived from home, animal or food processing plants and includes human excreta, soaps, organic materials, different types of solids, waste food, oil

Table 3.1 Characteristics of the domestic wastewater pollutants entering the marine environment of Bombay. Daily domestic wastewater output at 1200 mld (approximate) (Source: Zingde, 1985)

Parameter	µg/l	kg/day
Dissolved Solids	1450*	17.4**
Suspended Solids	245*	2.9**
BOD	258*	3.1**
Sulphate	75*	0.4**
Nitrogen	35*	0.9**
Phosphorus	9*	7200
Chloride	587*	7
Manganese	507*	608
Iron	2529	6035
Cobalt	30	36
Nickel	81	97
Copper	110	132
Zinc	251	301
Lead	11	13

*x one thousand
** x hundred thousand

detergents, paper and cloth. They are the largest group of water pollutants. Water pollution is caused by uncontrolled dumping of waste collected from villages, towns and cities into ponds, streams, lakes and rivers. A major ingredient of most detergents is phosphate. Phosphates support luxurious growth of algae. Algae withdraw large quantities of oxygen from water. It becomes detrimental to other organisms. Domestic waters are the primary sources of water pollution.

The domestic sewage contributes to the largest amount of waste and it has been estimated that approximately 20,000 million litres per day reach the coastal environment of the country. The characteristics of domestic sewage of Bombay is given below.

Bombay has the capacity to treat only 390 mld against 1200 mld of domestic sewage. Due to such partial sewage treatment, waste water retains the original characteristics: that result into severe damage to water quality The water quality at Bombay Harbour is mentioned below-

Table 3.2 Levels of various biological and chemical pollution parameters in the Bombay harbour water (Source NEERI 1985; ND - Not Detectable)

Parameter	Concentration in water	Safe limit
pH	7.0	6.5-8.5
BOD	110-225 mg/l	4.0 mg/l
COD	212-448 mg/l	180 mg/l
Oil & grease	4-15.6 mg/l	1.0 mg/l
Ammonical Nitrogen	0.7-3.5 mg/l	1.2 mg/l
Nickel	30-240 mg/l	0.300 µg/l
Lead	50-100 mg/l	0.100 µg/l
Chromium	ND -20 mg/l	0.200 µg/l
Cadmium	ND -45 mg/l	0.300 µg/l

The huge discharge of sewage allows following events that lead to water pollution.

i. **Depletion of Oxygen Contents**: The aerobic bacteria present in water are responsible for the decomposition of organic matter. The quantity of oxygen utilized by the bacteria for the degradation of organic substance is called biological oxygen demand (BOD). Thus BOD value can be used as an indicator of water pollution. Alongwith BOD, the quantity of oxygen dissolved in a water body (DO), indicates the quality of bio-life in a water system. DO below 4 to 5 ppm is detrimental to the system.

ii. **Promotion of Algal Growth**: Stimulation of heavy algal growth and shift in the algal flora to the blue green algae, leading to the formation of obnoxious blooms, floating scums or blankets of algae results into eutrophication. Most of the algal bloom do not seem to be utilized as food by the invertebrates or zooplankton, thereby, minimizing the predatory control. Biological decomposition of such algal masses in turn leads to oxygen depletion. In a poorly oxygenated condition, fish and other animal die and clean river is turned into a stinking drain.

iii. **Spread of Infections/Diseases:** Micro-organisms, usually viruses, bacteria, some protozoans and helminthes occur in water bodies as a result of sewage disposal. Consumption of contaminated water causes water borne infectious diseases.

Table 3.3: Water borne infectious diseases in man.

Organism	Diseases
Viruses	Viral hepatitis, poliomyletitis
Bacteria	Cholera, typhoid, paratyphoid, dysentery, diarrhoea
Protozoa	Amoebiasis, giardiasis
Helminthes	Roundworm, hookworm, threadworm

Presence of *Escherichia coli* in water indicates the degree of water   pollution as describe below –

Heavily polluted	$10,000\ l^{-1}$
Polluted	$1,000\ l^{-1}$
Slightly polluted	$100\ l^{-1}$
Satisfactory	$10\ l^{-1}$
Drinking water	$3\ l^{-1}$

3.3 Industrial Water Pollution

Pollution mainly caused by the discharge of industrial effluents into the water body is known as industrial pollution. These effluents contain a wide variety of inorganic and organic substances such as oils, greases, plastics, plasticizers, methylic wastes, suspended solids, phenols, pesticides, heavy metals and acids. Major industries of the country are located on or near the coastline or riversides.

Table 3.4 Industrial pollution in representative states.

State	No. of coastal industrial complexes/industries	Approx. quantity of wastes discharged in kiloliters/day
West Bengal	10 major industries	20000
Orissa	3 major industries	180
Tamil Nadu	17 major industries	1237160
Pondicherry	4 major industries	10710
Kerala	20 major industries	24630
Maharashtra	20 industrial complexes	240065
Gujarat	2 industrial complexes	219880

* includes 11,50,000 KLD of seawater discharged from Cooling Plant of Tuticorin Thermal Power Station.

Most of the Indian rivers and freshwater streams have been seriously polluted by industrial waste.

Table 3.5 Some Indian rivers and their major sources of pollution

Name of the river	Source of pollution
Kali at Meerut (UP)	Sugar mills; distilleries; paint, soap, rayon, silk, yarn, tin and glycerine industries.
Jamuna near Delhi	DDT factory sewage, Indraprastha Power Station, Delhi
Ganga at Kanpur	Jute, chemical, metal and surgical industries; tanneries, textile mills and great bulk of domestic sewage of highly organic nature.
Gomti near Lucknow (UP)	Paper and pulp mills; sewage.
Dajora in Bareilly (UP)	Synthetic rubber factories.
Damodar between Bokaro and Panchet	Fertilizers, fly ash from steel mills, suspended coal particles from washeries, and thermal power station.
Hoogly near Calcutta	Power station; paper pulp, jute, textiles, chemical, paint, varnishes, metal, steel, hydrogenated vegetable oils, rayon, and soap, match, shellac, and polythene industries and sewage.
Sone at Dalmianagar (Bihar)	Cement, pulp and paper mills.
Bhadra (Karnataka)	Pulp, paper and steel industries.
Cooum, Adyar and Buckinghum canal (Madras)	Domestic sewage, automobile work shops.
Cauvery (Tamil Nadu)	Sewage, tanneries, distilleries, paper and rayon mills.
Godavari	Paper mills.
Siwan (Bihar)	Paper, sulphur, cement, sugar mills.
Kulu (between Bombay and Kalyan)	Chemical factories, rayon mills and tanneries. Sugar industries.
Suwao (in Balampur)	Sugar industries.

The discharge of industrial wastes results into the following –
  i.   Organic substances deplete the oxygen content.
  ii. Inorganic substances render the water unfit for drinking and other purposes.
  iii. Acids and alkalies adversely effect the growth of fish and other aquatic organisms.
  iv. Dye change the colour of water and affect the aquatic life.
  v.   Toxic substances cause serious damage to flora and fauna.
  vi. Oil and other greasy substances interfere with the self purification mechanism of water.

3.4 Agricultural Water Pollution

Water pollution can be caused by agricultural wastes such as fertilizers, pesticides, soil additives and animal wastes that are washed off from the land to the aquatic system through irrigation, rainfall and leaching. India uses about 16 kg/ha of fertilizers on an average. However, average all over the world is 54 kg/ha. Netherlands alone uses 709 kg/ha. However, an increased use of pesticides has been observed in India from 2.8 Mt in 1975-76 to 6 Mt in 1994-95 and 9.7 Mt in 1994-95. The demand for fertilizers are non-biodegradable. In both fresh and marine systems they enter the food chain, accumulate in non-target organisms and increase in animal tissue to alarming concentrations. They may find entry into the drinking water supplies.

3.5 Solid Waste Pollution

Solid waste varies in composition with the socio-economic status of the generating community. The following materials could be classified as solid waste.
  i.   **Garbage** – Decomposable wastes from households, food, canning, freezing and meat processing operations that are not disposed off in wastewater.
  ii. **Rubbish** – All non-decomposable wastes, garden wastes, cloth, paper, glass, metals and chemicals.
  iii. **Sewage Sludge.**
  iv. **Miscellaneous Materials** – chemicals, paints, explosives and mining wastes.

River or ocean dumping of these materials leads to water pollution. This tends to contaminate surface as well as ground water. Because of a high organic nature of this material, a large BOD is placed on the receiving waters and the sediment becomes coated with a highly organic ooze. Non-combustible materials are generally disposed off by ocean dumping.

The National Environmental Engineering Research Institute (Nagpur, India) has categorized sea off Mumbai upto 5 km as one of the most polluted coastline where 1,800 millions liters of city's discharge is dumped everyday. In 1994, 800 truckloads of garbage was washed onto the Juhu beach in a single night. Others adding to the coastal pollution include foreign liners docking along the Bombay Port Trust (BPT) which openly dump their wastes into the sea and leave a huge trail of fuel along the coastal stretches while departing.

3.6 Thermal Pollution

An increase in the optimum water temperature by industrial processes (steel factories, electric power houses and atomic power plants) may be called as thermal pollution. Many industries generate their own power and use water to cool their generators. This hot water is released into the system from where it was drawn, causing a warming trend of surface waters. If the system is poorly flushed, a permanent increase in the temperature may result. However, if the water is released into the well flushed systems, permanent increase in temperature does not occur.

Many organisms are killed instantly by the hot water resulting into a high mortality. It may bring other disturbance in the ecosystem. The eggs of fish may hatch early or fail to hatch at all. It may change the diurnal and seasonal behaviour and metabolic responses of organisms. It may lead to unplanned migration of aquatic animals. Macrophytic population may also be changed. As temperature is an important limiting factor, serious changes may be brought about even by a slight increase in temperature in a population.

3.7 Oil Pollution

There are about 15 million water crafts on navigable waters throughout the world. Their combined waste discharges are equivalent to a city with a population of 2,000,000. Thus oil pollution, an oxygen demanding waste, is of concern not only from sensational major spills from ships and offshore drilling rigs but also from small spills and cleaning operations.

The information available on oil pollution in Indian ocean is fragmentary. In all, 6,689 observations were made on oil sliks and other floating pollutions along the tanker and trade routes across the northern Indian Ocean, of these, oil was sighted on 5582 occasions (Qasim, 1991). The percentage of oil sightings ranged from 51 to 96. The number increased away from the source of oil, that is from the ballast and bilge washings and northern Indian Ocean is well known for oil slicks.

Observations on the floating petroleum residues from the Indian Ocean region show variations in time and are fairly high occasionally along the tanker routes. In the Arabian Sea, the concentration ranged from 0 to 6.0 mg/m^2 with a mean of 0.59 mg/m^2. The range in the Bay of Bengal tanker route varied from 0 to 69.75 mg/m^2. These figures indicate that the tanker route in the Bay of Bengal is relatively more polluted than the Arabian Sea.

A layer of oil floating on the ocean surface can interfere with the exchange of oxygen and carbondioxide and reduce the rate of photosynthesis of marine plankton and the respiration of marine animals. Fuel oil added to sea water in very low concentration (199 ppb) depresses photosynthesis.

The death of birds from oil spills has attracted much attention. In the Torrey Canyon incident in 1967, an estimated 40,000 to 100,000 birds died. The Fort Mercer and Pendleton Collision in 1952 reduced the wintering population of eider ducks from 5,00,000 to 1,50,000. Some believe that the jackars penguin, which lives in South Africa, is endangered because floating oil from tankers rounding the Cape of Good Hope is killing hundreds of thousands of these birds each year. The chronic effects of oil spills appear less serious than the acute effects.

3.8 Toxic Water Pollutants and Their Effects

Two main categories of water pollutants have been identified from environmental health point of view. The first group comprises conventional water pollutants that come primarily from non-industrial sources and are highly bio-degradable e.g. faecal coliform bacteria, nutrients, nitrates and sediments.

Unlike conventional pollutants the second group comprises toxic water pollutants e.g. pesticides, heavy metals, chlorinated hydrocarbons and a variety of inorganic and organic compounds. They are not easily biodegraded and are a serious health problem of highly industrialized countries.

3.8.1 Water Borne Diseases

Table 3.6 Burden of water-related diseases in India (in millions of DALYs)

Disease	Female	Male	Total
Diarrhoeal disease	14.39	13.64	28.03
Intestinal helminthes	1.0	1.06	2.06
Trachoma	0.07	0.04	0.11
Hepatitis	0.17	0.14	0.31
Total	15.63	14.88	30.51

DALYs = disability adjusted life-years

3.8.2 Coliform Bacteria

The first causal epidemiological connection between human faeces and disease was made by Dr. John Snow in 1854 when he unrevealed the mystery of *London chlorea epidemic* by observing that the common link between all the victims in the Broad Street area was that they obtained their drinking water from the same well. Today we know that coliform bacteria make up something like one-fifth to one-third of the average person's wastes and billions of them are released into the sewage systems every day. They cause cholera, typhoid fever and gestro-enteritis. Chlorination of water is the only remedy but we should not forget that chlorination may pose health hazards of its own.

3.8.3 Nutrients

Nutrients like phosphorus, nitrogen, and carbon are essential to aquatic life. However, excess of these nutrients in a water body causes paradoxical kind of pollution known as *"eutrophication"* the term derived from Greek word meaning *"well nourished"*. Lake Erie in U.S.A. and Dal Lake in India are facing severe eutrophication problems.

3.8.4 Nitrates

Nitrates occur naturally in water at low levels. However, heavy application of artificial fertilizers most of which contain nitrogen in the form of urea may become a health problem. Nitrate levels above 10 mg/l in water can cause *methemoglobinemia* known as *"blue body"*. Nitrate levels as high as 40 mg/l have been found in some parts of Midwest.

3.8.5 Sediments

Sediments are the particles of sand, grit and other inorganic matter that flow into waterways from mining sites, construction sites, logging sites and urban run-off. Severe sedimentation can physically fill small ponds and lakes. They may cause turbidity. Mining activities are one major cause of sedimentation problems.

3.8.6 Acid Rain

The problem of acid rain first came to the attention of scientist in Northern Europe more than thirty years ago when lakes of southern Sweden and Norway were found losing their fish populations. In 70 percent of some 1500 lakes analyzed in Norway, pH values were below 4.3 and they contained no fish at all. Acid mists caused by coal-fired plants in great Britain across the North Sea were carried by prevailing atmospheric currents to Scandinavia. Soon, scientists from other countries reported increased acidity in their waterways. Nearly 200 lakes have been found dead in North America and Canada.

3.8.7 Mercury (Minamata disease)

In 1956, an epidemic of organic mercury poisoning broke out in Minamata, a small town of 50,000 residents located on the coast of Kyushu, Japan's southernmost island. It was caused by dumping of mercury wastes dumped into Minamata Bay by Chisso Corporation. Samples taken from the mud of Minamata Bay near Chisso's effluent outlet were found to be contaminated with over 2,000 ppm of mercury. Fish and shellfish samples also contained high methyl-mercury levels ranging from 10 to 40 ppm. Autopsies of the victims revealed levels as high as 70 ppm in liver and 144 ppm in kidney, and 24 ppm in brain which is acutely sensitive to methyl-mercury. Even residents of Minamata town with no outward sign of disorder had levels ranging from 100-150 ppm in hair samples, normal levels run about 8 to 90 ppm.

Until the early 1960s, it was widely assumed that elemental mercury emitted or dumped into the biosphere was not a hazard because it does not react easily with other substances, is quite stable, and above all, is only very slightly soluble in water. However, Swedish environmental scientists began to wonder that mercury dumped into the water was in an inorganic elemental from but the fish that inhabited these waters contained more toxic alkyl from, the methylmercury. By 1972, it had been repeatedly demonstrated that aquatic micro-organisms living in the bottom sediments of natural waterways had the power to transform the relatively insoluble form of elemental mercury into highly soluble and highly toxic form of methylmercury. Once it has been "biotransformed" in this way, methyhnercury was easily absorbed by the tissues of fish and from these it easily passed to man and other predators that consumed the contaminated fish.

Shortly after the 1956 outbreak, Chisso's own physician Dr. Hajime Hosokawa had quietly begun to test Chisso"s wastes by feeding them to laboratory cats and observing the effects. On Oct, 7, 1959 he found that cats convulsed, salivated and suddenly died while crashing into laboratory walls. However, Hosokawa did not publish his results and lull followed for almost ten years. In 1966, an identical case of mercury poisoning broke out in the town of Niigata in Central Japan. In this case Shown Denko

Fig. 3.1: Location of island of Kyushu (top), Minamata (center), and the Chisso factory and Minamata Bay (bottom).

Company was proved guilty of dumping toxic substances into the Agano River. The victims promptly sued the suspected source. Their success united anger in Minamata and suits were filed against Chisso. Finally Chisso stopped dumping mercury wastes into Minamata Bay in 1968. On March 20, 1973, a Japanese court decided the case in favour of plantiffs and ordered Chisso to pay each one of them including the heirs of those who had died from sixty thousand to eighty thousand dollars. As of the end of 1979, 1401 individuals had been certified as victims of Mianamat disease. Of this number 353 had died. Chisso had paid over $ 200 million in damages.

The afflicting symptoms of Minamata disease are madness, muscle spasms, convulsions that arch the body like a taut bow, paralysis, loss of speech and vision, and in many cases the crippling of arms and legs. In the final stages, the victims muscles waste away and they become comatose. They loose all emotional control and eventually pass from coma to death.

3.8.8 Mercury Pollution in India

In India, most of our rivers have become noxious sewers and veritable death traps of fish. Chaliyar river in Kerala has been reported to contain high level of mercury discharged by a nearby rayon factory. It has contaminated the fish that forms the popular diet of people living in surrounding villages.

Mercury poisoning is also evident in Orissa's Rushkulya river and in the Thane creek near Bombay.

The Kalu river which flows through Bombay's industrial suburbs of Ambernath and Ulhasnagar has been found to contain high concentrations of mercury.

Incidences of mercury poisoning have also been found in other countries like Iran and Sweden where mercury contaminated seeds were eaten by farmers and birds.

3.8.9 Detergents

The word detergent has been derived from the Latin term "detergene" meaning to cleanse. Substances like soaps that possess considerable cleansing property are known as detergents. Soaps differ greatly physically and chemically, but they have one property in common-they lower surface tension of liquids in which they are dissolved. That is why they are known as surface active agents or in short "surfactants".

The detergents enter into the water from various sources i.e. industry (textile, leather, dyeing and finishing), agriculture and refuse of cities and towns. The increased personal and domestic consumption of detergents is certainly one of the main causes of pollution of surface waters.

The detergents are responsible for nuisances like foam. They affect the oxidation of dissolved matter and transfer of oxygen. They affect the bacterial and algal flora and influence the permeability of soils.

In animals the lethal dose varies between 1 and 223 of anionic detergent per kg of their weight, depending on the species. In one hour, 20 ppm will cause the death of 65-day old trout, while 15 ppm will cause the death of a 15 cm young through in the same time. The amount of dissolved oxygen in the water is also very important. When the oxygen content is reduced, the affect of toxic substances is proportionately increased, and in polluted streams with a low oxygen content small concentrations of surface active agents can have a toxic effect.

In man, detergents may cause dermatitis. 131.1 percent of dermatitis cases in United States have been attributed to detergents. Owing to their peculiar physical properties the surface active agents are capable of penetrating into the skin either directly or through sweat glands. Their ability to denature proteins or reaction with keratin may trigger serious disturbances.

3.8.10 Pesticides

Pesticides are biologically active chemicals used for killing plants or animal pests. It is a general term embracing insecticides, herbicides, fungicides, nematocides etc. (Table 3.7). Sometimes a collective term *biocide* is used to include all these categories. Pesticides have been used in one form or another for more than 2,000 years, but the use became widespread only in the last century. The primary problem with pesticides is that they are intentionally toxic and if they are toxic to undesirable organisms, they may also be toxic to desirable one including ourselves. The danger from pesticides was brought to public attention in 1962 with the publication of SILENT SPRINGS, a book by Rachel Carson. During the last two decades the use of pesticides has increased 12 times (Freed et al. 1977) and millions of kilograms are produced every year.

Dichloro diphenyl trichloroethane (DDT), benzene hexachloride (BHC), chlordane, haptachlor, methoxychlor, toxaphene, aldrin, endrin and polychlorinated biphenyls (PCB) are the commonly used pesticides. Extensive and indiscriminate use of these biocides has made them an integral part of the biological, geological and chemical cycles of the earth. Today they are present in air, soil, water and each ecosystem (Table 3.8).

DDT is a classic case of a pesticide thought to be safe and affective. Its discoverer Paul Muller was awarded a Nobel Prize. After world war II, DDT was found to be an

Table 3.7 Classes of pesticides

Class	Purpose
Acaricide	Used against mites and ticks, members of the Acaridae
Alganecide	Used against algae
Antifouling agent	A paint additive to protect against organisms that grow on moist or wet surfaces including those underwater
Attractant	A chemical that attract insects or birds or other animal pests to a trap or a poison
Avicide	Kills or discourages birds
Chemosterilant	A chemical that will lower or stop reproduction by a pest
Defoliant	A chemical that induces plants to drop their leaves prior to harvest (sometimes used as a herbicide or silvicide)
Dessicant	A chemical that dries plants out before harvest
Fumigant	Used to kill soil pests and sometimes weeds
Fungicide	Used against fungi
Herbicide	A weedkiller, strictly speaking, but defoliants and silvicides are often called herbicides
Insecticide	Used against insects
Molluscide	Used to kill or control mollusks and other invertebrates (e.g. predatory snails, slugs, oyster drill, lamprey)
Nematicide	Used against nematodes, tiny round worms that feed on decaying matter, roots or other parts of plants
Ovicide	A chemical used against the eggs of insects, mites, and nematodes
Piscicide	Used to reduce the population of rough fish in a body of water
Plant regulator	A chemical that alters the normal pattern of growth and development of a plant; usually used for herbicidal purpose
Repellent	A chemical that repels insects or other animals
Rodenticide	Used against mice, rats, and other rodents
Silvicide	Used against bush, trees, and other woody plants and shrubs

amazing chemical. It controlled atleast 27 diseases, saved 5 million lives and prevented 100 million people from various illnesses. Although, it was recognized as early as 1948, that it is stored in body fat, no direct effect of DDT on man were noticed. Its widespread distribution in the biosphere was evident when it was detected in Antarctic penguins, in cow's milk and human milk. The drainage of DDT into rivers, swamps and coastal waters killed crabs, and fish. When these aquatic organisms were eaten by predators- their metabolism tended to favour the retention in body fat. Each step in the food chain resulted in an increased concentration of DDT. Finally the carnivorous birds contained sufficient DDT to cause toxic effects.

As the history of DDT illustrates, fat soluble pesticides can have subtle, hidden but important environmental effects. Because the compounds poorly soluble in water appear rare in the nonbiological environment, they are easily ignored. However, their biomagnification and negative effects can be serious though the compound may be present in traces.

After decades of discussion, DDT still remains controversial as far as its actual effects on human health are concerned. However, many countries like U.K., U.S.A., Sweden Germany and Denmark had banned the use of DDT and other organochlorine pesticides.

Table 3.8 Annual pesticide production

Class	Common Name	Production (millions of kg A.I.)
Insecticide	Methyl parathion	20
	Toxaphene	23
	Carbaryl	20
	Malathion	16
	Chlordane	11
	Parathion	7
	Methoxyclor	4.5
	Diazinon	4.5
	Carbofuran	3.6
	Disulfoton	3.6
	Phorate	3.6
Herbicide	Atrazine	41
	2,4-D	20
	MSMA-DSMA	16
	Sodium chlorate	14
	Trifluralin	11
	Propachlor	10
	Chloramben	9
	Alachlor	9
	CDAA	4.5
	2,4, 5-T	2.7
Fungicide	PCP and salts	21[b]
	TCP and salts	21[c]
	Captan	8
	PCNB	4
	Dodine	4

Source: From Freed et. al. 1977
Active ingredient.
Includes use as herbicide, desiccant, molluscicide, and for termite control.
Includes CDEC, Ditane M-45, Ditane S-31, Ferbam, Maneb, Metham, Nabam, Niacide, Polyram, Zineb, and Ziram.

According to World Health Organization (WHO), the level of DDT in human milk in India is the highest in the world. Monitoring of contamination of dairy milk within the country reveals that the residual level is the highest in Delhi, followed by Gujarat, Punjab, U.P. Haryana, Maharashtra, Bihar, Andhra Pradesh and Karnataka.

3.8.11 Handigod Syndrome

The indiscriminate use of pesticides in agriculture and public health operations has increased the scope of disruption of ecological balance as many of the "non target" organisms that are important links of food chain perish in the process thus adversely effecting the secondary and tertiary productivity of fresh water ecosystems. The tragic incident occurred in Karnataka. The incident has been reported in Cochin area in 1955 and Sitapur-Hardoi (U.P.) area in 1977.

Among various chlorinated insecticides viz. DDT, lindane, dieldrin, aldrin, endrin, endosulfan and chlordane examined for their toxicity in fish, endrin has been found to be highly toxic and endosulfan the least toxic. In catfish, endrin was reported to be 70,000 times more toxic than carbaryl.

3.8.12 Pollution Problems of Leather Industries

The leather industry occupies an important place in the National economy of every country. India occupies a predominant position in the world production of hides and skins. Tanneries in Tamil Nadu alone fetch the exchequer over Rs. 1,400 crore through exports but bring multifaceted environmental problems.

Tannery effluents contain vegetable tannins and non-tannins which cause chemical oxygen demand. They also contain high amounts of proteins. These proteins are biodegradable and exert high BOD where water from tanneries contain about 46% proteins, 10% fatty acids, 4% tannins, 20% inorganic volatile solids and about 21 % total dissolved volatile solids.

Chlorides, trivalent chromium, nitrogen, phosphorus, sulphate, ammonium salts, lime etc. are the inorganic pollutants present in significant quantities. For each kg salted hide 35 litres of water is required. The discharge of untreated waste waters in a nearby water body may effect the physical, chemical and biological characteristic of water. High oxygen demand, high pH, excessive alkalinity, suspended matter, sulphides are injurious to fish and other aquatic organisms. The chrome process involves 250 chemicals including cadmium, arsenic and chromium. Chromium in particular is used in large quantities.

Tanneries make the ground water unfit for drinking, and irrigation. Groundwater of a few districts vlz. Pallavaram, Chrampet, Ambut, Ranipet, Pernabet and Vaniyambadi in Tamil Nadu have been affected by tanneries. A 100 km stretch of river Polar in T.N. receives the effluents of about 300 tanneries. Therefore, subsurface water of river Polar have been badly polluted. Another river Vaniyambadi faces similar problems. A tannery waste can contaminate the groundwater in a radius of 8 kms. The effluent from vegetable tanning is coloured which makes the water turbid. Turbidity reduces light penetration, thereby reducing the photosynthetic activity of aquatic plants. Pathogenic micro-organisms such as, *Bacillus anthrasis* may increase in these waters and may be transmitted to human beings.

3.8.13 Dyes and Pigments

As many as 139 organic chemicals, heavy metals (Zn, Pb, Cr, Cu, Hg, Me, Ba) and their salts, acids and alkalies are used in the manufacture of dyes. Most of these compounds, besides small quantities of intermediate compounds along with final products are discharged into waste waters causing serious pollution problems. Total dye-stuff production in the organized sector alone was 187440 tonnes during 1982-83. Beside this a large number of small units are in operation in India. The estimated demand for dyes for 1983-84 was 36,100 tonnes. The use of dyes has enormously increased during the last decade. The main intermediate used are benzoic acid, nitrobenzene, chlorobenzene and aniline all ranked as highly toxic and suspected carcinogens.

Heavy metals such as mercury, lead, cadmiun and zinc are dangerous pollutants and are often deposited with natural sediment in water systems. Then they are incorporated in plants, food crops and animals. Heavy metal poisoning occurs when this water is used for agricultural and human consumption.

3.8.14 Organic Compounds

Artificial organic compounds are used primarily in industrial processes, pest control, pharmaceuticals, food additives and other consumer products. Many of them escape into the environment and spread widely throughout the biosphere and can be found in many species. Their presence in non-target marine organisms far from the sources of their production confirms their presence in water bodies. The dissolved organic matter (DOM), suspended organic matter (SOM) and bottom organic matter (BOM) cause a steep rise in BOD, killing off zooplankton and fish. SOM is more hazardous to the fish as the organic matter gathers round each silt particle and this silt organic mass clogs the gills of fish causing asphyxiation. The water polluted by organic compounds is unfit for human consumption.

3.9 Groundwater Pollution

We have long believed that ground water in general is quite pure and safe to drink. Therefore, it may be alarming for some people to learn that groundwater infact may easily be polluted by any one of the following sources (Table 3.9). Problem of groundwater pollution can be understood by two examples i.e. Love Canal episode (1976-77) and arsenic poisoning in West Bengal (India).

Table 3.9 Classification of sources of groundwater pollution and/or contamination.

Wastes		Non-wastes
Sources designed to discharge to the land and/or Groundwater	Sources that may discharge waste to the land and ground water unintentionally	Sources that may discharge a contaminant (not a waste) to the land and groundwater
Spray irrigation	Surface impoundments	Buried product storage tanks and pipelines
Septic systems, cesspools etc	Landfills	Accidental spills
Land disposal of sludge	Animal feed lots	Highly deicing salt stockpiles
Infiltration or percolation basins	Acid water from mines	Ore stockpiles
Disposal wells	Mine spoil piles and tailings	Application of highway salt
Brine injection wells	Waste disposal sites for hazardous chemicals	Product storage ponds Agricultural activities

3.9.1 Love Canal Episode

One of best examples of ground water pollution is the Love Canal near Niagara Falls, New York where burial of chemical wastes caused serious water pollution and health problems. From 1920s to the 1950s, more than 80 different chemicals were dumped there. In 1953 the company donated the land to the city of Niagara falls. Eventually township developed. Years later in 1975-76, heavy rainfall and heavy snowfall set off the events that led to Love Canal catastrophe. Trees and garden began to die. Bicycle tires and rubber shoes disintegrated. The old dumpsite contained a number of substances that were suspected to be carcinogens including benzene, dioxin, dichloroethylene and chloroform. Residents of the area had higher miscarriages, birth defects, liver and blood abnormalities and chromosome damage. More than $ 100 million were spent to clean up the Love Canal site and rehabilitate the residents. In addition, approximately $ 3 million/year may be needed to monitor the area for future.

3.9.2 TCE in Ground Water

Synthetic organic wastes like trichloroethylene (TCE) are also known to aggravate the groundwater pollution problems. In late 1979 a routine water analysis in Southern California showed that water of wells from San Gabrid Valley contained poisonous halogenated hydrocarbon TCE, ranging as high as 770 mg/liter. No one is sure how these wells became contaminated with TCE. However, local officials surmise that some fifteen to thirty years ago, industries might have disposed of TCE from degreasing processes by simply dumping it on the ground. The bed of the prehistoric river that formed the valley composed of an alluvial gravel allowed TCE to escape into the soil and then to water. TCE causes skin and mucous membrane irritation, toxic hepatitis and kidney damage. Chronic skin contact with water containing TCE can cause cancer. National Academy of Sciences USA estimated that a population drinking a typical two liters of water per day for an average span of 70-80 years will experience one additional case of cancer per million people. If the water they drink is contaminated with as little as 4.5 µg/liter of TCE. Environmental Protection Agency (EPA) has found the concentration of 2.7 µg/liter to be safe.

3.9.3 Arsenic Poisoning in West Bengal (India)

In a survey made from 1987 to 1996 (Tab le 3.10), seven districts of west Bengal-Malda, Murshidabad, Bardhaman, Nadia, Hooghli, 24-Parganas (North) and 24-Parganas (South) - arsenic has been found in groundwater above maximum permissible limit (0.05 mg/l) recommended by WHO (Mandal et at. 1996). According to the estimates of School of Environmental Studies, Jadavpur University, more than 1.0 million people are drinking arsenic contaminated water and more than 2,00,000 people have arsenical skin lesions.

Symptoms of arsenical toxicity develop insidiously after six months to two years or more depending upon the amount of intake of arsenic laden ground-water and arsenic concentration in the water sample. Darkening of skin (diffuse melanosis) in the body or in palm is the earliest symptom. Spotted pigmentation (spotted melanosis) is usually observed on chest, back or limbs. Leucomelanosis as white and black spots on side by

Table 3.10 Survey report of arsenic-affected seven districts, upto January 1996

Total area	37493 km
Total population	34,632,0.24
Total no. of blocks/PS	162
Total no. of arsenic-affected blocks/PS	50
Total population of arsenic-affected blocks	9,562,898
No. of arsenic-affected villages/wards	560
Approx. population drinking arsenic contaminated water above 0.05 mg/l	>1,100,000
Approx. population drinking contaminated water above recommended	>1,500,000
Approx. population showing arsenic related skin Manifestation on their body	>220,000

side also develops in many patients. Diffuse with nodular keratosis on palm and sole is another sign of moderately severe toxicity. Rough and dry skin often with palpable nodules (spotted keratosis) on dorsal surface of hands, feet and legs are the symptoms of severe arsenic poisoning. Complications like liver enlargement (hepatomegaly), spleen enlargement (splenomegaly) and fluid in abdomen (ascitis) are seen in several cases. Squamous cell carcinoma, basal cell carcinoma, carcinoma of lung, uterus, bladder, genitourinary tract have also been observed in advanced neglected cases suffering for more than ten years.

A simple calculation of water withdrawl and the amount of arsenic in that water from a single Rural Water Supply Scheme (RWSS), supplying water to few villages in Jothgopal, Malda indicates that 147.825 kg of arsenic came out in one year,

Not only India groundwater arsenic contamination has been reported from Taiwan, Chile, Argentina, Mexico and Thailand (Table 3.11). However, India faced the biggest arsenic calamity in the world.

Table 3.11 Some major groundwater arsenic contamination incidents all over the world.

Location	Year	No. of people exposed	People showing arsenical skin
West Bengal, India	1978 to Jan. 1996	>1,000,000	20%
Taiwan	1961-1985	103,154	13%
Antofagasta, Chile	1958-1970	130,000	16%
Monte Quemado, Argentina	1938-1981	10,000	Many
Region Langunera, Mexico	1963-1983	200,0000	21%
Ronpiboon, Thailand	1987-1988	-	18%

Possible source of arsenic contamination has been found to be geological. Arsenic rich sediments are present in layers. Arsenic rich sediments have been characterized by using electron probe micro analyzer (EPMA), laser microprobe mass analyzer (LAMMA) and X-ray diffraction. All these techniques indicate the existence of arsenic in pyrite (FeS_2). Why and how pyrite decomposed. Kinniburghs and associates (1994) have mentioned that high ground-water withdrawl (FeS_2) causes its oxidation. Acid released during oxidation of pyrite reaches with minerals leading to high concentrations of cations in the pure water. Further while explaining the presence of arsenic in groundwater of Western United States, Welch et al. (1988) reported that mobilization of arsenic in sedimentary aquifers may be caused by changes in geochemical environment brought by excessive withdraw of groundwater for irrigation. The world should learn a lesson from this arsenic problem

3.9.4 Water (Prevention and Control of Pollution) Act, 1974

In India, the problem of water pollution was officially recognized in early sixties. The Ministry of Health appointed an Expert Committee in October, 1962 to study the question and also to prepare a draft legislation to deal with water pollution resulting from domestic and industrial wastes. The Committee, after a comprehensive study of all aspects of pollution, recommended that Central as well as State legislations on the subject may be enacted. The Central Council of Local Self Government which considered the recommendations of the Expert Committee on 7th September, 1963 resolved to recommend the enactment of a single law by the Parliament, so that there may be uniform measures throughout the country to control water pollution. The Government of India considered the recommendations of the expert Committee and the resolution of the Central Council and decided to have a central legislation on water pollution. Accordingly, a draft bill was circulated to all the State Governments in December, 1965 with the request to pass resolutions authorizing the Parliament to enact the law on their behalf as required in Article 252(1) of the Constitution.

In December, 1969, the Central Government introduced the "Water (prevention and I Control of Pollution) Bill, 1969 in the Rajya Sabha after 6 States had passed enabling resolutions authorizing the Parliament to enact on their behalf. In August, 1970, the Rajya Sabha decided to refer the Bill to a joint Committee of both the Houses. The joint Committee, after comprehensive examination, modified the Bill in several respects and presented its report along with the modified Bill to the Parliament on November, 13, 1972. The Parliament passed the Bill in early 1974, which received the assent of the President on March 13, 1974. By then, 6 more States had passed enabling resolutions. The Act came into force from this date in all the 12 States, viz: Assam, Bihar, Gujarat, Harayana Himachal Pradesh, Jammu & Kashmir, Karnataka, Madhya Pradesh, Kerala, Rajasthan, Tripura, and West Bengal and all Union Territories. With the intention of extending the Water (Prevention and Control of Pollution) Act, 1974 to the whole country certain provisions are made in the Act. Under Section 1(3) of the Act any State can adopt the Act and it will come into force in that State from the date of adoption.

3.9.5 Objectives of the Act

The objective of the Act is to prevent and control water pollution and also to maintain and restore the wholesomeness of water. Prevention refers to the new sources of pollution while control refers to the existing source of pollution. It may be appreciated that an existing industry cannot be asked all of a sudden to stop its discharge of effluents into a water course, which the industry might have been doing for years, without seriously dislocating the industrial activity. Such a drastic action would otherwise affect industrial production and would also create several other social problems, including unemployment. Therefore, in respect of the existing source of pollution, the remedy lies in a gradual control of pollution without causing any serious dislocation to the industries. It is with this view a special provision has been made in Section 26 of the Act, that the existing industries discharging effluents in the water course, should apply for consent within three months of the constitution of the respective State Boards.

In respect of new industries, standards can be laid down and enforced strictly so that the arrangements are made, well in advance, for the treatment of the effluents to the required standards before they are let out into the water course. In the case of new industries the concept of "prevention", thus, can very well be applied. Section 25(1) of the Act lays down that no person shall, without the previous consent of the State Board, bring into use any new or altered outlet for discharge of sewage or trade effluent into a stream or well, etc.

SUGGESTED FURTHER READINGS

Moss, B., 1988. Ecology of fresh waters. Man and medium (2nd edn.) Oxford: Blackwell.

OECD, 1986. Water pollution by fertilizers and pesticides. Paris: Organization for economic Co-operation and development.

Speidel, D., Ruedisili, L. and Agnew, A. 1988. Perspectives on water uses and abuses. Oxford University Press.

STUDY QUESTIONS

1. Define water pollution. Classify different types of water pollution with examples.
2. Enumerate toxic water pollutant and discuss their health effects in man.
3. Write short notes on the following-
   a. Mina-mata disease
   b. Eutrophication
   c. Water pollution by pesticides
   d. Water pollution by leather industries
   e. Ground water pollution
   f. Water (prevention & control of pollution) Act 1974.

Chapter 4

MARINE POLLUTION

Water, the most abundant natural resource *(q. v.)*, is mostly available in nature in the form of sea and ocean water. This water has a specific composition but by the discharge of wastewater, by dumping of different pollutants or by other means the natural composition alters, giving rise to the phenomenon known as 'marine pollution. Although description of marine pollution is simple, no definition is commonly accepted by everybody. Ecologists, for example, have a tendency to define marine pollution as 'any activity that causes a change in the marine ecology and disappearance or reduction of any aquatic species.' On the other hand, according to engineers, marine pollution is a change in the quality of the sea water that prevents its use for the benefit of human beings. The UN Food and Agriculture Organization defines marine pollution as the discharge of contaminants into sea water, which changes its quality, threatens public health, constitutes danger for aquatic life, and limits fishing activities.

As can be seen, the definition of marine pollution is subjective. Because of that, marine pollution is described as failure to satisfy predetermined standards. These standards vary according to the intended use of the water body.

4.1 Sources of Marine Pollution

The sources of marine pollution can be classified in two main groups: *point sources* and *non-point* or *diffused* sources. At the point sources, wastewater is flowing into the marine environment through a definite point. Control of wastes generated from a point source is relatively simple because their collection is easy. Non-point sources are more troublesome. As these sources are diffused, it is very difficult to disperse or treat them. They flow freely as surface or underground runoff to the marine environment.

4.1.1 Point Sources

There are many different point sources of marine pollution. Among them the ones discussed briefly below can be considered as the major examples:

Domestic areas. Mainly in developing and less developed parts of the world, many cities discharge the wastes which are generated from residential areas directly into the sea. The modern tendency to consider septic tanks as out of date and primitive has resulted in a considerable increase in the amount of domes-tic wastewater discharged in the marine environment. These wastes are mainly rich in organic materials and have unpleasant effects on the receiving body such as microbial pollution, increase in the concentration of nitrogen, phosphorus and sometimes solid materials, heavy metals and toxic elements, and depletion of oxygen levels. Rapid increases in population and growth of urban areas have also contributed to the increase of the amount of domestic wastewater generated.

Industries. Many industries discharge most of their wastes directly or indirectly through rivers into the marine environment These industrial wastes are generally toxic and cause the depletion of the oxygen level and increase of the concentration of suspended solids, oils, heavy metals and so on. Large power plants discharging their cooling waters are also hazardous to aquatic life (see *Thermal Pollution* page 51).

Stormwater. Rain water collected by sewers and carried into the marine environment constitutes a point source. These waters are significant sources of pollution because they carry almost all kinds of impurities which can be found on the surface of the Earth, such as solid wastes, leaves, soil, and even sometimes lead generated from the exhaust gases of vehicles.

4.1.2 Non-point Sources

The main sources of these kinds of wastes are as follows.

Urban areas. Pollutants generated from urban areas may occur in liquid or solid form. Rain water not collected by sewers, leachate generated by open dumps or landfills, the contents of septic tanks which overflow accidentally, and oils are examples of the liquid wastes. On the other hand, particulate matter, such as dust, generated by air pollution and precipitating on the marine environment, constitutes an example of solid pollu-tants which may be discharged into the marine environment. The wastes generated from the sources mentioned above con-tain all kinds of pollutants, such as toxic materials, heavy metals, bacteria and nutrients.

Agricultural areas. Wastewater originating from agricultural areas may contain excess amounts of nutrients such as nitrogen and phosphorus generated from natural and synthetic fertiliz-ers. The concentration of bacteria, suspended solids and pesti-cides in the marine environment is also increased by the discharge of this wastewater. Also runoff from the areas con-taminated by livestock and poultry wastes, particularly from the feedlots, may contribute to marine pollution.

Mines. Wastewater generated from mining activities can be rich in toxic metals, such as mercury and cadmium, which may be harmful to aquatic life.

Forests. Mainly solid materials, such as leaves carried to the marine environment by storms, cause an increase in the concen-tration of solid materials of the marine water.

Ships and other vehicles. Commercial passenger and transport ships, private boats, and yachts many times discharge their wastes (sewage, bilge water, solid waste, litter, etc.) into the marine environment, thus contributing to its pollution.

Petroleum wells, tanker accidents. The cooling water of refiner-ies, seepage of petroleum from ships, boats, and petroleum wells, research conducted in the seas and accidents such as *Torrey Canyon* and *Exxon Valdez* disasters are the main sources of petroleum pollution of the marine environment. Because of its toxicity, petroleum pollution is hazardous to marine aquatic life.

4.2 Types of Marine Pollution

Marine pollution can be classified in the following groups:

Microbial pollution. The presence of pathogenic (diseasecaus-ing) microorganisms in the marine environment is the source of microbial pollution. This type of pollution is a real threat to public health due to the fact that it may cause the spread of waterborne diseases, and the marine environment is consid-ered a hazard to public health until these microorganisms disappear. The origin of these microorganisms are wastes (mostly sewage) discharged into the marine environment. Since it is very difficult to determine the types of all microorganisms one by one that cause microbiological pollution, a few species of microorganisms of enteric origin are used as indicators of fecal contamination. Coliform bacteria, *Streptococcus fecalis* and *Salmonella* are more resistant to environmental conditions than other pathogenic microorganisms, more numerous than associated enteric pathogens and can be detected and estimated numerically by relatively simple bacteriological procedures.

Death of bacteria. The number of pathogenic microorganisms decreases in the marine environment because they cannot adapt to the new environmental conditions. The increase in the temperature of seawater, radiation from the sun, and the salinity of seawater accelerate the dieoff rate of pathogenic microorganisms, while the turbidity of the seawater decreases their dieoff rate. The dieoff rate of microorganisms is expressed by *T90*, a variable that expresses the time required for the removal of 90 per cent of the microorganisms.

Organic pollution. The increase in the concentration of organic material in the receiving media causes organic pollution. These organic materials serve as food for indigenous microorganisms that carry out oxidation processes in which organic materials are decomposed and carbon dioxide and water are produced. During that process the available oxygen in the marine environment is depleted and, if there is an excess amount of organic materials, anaerobic conditions which cause odor and nuisance occur. Then anaerobic bacteria continue to decompose the organic material, producing carbon dioxide, methane, hydrogen sulfide gases and some stable compounds. The excess amount of organic material not only disturbs the marine ecology, it also causes noxious odors. Decrease in the dissolved oxygen of the marine environment has detrimental effects on the fauna. When the dissolved oxygen value falls below 4 mg/l certain fish species disappear.

The concentration of organic materials is ascertained by the test for biochemical oxygen demand *(q. v.)* which is the amount of oxygen needed to decompose and stabilize the organic materials biologically under aerobic conditions at 20°C. Other methods in the determination of the organic pollution level are the chemical oxygen demand (COD) and total organic carbon (TOC) tests.

Inorganic pollution. Industrial wastewater discharged into the marine environment may contain numerous metallic salts and toxic, corrosive, colored and tasteproducing materials. Iron, manganese, chlorides, heavy metals, nitrogen and phosphorus are also among the pollutants. The effects of each of these pollutants on the marine environment are different. They are toxic to aquatic life. They are diluted in the receiving media, but although they are not appreciably changed in total quantity they accumulate on sediment. Self purification mechanisms for this type of pollution are dilution and sedimentation.

Oil pollution. Although oil pollution in the marine environment is caused by many sources, it attracts attention when oil (petroleum and petroleum products) is discharged into the marine environment after a tanker accident, because in such cases a huge volume of oil is discharged into a relatively small area at a very short time. Since the specific gravity of oil is 10 per cent less than seawater, it may be imagined that oil would stay on the surface of seawater, but this is not so. The volume of discharged oil is decreased over time because of the evaporation of the components which are volatilized at low temperatures, while the remaining part is emulsified and decomposed by microorganisms by photooxidation and oxidation processes. Thus the volume of the discharged oil is decreased by 85 per cent after a few months, and the remaining tarlike material either settles on the bottom of the sea or reaches coastal areas. The petroleum and petroleum products are toxic and have anesthetic, narcotic and carcinogenic effects. The oil layer on the surface of the seawater prevents transfer of oxygen and the penetration of the sun's rays, thus hindering photosynthesis. As a result, oxygen concentration, which is essential for aquatic life, is depleted. This also causes the death of diving birds. The remaining part, tar, settles on the bottom, changes the benthos composition and affects benthic life adversely.

Thermal pollution. This is a change in the temperature of the marine environment, which causes disturbance to marine ecology. Power plants and the cooling water of industries are the main sources. As a result of thermal pollution (increase in the temperature of seawater) the rate of increase of plankton and benthic life is accelerated, causing a change in aquatic populations. The change of temperature negatively affects fish which are accustomed to live at a certain temperature range. With an increase in seawater temperature, the oxygen saturation level of the water is also decreased. Thus the solubility of oxygen in the seawater is decreased. Also, the metabolic activity of bacteria and aquatic life increases, the oxygen concentration of seawater decreases and then anaerobic conditions are rapidly reached. At high temperature the settling rate of solid particles is also increased and thus benthic composition is changed.

Several comprehensive texts cover the topic of marine pollu-tion, including Clark (1989), Goldberg (1976), and Gorman (1993). Specific aspects of coastal pollution by garbage have been dealt with by the US National Research Council (1995), while Bishop (1983) and Williams (1979) have reviewed techniques of marine pollution control. Lastly, Tippie and Kester (1982) discussed the impact of marine pollution on socioeco-nomic systems.

SUGGESTED FURTHER READINGS

Bishop, P.L., 1983. *Marine Pollution and Its Control.* New York: McGraw-Hill.
Clark, R.B., 1989. *Marine Pollution* (2nd edn). Oxford: Clarendon Press.
Goldberg, E.D., 1976. *The Health of the Oceans.* Paris: Unesco Press.

STUDY QUESTIONS

1. What are major sources of marine pollution?
2. Enumerate different types of marine pollution.

3. Write short notes on the following –
    a.   Thermal pollution
    b.   Oil pollution
    c.   Microbial pollution

Chapter 5

NOISE POLLUTION

Noise is unwanted sound. The world has been derived from a Latin word "nausea" meaning noise. The sound is a form of energy. The environmental effects of noise depend not only on total energy but on the sound's pitch or frequency and its time pattern. The apparent noise that is perceived by human ear depends on both the frequency and intensity of sound. The intensity of the sound is measured in decibels (d = deci meaning 10, B = Bell after the name of scientist Alfred Grahm Bell). Technically, a decibel (dB) is the amount of sound pressure that equals to 0.0002 microbars (energy of about 10^{-15} Watts). Since loudness is measured in logarithmic scale dB shall be --

$$dB = 10 \times \log_{10} \frac{I}{I_0}$$

Where I is the measured sound intensity and I_0 is the softest audible sound intensity. One decibel is the faintest sound intensity that can be perceived by human ear, and it can tolerate upto 180 dB. The frequency is defined as the number of vibrations. It is denoted by Hertz (hz). One hertz is equal to 1 vibration second^{-1}. People can hear sound from 16 to 20,000 hz. The range of vibrations below 16 is infra-audible and those above 20,000 hz are ultrasonic. Many animals can hear sounds that are inaudible to human ear. The loudness can also be expressed in sones. One sone is equal to 40 dB at a pressure of 1,000 hz. Sometimes the sound is expressed in psychoacoustic terms known as phon. The relationship between sone and phon can be expressed by the following formula.

$$\log 10\, S = 0.03\,(P - 40)$$

Where "S" denotes the loudness in sones and 'P' denotes the corrected sound level in phons. The sound intensity is measured by instruments such a Lam barometer or sonometer.

5.1 Sources of Noise Pollution

Noise may be either natural such as wind, thunder and movement of water or man made. If all symbols of civilization - from jet planes, vehicles, railway engines to factories generators, construction, machinery, television and radio sets and public address systems have something common - it is the noise.

65

5.2 Industrial Noise Pollution

The most offending noise sources are compressors, generators, furnaces, looms, grinding mills, releasing valves etc. that are used in many industrial processes- and are installed in partially closed or open sheds. 80-120 dB noise level is common to most units which is hazardous.

Table 5.1 Noise level in some important industries

Industry	Noise level in work area (dB(A))
Ceramics glass	90-100
Glass	82-95
Food processing	80-90
Chemicals	85-96
Papers	88-96
Foundry	90-100
Fabrication	82-95
Machinery	85-96
Leather	85-95

Industrial noise pollution in India has been studied by a number of workers. The results are summarized in the following table

Table 5.2 Noise levels is some Indian industries

Industry	Noise levels (dB(A))	References
Naval ships (Control room)	100-200	Ahluwalia and Kesar, 1978
Naval ships (engine room)	93-94	-do--
Aircraft repair shop	100-110	-do--
Heavy vehicle factory	110-115	-do--
Glass blowing shop	70-108	Ambasankaran et. al., 1978
Ammunition factory	85-101	Chandha and Singh, 1971
Saw mills	90-112	Lal,1984
Brass band units	90-112	Lal,1984
Silver foil manufacture	90-112	Lal,1984
Vanaspati (filling section)	76-87	Pant, 1972
Auto vehicles (engine testing)		
Hindustan Petroleum Corporation	74-98	Pradhan, 1985
Synthetic power industry	90-116	Singh, 1973
Power plant	90-100	Singh, 1975
Sugar industry	81-104	Singh, et. al., 1986
Light engineering unit	78-82	Ali and Tiwari, 1989A
Plastic industry	87-94	-do -
Leather industry	75-80	-do-
Heavy engineering	84-93	-do-
Fabrication unit	75-80	-do-

(Source: Adopted from Tiwari, 1989)

5.3 Domestic Noise Pollution

Transistors, radio, T.V., other musical instruments, air-conditioners, washing machines, kitchen appliances are common sources of noise at home. They not only do affect the users but the neighbours too.

5.4 Traffic Noise

Continuous movement of vehicles causes traffic noise. It affects not only those who are moving but those too who live near the roads, railway lines, stations and airports. There had been a gradual increase in the traffic noise in recent years due to increased density of vehicles. According to some estimates (Singh, 1984), traffic noise level in Delhi is 90 dB and in Bombay it is 95 dB. Average noise level in Delhi, Bombay and Calcutta is about 95 dB.

Near the air ports, noise levels between 85 dB and 95 dB were recorded with an increase of 20-25 dB during landing and takeoffs.

Near railway tracks too, ambient noise level increases upto 10-20 dB during train movement. Relative noise levels of different vehicles are given below-

Table 5.3 Relative noise levels of different vehicles

Type of vehicle	Noise level (in dB)
Luxury limousine	77
Small passenger car	79
Miniature passenger car	84
Sports car	91
Motor scooter (I-cylinder 2-stroke)	80
Motor-cycle (2-cylinder 4-stoke)	94

5.5 Other Sources of Noise Pollution

In metropolitan cities like Bombay, Calcutta, Delhi and Madras, the crowded markets have become great source of noise pollution. Theatres, dance halls, circus carnivals, religious activities, fairs, and festivals are other sources of noise pollution. Aircraft noise pollution is a matter of serious concern specially in areas close to international air ports. Satellite launching operations and atomic explosions emit lot of noise.

During last 100 years, there has been substantial increase in noise from man made sources. According to a survey made by EEC, 48% of population suffers from traffic noise, 30% faces industrial noise and 2% population bears aircraft noise. Noise levels are known to double in every ten years.

5.6 Effects of Noise Pollution in Man

Ambient Air Quality Standards, as prescribed in Environment Protection Third Amendment Rules (1989), noise level should lie between 50 to 75 dB in day time and 40 to 70 dB in night time.

Table 5.4 Ambient air quality standards

Area	Limits in dB	
	Day time	Night time
Industrial area	75	70
Commercial area	65	55
Residential area	55	45
Silent zone	50	40

However, in many industries around the world workers have to endure sound levels as high as 110 to 115 dB. Sound is uncomfortable beyond 100 dB. Noise levels generally exceed 85 dB in metropolices allover the world. The effects of noise in man are described below.

5.6.1 Auditory Effects

The first organ to be effected by noise is the ear. More than 40% cobblers, fruitsellers, shopkeepers and drivers complain of tinnitus in the ears. Continuous exposure may lead to deafness/permanent loss of hearing. It is due to complete destruction of the organ of corti that transmits sound from ear to brain. WHO statistics suggest that around 5% of the school children suffer from varying degree of deafness. According to US Public Health Service, more than 7 million people are working where noise levels are high enough to damage hearing. Noise is a major health hazard in industry.

Hearing loss due to noise exposure is a complex func-tion of noise parameters and individual susceptibility. When noise does cause damage to the ear, the effects are pervasive and can involve damage to all the major cell systems of the cochlea. Figure 5.1 shows damaged cochlear tissue after noise exposure. Perhaps the most sensitive structures are the stereocilia, which are the transduction links between the sensory hair cells and the tectorial membrane. One of the first pathologies seen is disruption of the mechanical cross-links between individual stereocilia on a given hair cell. Stereocilia can also develop "blebs" or growths and can fuse to form "giant stereocilia."

The outer hair cells (the motile sensory cells responsible for the exquisite sensitivity and frequency selectivity of the cochlea) appear to be more vulnerable to noise damage than the inner hair cells (the sensory cells that are responsible for conveying electrochemical information to the brain via auditory nerve fibers). In most of the cochlea's examined after acute exposures in experi-mental animals or after a lifetime of exposures in humans, the loss of OHCs is usually pervasive. When inner hair cells are missing there is usually secondary degeneration of the afferent fibers that innervated them.

The relation between cochlear blood flow and hear-ing loss from noise is not well understood, but there is relatively strong evidence that significant hearing losses are usually accompanied by capillary loss in stria vascularis and the spiral ligament.

Gunfire and other short-duration, high-level impulse or impact noises constitute a special hazard to the ear. The high sound pressures associated with impulse noise cause the basilar membrane to be displaced a relatively large distance in a short time, thereby developing large inertial forces that literally rip the cochlea apart. In addition, exposure to high level impulse noise can lead to ripping of tight cell junctions and ejection of sensory cells from the cuticular plate.

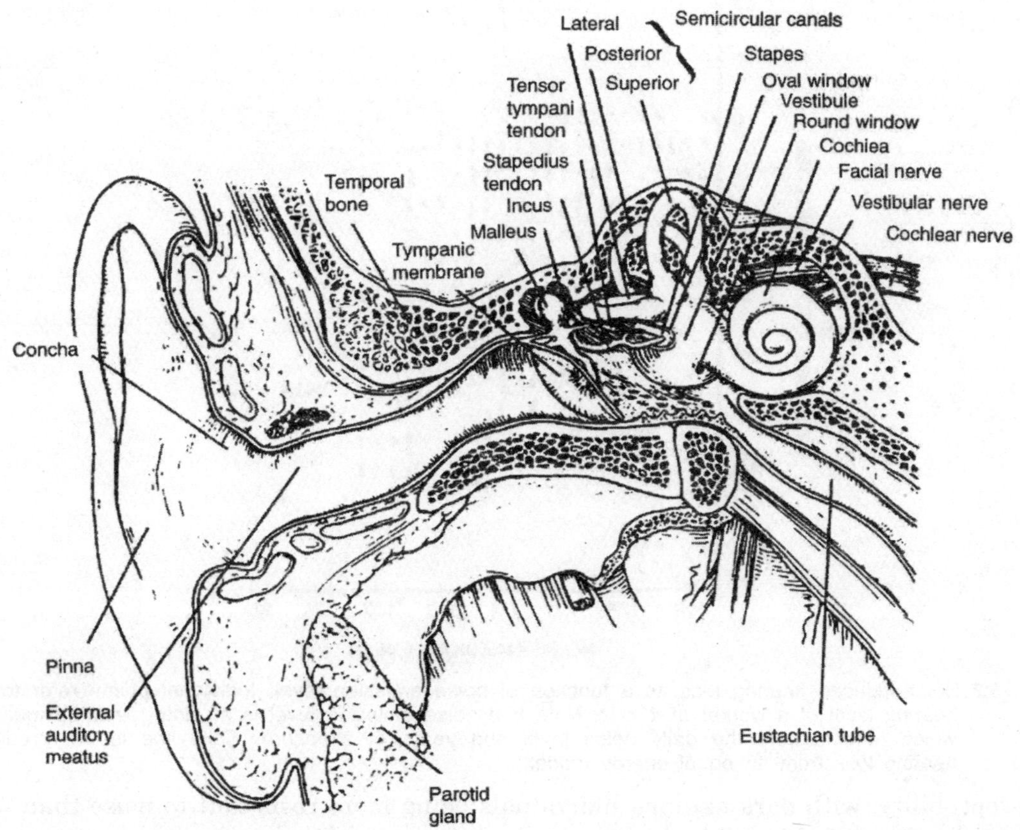

Fig. 5.1 A cross section of the human temporal bone showing the major components of the External, middle, and inner ear.

5.6.1.1 Variability and Susceptibility

Individual variability in susceptibility to noise induced hearing loss (NIHL) can be impressively large. In a particularly well-controlled epi-demiological study of hearing loss in a foundry and forging plant, workers with ostensibly the same exposure to noise developed markedly different hearing losses. Figure 5.2 shows that individuals who received the same overall "noise dose" (daily noise power X years of exposure) ranged from having absolutely normal hearing to having 60 dB of hearing loss. Despite careful attention to experimental details, ultimately it is difficult to control for all relevant factors that influence the amount of hearing loss a given individual will develop. Surprisingly, even when subjects are carefully screened and their noise history is known, there is still large intersubject variability in. The large degree of intersubject variability in the amount of hearing loss caused by exposure to noise highlights the need for developing predictive measures of susceptibility to noise.

A number of factors that correlate with susceptibility to NIHL have been identified. For instance, an individual's concentration of melanin can be related to

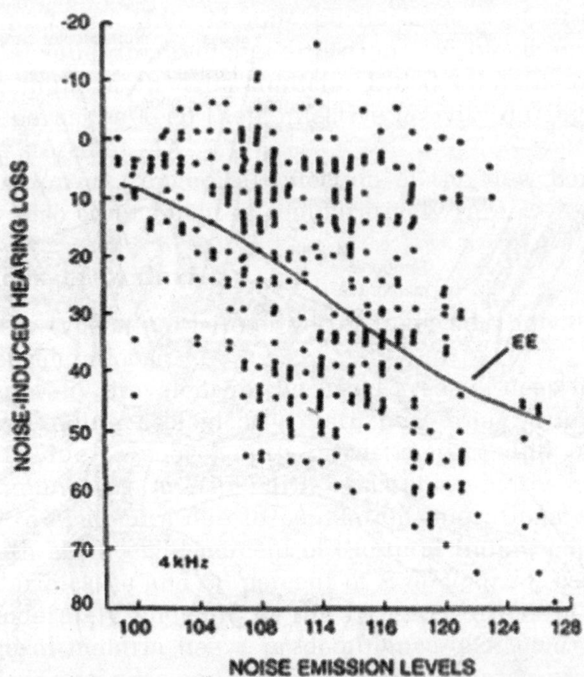

Fig. 5.2 Noise-induced hearing loss as a function of noise emission level. Individual points refer to the hearing level of a worker at 4 kHz. Note that noise emission level is an energy-related measure which reflects both the daily noise level and years of exposure. Solid line is the predicted hearing loss from an equal energy model.

suscepti-bility, with dark-skinned individuals being more resis-tant to noise than fair-skinned or blue-eyed individuals.

Research has suggested that melanin plays a role in counteracting the toxic reactions produced by high level noise exposure. Another factor related to susceptibility is acoustic reflex contraction of the middle ear muscles. Individuals with efficient acoustic reflexes (fast acting, strong contracting, and resistant to adaptation) develop substantially less NIHL than individuals with less effective reflexes. Although factors such as these have been shown to be correlated with susceptibility to NIHL in large-scale studies, there are still no practical means of assessing an individual's susceptibility to NIHL prior to noise exposure.

The high degree of variability seen in industrial survey research may be partially attributable to interactions between continuous noise and other agents, including exposure to impulse/impact noise, ototoxic drugs, and organic solvents.

5.6.2 Non-auditory Effects

Noise promotes the development of several non-auditory health effects including speech disturbances, sleep problem, insomnia, annoyance, hypertension and physiological disorders. These are neurosis, anxiety, insomnia, hypertension, sweating, giddiness, nausea and fatigue. Cardiovascular system is specially vulnerable to high levels of noise. Chronic noise may lead to abortions and congenital defects.

Reactions to noise pollution could be listed almost endlessly when one hears a loud noise, the blood vessels constrict, the skin pales, the pupils dilate, the eyes close, one winces, holds the breath and the voluntary and involuntary muscles become tense. Gastric secretion diminishes and blood pressure increases. Adrenaline is suddenly injected into the bloodstream which increases tension, nervousness, irritability and anxiety. Prolonged exposure to such noise can lead to severe mental disorientation and in some cases the violence. According to *Nobel Laureate Robert Koch noise is the worst enemy of health.*

5.7 Effects of Noise Pollution on Wild Life

Wild life too has been affected by noise. Health status of several Zoo-animals viz: deers, lions, rhinos are known to be affected by noise. They appear dull and inactive. Several migratory birds have stopped resting in a habitat close to noisy cities. Grizzly bears, muskoxen and kangaroo rats in Malaysia have been affected by noise. They leave habitat and move to calm places. This change in habitat alters their food habits, health and mating behaviour.

SUGGESTED FURTHER READINGS

Henderson, D., and Hamernik, R.P. (1995). Biologic bases of noise-induced hearing loss. In *Occupational Medicine: State of the Art Reviews,* Vol. 10. Hanley & Belfus, Philadelphia.

STUDY QUESTIONS

1. Define noise pollution. Describe different source of noise pollution.
2. Give a detailed account of the effects of noise pollution on human health.

Chapter 6

RADIOACTIVE POLLUTION

Low levels of radiation occur naturally from several sources. Natural radiation can come from earth materials, originate in outer space or result from the action of radiation on materials in the biosphere. Some radioactive elements occur as natural minerals in the Earth's crust. Uranium which is mined for use in atomic weapons and in nuclear power plants is one such element. Potassium 40 and radioactive carbon are among the more common radioisotopes. An important characteristic of radio-isotope is its *half-life*. It is the time required for one half of a given amount of the isotope to decay to another form. Every radio-isotope has unique characteristic half-life. For example radioactive iodine has a short half life of 8 days, Radioactive carbon has an intermediate half life of 5570 years. Uranium 235 has a half life of 700 million years.

A radioisotope is a form of a chemical element that spontaneously undergoes radioactive decay. It changes from one isotope to another and during this process emits one or more forms of radiation. Some radioisotopes undergo a series of decay finally reaching a stable nonradioactive isotope.

There are four major kinds of radiations: alpha and beta rays, are both particulate emissions. Gamma rays having no mass contain pure energy. The fourth type of radioactive emissions are neutron rays- produced in atomic bombs and nuclear reactors. They are all harmful because they can "ionize" living tissue that alters its fundamental atomic structure. Ionizing radiations differ in their power of penetration and ionization. Alpha radiation is highly ionizing but lacks penetrating power. Beta radiation is somewhat less ionizing, but somewhat more penetrating as well. X-rays and gamma rays are still less ionizing but have extremely high penetrating powers. Alpha and beta rays are generally harmful only if swallowed, inhaled or absorbed through direct skin contact. Major radiation hazards are the result of exposure to gamma, X or neutron rays.

Curie (Ci), named after Madam Curie is the basic unit used to express the intensity of radioactivity. One curie is equal to 3.7×10^{10} nuclear disintegrations g^{-1} second^{-1}. Potency of radiation may be expressed as Roentgen (R). One R is equal to 2.083×10^{10} ion pairs produced in 1 ml of air. Rad is the amount of radiation that gives an energy absorption of 100 ergs g^{-1} of tissue. Rem is the unit dose of radiation that gives the same biological effects as given by one roentogen of X-rays.

6.1 Sources of Exposure to Radiation

6.1.1 Natural

Exposure to radiation may occur through natural and/or anthropogenic sources. Natural sources of radiation are the cosmic rays and the radioactive compounds in the soil and the residues they emit into the air, water and food crops. Even the very cells of our bodies contain some natural radiation, in the form of potassium K^{40}; most of us probably receive more internal radiation from this source than any other, under normal conditions.

Although levels of background radiation do exceed 1000 mR per year in a few parts of the world, the average American is exposed to about 100 to 125 mR/year at sea level. Thinning of the atmosphere provides less and less protection against cosmic rays at higher altitudes. Risk of exposure increases with increase in duration. Due to this reason, residents of states like Utah, Colorado, Wyoming and New Mexico receive significantly more natural radiation from cosmic rays than residents of other states. The crews of jet-air crafts receive even higher exposure, high enough to make it necessary for them to limit the amount of time they spend above 30,000 feat each year. It should not exceed 5,000 mR/year, the legal prescribed limit of exposure.

Background radiation probably does contribute in a small way to cancer, birth defects and other health problems.

6.1.2 Man Made

Man has been using radiation technology for health care and energy demands.

6.1.3 Health Care

Since the advent of X-ray technology, these rays are routinely used for the examination of teeth, bones, lungs and other organs. Different parts of the body can pick up different doses emitted form X-rays. A single chest X-ray increases our gonadal exposure by some 5 to 8 millirads, a dental X-ray by 9 millirads, while a full X-rays of the lower back may be as high as 2,500 millirads. Patients and radiologists both are exposed. Infants and pregnant woman are more liable to their effects than adults.

Much higher exposure occurs in subjects who use plutonium powered pace-makers. Professional radiologists are likely to suffer high risks of aplastic anemia, leukemia and certain types of cancer.

6.1.4 Nuclear Fallout

Since 1945, the desire by nations to add nuclear weapons to their military arsenals has introduced a unique kind of man made exposure to radiation. Explosion of a nuclear device in the atmosphere releases an intense shower of many kinds of radiation, most of which fall back to earth in the immediate vicinity of the blast within hours. Within 80 to 160 km of the blast, alpha and beta emitters can contaminate air, water and food supplies. Some fallout is also pulled aloft into the upper atmosphere by wind currents and carried around the earth for years, even decades, constantly subjecting populations to a slow, steady drizzle of radioactivity. Like all serious human questions, the use and non-use of nuclear weapons goes beyond political and military considerations. It is a moral and philosophical question as well. Exposure from nuclear fallout has been predicted to be 5 mR/year/person by the year 2,000.

6.1.5 Nuclear Power

As a result of growing pressure against the expanding use of fuels (coal, oil and natural gas), nuclear-generated energy has been relied by many nations to meet the burgeoning energy demands. Power needs are doubling every decade in the U.S. The Atomic Energy Commission predicts that 50% of the power generated by U.S. will be nuclear by 2,000. First nuclear power generator was installed in 1958 in U.S. By 1981, seventy three commercial reactors were generating more than 10% of electricity.

The nuclear fuel cycle begins with the mining, extraction, purification and other processing steps of uranium ore. Processed uranium is known as, U^{238}. It also contains a small amount of U^{235}. For efficient fission, the percentage of U^{235} must be brought upto 3 to 4 percent, a step known as enrichment. Power is generated from enriched uranium, as well as releasing hundreds of highly radioactive by-products, fission also produces heat, and this is what is used to create electricity. It takes 30 trillion fissioned uranium atoms to generate 1 Kilowatt of electrical power.

The reactors are so designed that they can not explode like atom bomb. However, no nuclear power plant is contaminated safe. The main danger posed by commercial reactors is the release of intense radioactivity generated by fission into the ambient environment in the form of gases, contamination steam, or radioactive water. In U.S. Three Mile Island nuclear power plant leaked in 1979. Chernobyl nuclear power plant leaked in 1986 in USSR.

6.1.6 Routine Emissions

Nuclear reactors routinely release small amounts of radio-active gases, primarily xenon, krypton-85 and radioactive iodine into the air and flush small amounts of liquids contaminated with radioactive tritium and cesium into local water ways.

6.1.7 Meltdown

The heat generated by fission is so intense that, if not carefully controlled, it would soon melt the contents of the core and sent tons of molten slag through the floor of the containment building. The fanciful possibility that the extremely radioactive lump of hot (over 2875^0C) metal would continue burning a hole straight through the Earth has given rise to the "China syndrome". A full meltdown would be an unparalleled' environmental catastrophe, releasing high levels of radioactivity into the air, soil and both surface and groundwater supplies.

6.2 Biological Effects of Radiation

Uranium mine workers, radiologists, radium dial painters are among those who suffer seriously from these radiation hazards. Most serious disorders are caused by radiation escaped from nuclear power plants and nuclear explosions. These include miscarriages, stillbirths, mental retardations and reduced fertility. Other effects include damage to brain cells, nausea and death from extraordinary agony. Genetic malformations could occur in plants, animals and birds destroying the whole biosphere.

6.2.1 Acute Effects

Accidental industrial radiation of the whole body by gamma or neutron rays causes a distinct pattern of illness known as acute radiation syndrome (ARS), or radiation sickness. At 200 to 400 rems, ARS initially manifests itself as a flu like condition characterized by chills, headache, fever, sore throat and lassitude. As it progresses, these symptoms are succeeded by weight loss, hair loss and bleeding gums. Upto about 500 to 600 rems, the major effects of ARS is on the organs of the body that make blood-bone marrow, spleen and lymphnodes. From 600-1000 rems, the gastrointestinal tract becomes the major target of ARS. At doses above 700 rems few if any exposure victims will survive.

6.2.2 Chronic Effects

The story of radiation research and its use in the early years of this century is drawn with endless examples of how a lack of appreciation for the hazards of chronic exposure to radiation could lead to tragedy. Within the decade of Roentgen's discovery of X-rays in 1895, the first case of cancer caused by excessive X-ray exposure had been reported in medical journals. Marie Curie, the discoverer of radium succumbed to leukemia, as did her daughter, undoubtedly because of the radiation exposures. In many cases the hands were the main targets of these excessive exposures. Effects varied from cracking and roughening of the skin to blistering, ulceration and gangrene, causing intense pain, scar formation and deformity.

6.2.3 Genetic Effects

Genetic effects are caused by mutation. Ultraviolet radiations and ionizing radiations and a variety of chemicals can cause spontaneous mutation. They are called as mutagenic agents. We can detect a new spontaneous mutation about once in a million gene replications. Probability of making a mistake during replication is about 10.7 per nucleotide per replication. There are several ways in which mistakes can occur. The wrong base might be inserted in the daughter strand because of the tendency of bases to spend little time in alternate tautomeric forms. Another way in which replication errors can occur is for single stranded loops to form in either the parent or daughter strand. A loop in the parent would cause bases to be missed, yielding a deletion mutant. A loop in the daughter would produce an insertion mutant. At the next replication, in either case, a fully mutant double stranded molecule would be produced. These mutations induce the risk of cancer.

6.3 Famous Incidents of Radioactive Pollution

6.3.1 Hiroshima Catastrophe

At 8:15 AM on August 6, 1945, an American B-29 bomber nicknamed Enola Gay appeared on skyline over Hiroshima. It was a bright summer morning and the city lay in the sloping rays of rising sun, unaware of the doom that was to befall it. Forty three seconds later an 10,000 lb atomic bomb exploded on human civilization for the first time in the history of man. A ball of fire with a temperature of around 100 million degree centigrade at its core spread the landscape and much of Hiroshima was turned

Fig. 6.1: Remains of the city office of Hiroshima.

into an oven. After three days, it was the fate of Nagasaki to suffer the ravages of a nuclear attack. Over 100,000 people died in the immediate aftermath of the two bombings. In the years that followed many more perished from the effects of radiation.

6.3.2 Chernobyl Accident

The devastating nuclear explosion occurred in Chernobyl nuclear power plant in the Ukraine on April 26, 1986. Four million people were directly or indirectly affected by large scale contamination that led to evacuation and resettlement of people. Radioactivity released in the accident continued to spread via ground water and surface waters, affecting the ecosystems of large areas of Europe.

^{131}Iodine and ^{137}caesium were the most serious amongst the 20-odd radioactive elements released during Chernobyl disaster. According to Soviet Health Ministry, 31 persons died shortly after the disaster of the 276,614 people who took part in cleaning work, a total of 1065 died by the end of 1990. UNESCO organized a conference on "hydrological" impact of nuclear power plant system" in Paris to study the lessons to be learnt from Chernobyl. Another Chernobyl must not be allowed to happen.

New regulations have been adopted restricting access to 30 km exclusion zone around the Chernobyl nuclear power plant. They deny access to any one under 18 in an attempt to end visits by children to parents and grandparents who have refused to leave. After the accident some 116,000 people were hastily resettled but some have since returned preferring the risk of ill health to the problems of making a new life away from their homes.

There is no longer any doubt that the massive increase in thyroid cancers in Ukraine, Belarus Russia are radiation induced, as a result of the Chernobyl accident, but the mechanism still requires more research. This was one of the conclusions of the World Health Organization's international conference, held in Geneva, on the health consequences of the Chernobyl and other radiological accidents. It looked mainly at the results of WHO's international programme on the health effects of the Chernobyl accident (IPHECA), the first phase of which ended last year, Wilfried Kreisel, executive director in charge of WHO's environment and health issues says: "WHO wants to see the volume and quality of medical assistance and scientific research increased. We will be doing a disservice if we fail to extract any benefit for mankind from this monumental human tragedy. If history is not to repeat itself we should learn the lessons of Chernobyl well".

6.3.3 Three Mile Island Episode

A few miles southeast of Pittsburg, Pennsylvania (USA) on an island (Fig. 6.2) in the Susquehanna River were situated two nuclear reactors designated as TMI-1 and TMI-2. On March 28, 1979, a local radio-station first broadcasted the story of TMI accident. High radiation levels (30,000 mR/hour) were recorded by the hovering helicopters. It was declared as the worst commercial nuclear accident in American history. Though

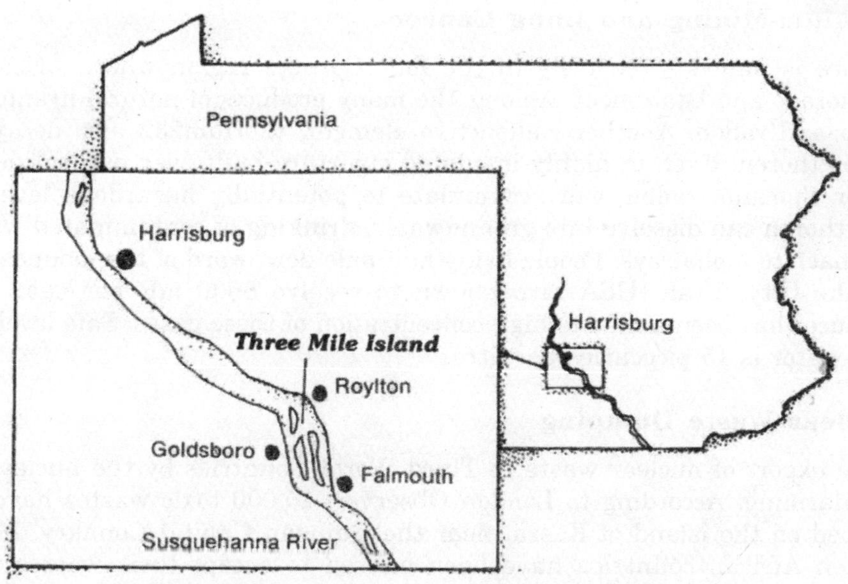

Fig. 6.2 Map showing the location of Three Mile Island.

the danger of melt down was contained within five days, radioactive emissions might have caused subtle, undetectable damage to infants such as birth defects and mental retardation. The nearly area of the plant was declared hazardous. The accident left the containment and auxiliary buildings filled with some 3 to 4 million litres of highly radioactive water, that might have taken years to decontaminate using special resin filters. The accident also left a sizable amount of krypton-85 gas trapped in the containment building which was vented through special filters in 1980. Air and groundwater samplings has been conducted more or less continuously by EPA and effluent monitors at the site had been set to sound an alarm anytime radioactivity exceeds certain levels.

6.3.4 The Radium Dial Painters Case

In the early 1920s dentists in the area of Essex county, New Jersey (USA) noticed that young female patients were developing strange complications. Later it was found that all of the women involved at one- time or the other have been employed in the local factory as painters of luminous dials on clocks and watches, using a paint made from water, zinc sulfide, manganese, copper, cadmium, acacia and two radioactive substances radium and mesothorium, both alpha emitters. By 1929, the tragedy had claimed fifteen lives. In most cases the cause of death was aplastic anaemia. However, several autopsies confirmed the presence of radioactive residues primarily mesothorium in amounts ranging from 50 to 150 micrograms. As known today, as little as 1.2 to 2 micrograms of these substances deposited in the bones may be fatal. Mesothorium has a biological half life of about seven years whereas radium has a biological half life of 1700 years.

6.3.5 Uranium-Mining and Lung Cancer

Uranium ore is mined extensively in the four Corners region where Arizona, New Mexico, Colorado and Utah meet. Among 1he many products of natural uranium decay are radium and radon. Another radioactive element, thorium232 also decays into a harmful gas thoron. Even in highly insulated structure built over natural deposits of uranium or thorium, radon, can accumulate to potentially hazardous levels. Since radon and thoron can dissolve into groundwater, drinking of contaminated water may expose stomach to alpha rays. People living half mile downward of the mounds of waste at Salt Lake City, Utah (USA) are known to receive 8000 mR per year. Lung & stomach cancer has been linked to high concentration of these gases. Safe level of alpha particle in water is 15 picocuries per litre.

6.3.6 Nuclear Waste Dumping

Large scale export of nuclear waste to Third World Countries by the nuclear powers has been alarming. According to London Observer, 15,000 toxic wastes have already been dumped on the island of Kassa, near the Guinean Capital Conakry. More than half a dozen African countries have been offered to accept toxic wastes from the industrialized world. According to Greenpeace (NGO), 5.4 million tones of hazardous waste have been shipped from Australia, Canada, Germany, U.K., and USA to 13 Asian countries between 1990-1993. They also plan to export nuclear reactor and radioactive wastes for disposal in Tibet and India. Third World countries must take up the issue with the Geneva based International Atomic Energy Agency (IAEA) on the movements of hazardous wastes.

6.3.7 Radioactive Contamination of River Periyar

At Indian Rare Earth (IRE) Limited in Kerala is a part of DAE ambitious plan to increase ten fold expansion of nuclear power generated by the year 2000. At IRE, monazite (separated and obtained from sand along the coastal belt of Kerala) is processed to produce thorium hydroxide, trisodium phosphate (TSP) and rare earth chloride. While rare earth chloride is exported, thorium hydroxide is stored for future use in the fast breeding reactors. TSP is a general purpose detergent supplied to a few soap manufacturers. 5000 RCC barrels containing thorium hydroxide are stored in a silo (godown) about 2 metres from the bank of river Periyar. Over 400 RCC barrels of radioactive waste have been buried within the IRE premises. Most of these barrels when broken, leak and contaminate the underground water. Moreover, radioactive effluents from the processing line as well as the water used to wash the floor, is regularly discharged into the river. According to estimates, some 1500 kgs of thorium and 15 kgs of uranium go into the river every year.

An epidemiological study has revealed that the incidence of cancer among IRE workers is abnormally high. Eleven workers have died of cancer so far. Seventeen cases of genetic disorders have been traced amongst the children of the workers. Lakhs of people whose daily lives are touched by the river are in a state of panic. Water and food are not safe for human consumption. A severe distortion of human, animal and plant life for scores of generations to come awaits at river Periyar.

6.3.8 Threat of Nuclear Summer

There is no doubt that "nuclear winter" will set in time following nuclear explosions. But at the same time with a drop in temperature, vast volumes of carbon will be released into the atmosphere in some areas of the globe where fire will be raging. According to experts, fire will effect an area of no less than one million km^2. There are scientific speculations that the northern hemisphere will plunge into darkness and dark carbon aerosol will settle on crops destroying them. The greenhouse effect will enhance sharply - the heat from the explosions and fire will lead to the onset of "nuclear summer". It may result into a global ecological catastrophe.

SUGGESTED FURTHER READINGS

Morgan, K.Z. and Turner, J.E. (editors). Principles of Radiation Protection, John Wiley and Sons, New York, 1967.

ICRP Publication 60, 1990 Recommendations of the International Commission of Radiological Protection. *Annals of the ICRP,* 21(4), (1991) 1-77.

UNSCEAR (United Nations Scientific Committee on the Effects of Atomic Radiation). *Sources and effects of Ionising radiation.* United Nations New York, 1993, pp. 221-373.

ICRP Publiaiton 62, Radiology protection in biomedical research. *Annals of the ICRP,* 22(3), (1992), pp. i to xxiv.

STUDY QUESTIONS

1. Define radioactive pollution. Enumerate the sources of exposure to radioactive pollution.
2. Discuss biological effects of radioactive pollution.
3. Describe famous incidents of radioactive pollution.
4. Write briefly on the following-
    a.   Hiroshima Catastrophe
    b.   Three mile island episode
    c.   Radium dial painters case
    d.   Chernobyl accident

Chapter 7

SOLID WASTE POLLUTION

Almost all processing and consumption activities of people, from agriculture, industry to commerce and health generate solid waste. According to an estimate a single person generates 1600 kg of waste in one year in developed countries. Several components of these wastes can be recycled whereas others remain to be non-biodegradable. Perhaps the largest waste disposal site in the world is located on a 1500-ha (3706-acre) site on Staten Island, New York. This facility known as fresh kills accepts approximately 1500 metric tons/day of municipal and commercial waste collected in the city of New York. The story of Love Canal episode (1976) that occurred in a residential area near Niagara falls demonstrates that waste management should be an important objective of planning to ensure a better environment.

7.1 Sources and Classification of Solid Waste

Solid waste, often called the third pollution after air and water pollution, is that material which arises from various human activities and which is normally discarded as useless or unwanted. It consists of the highly heterogeneous mass of discarded materials from the urban community as well as the more homogeneous accumulation of agricultural, industrial and mining wastes.

Solid wastes may be classified based partly on content and partly on moisture and heating value. A typical classification is as follows:

a. **Garbage:** Refers to the putriscible solid waste constituents produced during the preparation or storage of meat, wit, vegetables, etc. These wastes have a moisture content of about 70% and a heating value of around 6×10^6 J/kg.

b. **Rubbish:** Refers to non-putriscible solid waste constituents, either combustible or non-combustible. Combustible wastes would include paper, wood, scrap, rubber, leather etc. Non-combustible wastes are metals, glass ceramics etc. These wastes contain a moisture content of about 25% and the heating value of the waste is around 15×10^6 J/kg.

c. **Pathological Wastes:** Dead animals, human waste, and hospital waste constitute pathological masks etc. The moisture content is 85% and there are 5% non-combustible solids. The heating value is around 2.5×10^6 J/kg.

d. **Industrial Wastes:** Chemicals, paints, sand, metal ore processing, flyash, sewage treatment sludge, etc. are the examples of industrial waste.

81

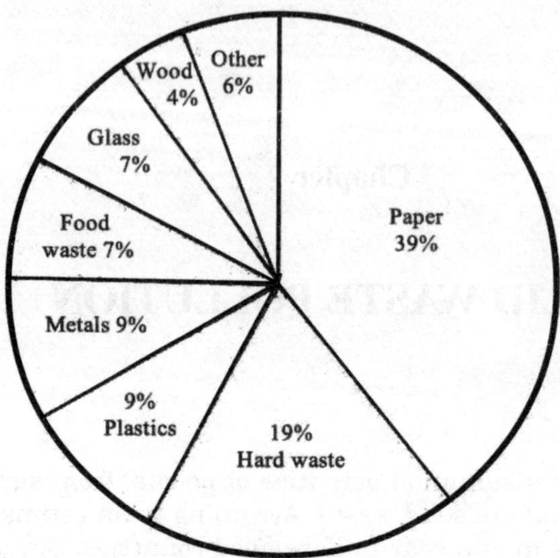

Fig. 7.1 Generalized composition of solid waste likely to end up at a disposal site. Paper is the most abundant of these solid wastes.

e. Agriculture Wastes: Farm animal manure, crop residues, etc.

The principle sources of solid wastes are domestic, commercial, industrial and agricultural activities. Many times domestic and commercial wastes are considered together as the so-called urban wastes. The main constituents of urban wastes are similar throughout the world, but the weight generated, the density and the proportion of constituents vary widely from country to country, and from town to town within a country according to the level of economic development, geographic location, weather and social conditions. In general, it has been found that as the personal income rises, kitchen wastes decline but the paper, metals and glass wastes increase; the total weight generated rises but the density of the wastes declines.

In India authentic information regarding the composition of the urban wastes is not generally available as regular analysis of the refuse is not carried out by the municipalities. In fact refuse is very heterogeneous in composition and the geographical, temporal and seasonal variations in its composition make it difficult to define a "typical refuse". The solid refuse generated in urban areas contains articles of various sizes and types and consists of dust, vegetables, waste paper, large paper-board cartons, glass bottles, worn out tyres, carcasses of animals and night soil. Table 7.1 gives the composition of refuse in various cities of India and a comparison is made with the urban refuse from a typical European city. As is seen from table, the average paper content in the refuse of Indian cities is about 2 to 3 percent as compared with about 27% for a typical European city. Similarly, the density of refuse in India is much higher than that of the refuse generated in the cities of Western countries because of the inclusion in it of the street sweepings. The amount of refuse collected from urban areas in India is of the order of 0.3 kg to 0.5 kg per person per day excluding night soil.

Table 7.1 Composition of city refuse (percentage by weight)

	Kanpur	Delhi	Calcutta	Banglore	Bombay	Typical European city
Paper	1.35	5.88	0.14	1.5	3.20	27
Vegetableputrescible matter	53.34	57.71	47.25	75.2	59.37	30
Dust, ash, etc.	25.93	22.95	33.58	12.0	15.90	16
Metals	0.18	0.59	0.66	0.1	0.13	7
Glass	0.38	0.31	0.24	0.2	0.52	11
Textiles	1.57	3.56	0.28	3.1	3.26	3
Plastics, leather, rubber, etc.	0.66	1.46	1.54	0.9	-	3
Other (stones, wooden matter, etc.)	18.59	6.4	16.98	18.9	16.4	3
Density, kg/m³ 500	-	578	-	-	-	132

Source: Report of the committee of urban wastes, Govt. of India Ministry of Works and Housing New Delhi. Dec. 1975.

Manufacturing industries produce wastes which are solid or semi-solid. These wastes can be pyrophoric (self-igniting), explosive, toxic or radioactive. Chemical process industries (CPI) generate a variety of wastes, both organic and inorganic, which are mixture with a wide range of component concentrations. Table 7.2 gives some of the physical and chemical characteristics of the wastes generated by the chemical process industry (CPI).

Table 7.2 Solid wastes generated in the CPI.

Origin	Physical and chemical characteristics	Comments
Coke manufacture	Coke and coal fires	Generally contained
Dyes and pigments	Reaction raw-material sludge, highly variable	in scrubber wastes
Pharmaceuticals and Fine-chemicals	Raw material solids, biological wastes	-
Inorganic chemicals	Insoluble salts, tailings, slimes	-
Metal processing (secondary)	Ash, scrubber wastes, metal hydroxide sludges	-
Petrochemicals	Oily, greasy, asphaltic	Usually float
Plastics and rubber	Latex or plastic crumbs, often Coagulated	-
Pulp and paper	Fibrous, often mixed with lime or alum	-

Source: R.W. Okey, D. Digregorio and E.G. Kominek, "Waste-sludge treatment in the CPI" *Chem. Engg.*, Vol. 86 (1979).

Most of the industrial wastes generated in cities come from small scale operations and these are usually disposed off along with the city refuse. Larger industries are often located outside the cities and the disposal of their wastes is primarily the responsibility of the industries themselves. Some of the industrial wastes are often recycled (scrap metal and paper) while others can be utilized as an energy source for specific processing plants in some regions. Energy can be recovered from solid wastes by numerous thermal routes as well as by biochemical conversion. The toxic and radioactive wastes, often classified as hazardous wastes, need special consideration before their disposal.

Agricultural wastes comprise both crop residues and animal wastes such as manure and urine. Whereas urban wastes amount to between 0.3 and 0.5 kg per person per day in India, agricultural wastes amount to around 2 kg per person per day. Animal and vegetable wastes contain valuable minerals and nutrients. Humus from agricultural wastes contains nitrogen, phosphorus, potash and trace elements which are vital to the fertility of the soil and optimum plant growth. Burning of wastes as fuel further, leads to loss of valuable nutrients.

7.2 Public Health Aspects

The relationship between solid wastes and human disease is difficult to prove. Nevertheless, improper handling of solid wastes is a health hazard and causes damage to the environment. The main risks to human health arise from the breeding of disease vectors, primarily flies and rats. A common transmission route of bacillary dysentery, amoebic dysentery and diarrhoeal disease in India is from human faeces by flies to food or water and thence to humans. It has been estimated that in warm climates, exposed garbage produces as many as 70,000 flies per 0.03 m^3 in a week.

The refuse dumps also serve as a source of food for rats and small rodents which quickly proliferate and spread to neighbouring areas. Rats destroy property, infect by direct bite and spread various diseases like plague, endemic typhus, salmonellosis, trichinosis, etc. Apart from diseases for which insects and rats are carriers, the handling and transfer of biological wastes poses a threat to the worker and to those he contacts. Disease transmission may occur through direct contact with the waste, through infection of open sores or through vectors.

The hazardous wastes are injurious to human health; some have acute effects while others pose a health hazard after prolonged period of exposure. Improper disposal of such wastes has resulted in the death of humans and animals through contamination of crops or water supplies.

The environmental damage caused by solid wastes is mostly aesthetic. Uncontrolled dumping of urban wastes destroys the beauty of the countryside; also, there is the danger of water pollution when the leachate from a refuse dump enters surface water or ground water resources. In addition, uncontrolled burning of open dumps can cause air pollution.

Solid waste disposal is at present limited to land and the ocean. Some of the wastes can be recovered and reprocessed, a procedure commonly known as recycling. However, before the wastes can be disposed of effectively, they must be colleted efficiently. These activities, i.e., collection, disposal and/or recovery form a part of the solid waste management system.

The old "dilute and disperse" concept of waste management no longer works and newer concept of "concentrate and contain" may give way to a concept called integrated waste management.

SUGGESTED FURTHER READINGS

Berry, R.S., and Makino, H., 1974. Energy thrift in packaging and marketing. *Technol. Rev.*, 76, 32-43.

Boustead, I., and Hancock, G.F., 1979. *Handbook of Industrial energy Analysis*. Chichester: Ellis Horwood.

Hocking, M.B., 1991. Relative merits of polystyrene foam and paper in hot drink cups: implications for packaging. *Environ. Manage.*, 15, 731-47.

STUDY QUESTIONS

1. Classify solid waste. Discuss the composition, physical and chemical characteristics of city refuse.
2. Discuss public health aspects related with solid waste pollution.

Chapter 8

SOIL AND LAND POLLUTION

The global weathered earth crust form soil over the centuries that supports the variety of microscopic and macroscopic life-forms (Fig.8.1). The process of soil formation is very complex involving a number of physical, chemical and biological transformations.

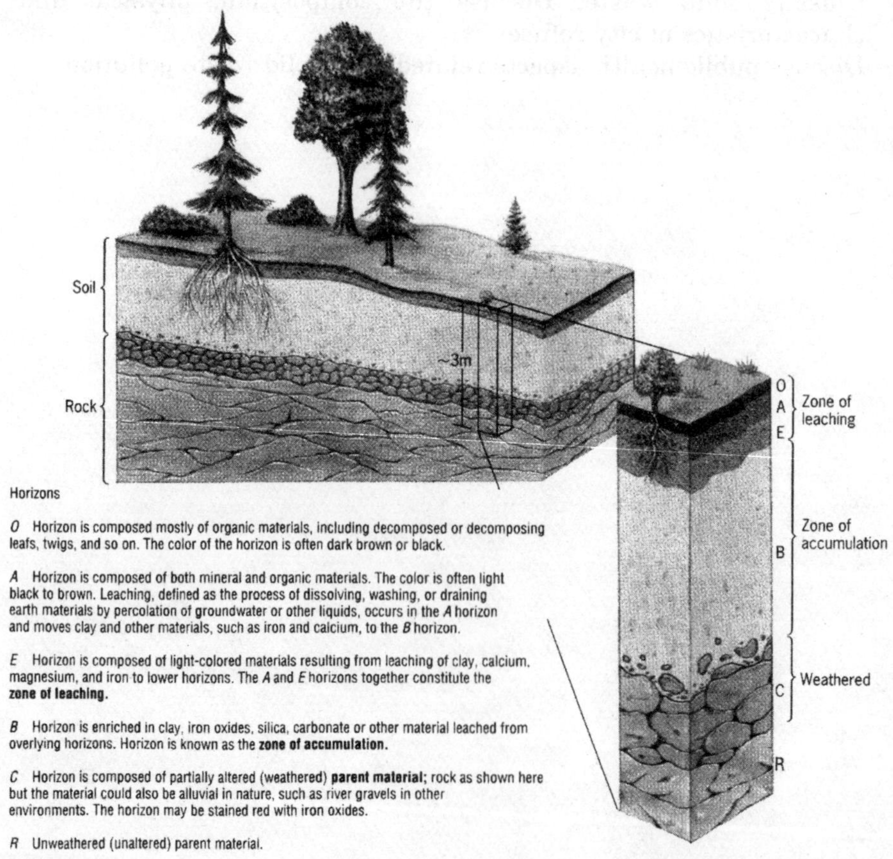

Horizons

O   Horizon is composed mostly of organic materials, including decomposed or decomposing leafs, twigs, and so on. The color of the horizon is often dark brown or black.

A   Horizon is composed of both mineral and organic materials. The color is often light black to brown. Leaching, defined as the process of dissolving, washing, or draining earth materials by percolation of groundwater or other liquids, occurs in the *A* horizon and moves clay and other materials, such as iron and calcium, to the *B* horizon.

E   Horizon is composed of light-colored materials resulting from leaching of clay, calcium, magnesium, and iron to lower horizons. The *A* and *E* horizons together constitute the **zone of leaching.**

B   Horizon is enriched in clay, iron oxides, silica, carbonate or other material leached from overlying horizons. Horizon is known as the **zone of accumulation.**

C   Horizon is composed of partially altered (weathered) **parent material**; rock as shown here but the material could also be alluvial in nature, such as river gravels in other environments. The horizon may be stained red with iron oxides.

R   Unweathered (unaltered) parent material.

Fig. 8.1: Idealized diagram showing a soil profile with soil horizons.

The topmost layer of soil is comparatively more rich in nutrients and supports maximum living forms. It is composed of minerals of various sizes and organic matter along with pore space filled with air and water. However, in deeper layers soil composition with respect to particles, nutrients and other associated properties varies distinctly. Thus, there are a number of zones one can not notice when a soil profile is made for analysis of soil depth. The profile character varies distinctly from place to place, particularly with respect to their depth, colour and composition.

There are different classes of soil types primarily based on the particle size distribution pattern. The details are given in textural triangle of soil classes (Fig.8.2).

There are various ways of addition and losses of nutrients that take place in soil. These nutrient cycles make the balance of organic and inorganic soil constituents (Fig.8.3).

Biological soil has a large number of biota living inside. They are categorized as algae, fungi, bacteria, actinomycetes, microarthropods, protozoans and nematodes etc. Many such biota help in decomposition of organic materials and nutrients. A typical microarthropod spectrum of fertile soil is shown in Fig. 8.3. It is further interesting to note that soil itself acts as an ecosystem with diverse forms of life, which could easily be arranged in various trophic levels. A food web exists (Fig. 8.3). More the complex food web vis-a-vis higher diversity of life forms in soil, it is better for soil, health.

However, over the years, soil is being contaminated by a number of ways which is popularly known as soil pollution/land pollution. The possibilities of soil contamination are shown by the Table 8.1.

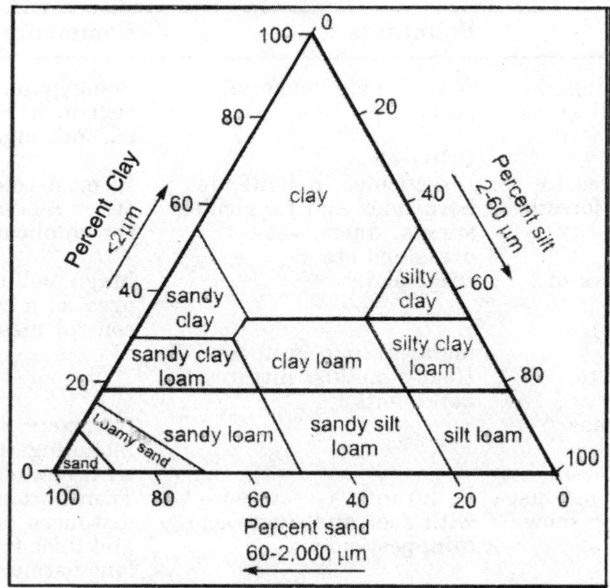

FIG. 8.2: Textural triangle of soil classes.

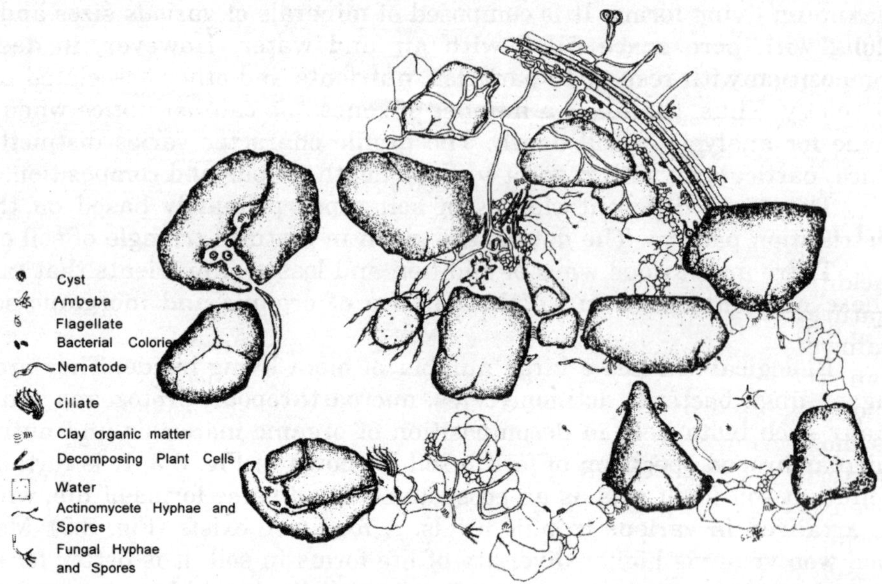

Fig. 8.3   Approximately 1 cm^2 of a highly structured section in the surface horizon of a grassland soil illustrating the trophic relationships among different groups of soil organisms. Interactions are controlled by accessibility of groups to their resources.

Table 8.1 Possible avenues of soil/land contamination

Route	Pollutants	Comments
Waste-dumping including rubbish dumps/landfall sites/ industrial dumps	A very wide range of pollutants PCBs etc.	Some industrial dumps are rich in particular pollutants e.g. oil, metal ore deposits,
Application of pesticides to agricultural land and forests	Insecticides, rodenticides, herbicides and fungicides sprays, dusts, seed dressings etc.	In most countries there are strict regulations controlling the application of pesticides
Control of insect vectors of disease	Insecticides malarial mosquito and fly	Major pollution over large area as a consequence of control measures against
Application of sewage to agricultural land flooding by rivers or seas	Heavy metals, nitrates, detergents	A variety of pollutants including those associated with sewage
Precipitation from air as dust or droplets or in rainor snow	Pollutants associated with soot and dust, acid rain, pesticides	Transport may be over short distances (spray drift, soot and dust from chimneys) or longdistance (brought down especially by rain and snow)

A continuous material cycle is operated in soil by microbes in addition to sustained addition and loss of nutrients from soil by various activities, resulting in the rapid transformation of soil nutrition in short and long-term basis.

8.1 Land/Soil Pollution

Soil gets polluted by a number of ways. The major kinds of soil pollution are:
- Acidification,
- Salinisation and sodification,
- Agrochemical pollution
- Contamination by metalliferous water

8.1.1. Acidification

Acidification has a number of natural and anthropogenic causes (table 8.3). The main natural causes are long-term leaching and microbial respiration. The acids found in rainwater (carbonic acid) and in decomposing organic material (humic and fulvic acids) can stimulate leaching by dissociating into H^+ ions and their component anions which then displace or attract base cations from the soil exchange complex. Leaching of bases is most common where precipitation exceeds evapotranspiration and includes many eastern parts of North America and northwestern parts of Europe, where soils have been subjected to leaching for 10,000-15,000 years of postglacial time. Microbial respiration also leads to soil acidification through the production of CO_2 which is dissolved in soil water to form carbonic acid. Other natural processes associated with soil acidification are plant growth and nitrification. During plant growth, nutrient base cations are obtained through root systems in exchange for H^+ ions, thus leading to increased soil acidity. Nitrification is an oxidative process of organic decomposition whereby NH_4^+ (ammonium) ions are converted to NO_3^- (nitrate) ions by nitrifying bacteria, with H^+ ions as a byproduct:

$$NH_4^+ + 1/2\ O_2 \Rightarrow NO_3^- + 4H^+$$

These ions are then available for displacing and attracting base cations from the soil exchange complex, as mentioned above, thus leading to soil acidification.

The main anthropogenic causes of acidification include certain landuse practices, such as needle leaf afforestation, excessive use of inorganic nitrogen fertilizers, land drainage and acid deposition resulting from urban and industrial pollution. Needle leaf afforestation has been associated with the acidification of soils and surface waters for a number of reasons. First, needle leaf trees produce litter which is very acidic in comparison with most broadleaf species. Second, because of their high canopy surface area, needle leaf trees are able to 'scavenge' acid pollutants from the atmosphere, later releasing them into the soil via fall and stem flow. Thirdly, due to modifications of the surface and soil hydrology by drainage channels and shallow root networks, water transfer is rapid and is concentrated either at the surface or in the uppermost layer of the soil. Under these conditions, the residence time of the water in the soil is limited, as is the depth to which it can percolate. Thus, the contributions of weathering and ion exchange reactions to the buffering process are limited (Bache 1983, Miller 1985).

Table 8.2 Causes of soil acidification

	Source	H^+ addition or equivalent (kg.H^+ ha^{-1} a^{-1})
Natural	CO_2 in soil pH> 6.5 (calcareous soils)	7.2-12.8
	Organic acids in acid soils and from vegetation	0.1-0.7
Acid "rain"	Wet deposition	0.3-1.0
	Dry deposition	0.3-2.4
	NH_3 and NH_4^+ oxidation	0.7
Land use	Cation excess in vegetation	0.2-2
	NH_4^+ oxidation (agricultural soils) and leaching	4-6
	Oxidation of N and S from organic matter and leaching	0-10

Source: Rowell and Wild 1985

Excessive use of inorganic nitrogen fertilizers in agricultural systems has also been associated with soil acidification, partly through the process of nitrification (see previous equation). If levels of NO_3^- ions in the soil are in excess of plant requirements, they will behave as mobile anions, thus encouraging the leaching process. The acidifying effect of nitrogen fertilizer mobilized in the form of soluble organo-metallic complexes. If mineral acids predominate, however, then aluminium is mobilized in its ionic, labile-monomeric form (Al^{3+}). This form of aluminium is particularly toxic to many :freshwater organisms including fish. In a similar way, the aluminium heavy metals, such as lead, zinc and cadmium, are more readily mobilized in acidic soils.

Acidification of soils and associated nutrient leaching has also been implicated in damage to trees in forested areas, particularly in Central Europe and Scandinavia. It is more likely, however, that nutrient leaching acts synergistically with other factor, including ozone pollution, acid deposition, NH_4^+ (ammonium) uptake, drought and frost to product stress in the tree. Signs of tree damage include needle discoloration, crown defoliation and deformed branch structures. Increased soil acidity and associated changes in aluminium mobility also have serious implications on the quality of surface waters which receive drainage from acidified soils.

Soils vary dramatically in their ability to buffer acidity, a characteristic known as the *buffering or acid neutralizing capacity*. The buffering capacity of a soil is defined as the amount *of* acid that needs to be added to cause a reduction in pH of one unit (Trudgill 1988). Soils which contain significant quantities of base-rich, weatherable minerals have a high buffering capacity whereas those which are dominated by quartz and similarly resistant minerals have a low buffering capacity. In 'Britain, for example, soils with the highest buffering capacities are found in eastern and southern areas where calcareous parent materials are particularly common. Soils with the lowest buffering capacities are found in areas with base-poor, often siliceous, parent materials which predominate in upland areas in the west and north.

There are a number of approaches to the management and remediation *of* soil and surface water acidity, including liming, sensible forestry management and reduction of acid emissions into the atmosphere. Liming has long been practiced on agricultural land to remedy the problems or modify, slow organic matter turnover, poor modulation in some legumes, calcium and molybdenum deficiency and aluminium and manganese toxicity. Liming materials most commonly used include ground limestone, chalk, marl and basic slag, the main active constituent being $CaCO_3$ (calcium carbonate); other minor components include quicklime (CaO), slaked lime ($Ca(OH)_2$) and $MgCO_3$ (magnesium carbonate). The lime requirement of a soil varies depending on its buffering capacity and is usually expressed as the amount of $CaCO_3$ (t ha^{-1}) required to raise the pH of the top 15 cm of soil to the desired value. In temperate areas, the ideal soil pH is about 6.5 for arable crops and about 6.0 for grasslands. In tropical areas, however, pH values of about 5.5 are often preferred particularly in soils with high exchangeable aluminium contents, where phosphorus availability may be restricted under more alkaline conditions.

8.1.2 Salinization and Sodificaiton

Soils that accumulate salts are known as saline or sodic soils. Growth of crops is adversely affected by soluble salts. Further, a saline soil is one that contains enough soluble salts to interfere with normal plant growth, while a sodic soil contains enough exchangeable sodium (ExNa) to also have an adverse effect on plant growth. A saline-sodic soil contains both soluble and exchangeable sodium at high levels.

Salt affected soils are a common feature of arid and semiarid geographic regions. Salinity can affect plant growth in three ways-
  i.   It can increase the osmotic potential and hence decrease water availability.
  ii. It can affect biological metabolism
  iii. It can diminish soil-water permeability and soil aeration by adversely affecting soil texture.

To reclaim salt affected soil, excess salt should be leached down so to that EC of the soil solution becomes lower than the critical threshold of the crop grown. The calcium loss must be replenished.

The soil pollutants which affect agricultural production and human habitat are: (a) heavy metals, (b) agri-chemicals, (c) radioactive materials, and (d) petroleum products.

8.2 Heavy metal Contamination

Heavy metals (*q. v.*) may be applied to the soil with pesticides, as plant nutrients, atmospheric fallout, and as a constituent of waste products. Heavy metals which tend to accumulate in soils include cadmium (Cd), chromium (Cr), copper (Cu), mercury (Hg), nickel (Ni), lead (Pb), and zinc (Zn). Sewage sludge (*q.v.*) contains all of these heavy metals and has been applied to agricultural soils for some years as a means of disposal of this material.

Considerable studies have been conducted over the years to determine the safe application rates of sewage sludge to agricul-tural soils in an effort not to exceed the recommended loading rates of each metal. If the above guidelines are followed, the application of sewage sludge to agricultural soils is considered beneficial, as it provides some of the essential plant nutrients and may improve soil physical properties and, in turn, plant growth and production. The mobility of heavy metals is very slow; thus they tend to accumulate if applied repeatedly. Heavy metals can occur in the following pools: (a) the exchange sites, (b) incorporated into or on the surface of crystalline or noncrystalline precipitates, (c) incorporated into organic com-pounds, or (d) present in soil solution.

Table 8.3 shows values of heavy metal content in soils for European countries, Canada, and the USA (Angelone and Bini, 1992). The concentrations of each of the trace elements considered as excessive are also shown for comparison. The concentration of some heavy metals is much greater than the recommended excessive levels in some countries. Substantial variation in concentrations of various heavy metals among the countries demonstrate varying degrees of accumulation of these elements in soils as a result of differences in land use.

Table 8.3 Mean of some total trace elements in soils in western Europe, in comparison with the contents of US, Canadian and world soils (mg/kg)

Country	Cu	Zn	Ni	Cr	Pb	Cd	Fe	Mn	B
Austria	17	65	20	20	150	0.20	13,300	310	.
Belgium	17	57	33	90	38	0.33	1,638	335	32
Denmark	11	7	7	21	16	0.24	1,236	315	
France	13	16	35	29	30	0.74	-	538	21
Germany	22	S3	15	55	56	0.52	1,147	806	-
Greece	1,588	1,038	101	94	398	7.4	-	1,815	-
Italy	51	89	46	100	21	0.53	37,000	900	
Netherlands	18.6	72.5	15.6	25.4	60.2	1.76	-	-	
Norway	19	60	61	110	61	0.95	-		-
Portugal	24.5	58.4	-	-				328	59
Spain	14	59	28	38	35	1.70	-	-	-
Sweden	8.5	182	4.4	2.3	69	1.20	6,300	770	-
England & Wales	15.6	78.2	22.1	44	48.7	0.70	3,141	1,405	
Scotland	23	58	37.7	150	19	0.47	-	830	-
Calculated average									
+ Greece	131.6	137	32.7	55.9	66.7	1.30	9,108	732	37
- Greece	19.5	68	27	52.7	39	0.79	633		
World soils	20	50	40	200	10	0.30	-	850	
US soils	25	54	20	53	20	0.50	-	560	
Canadian soils	22	74	20	43	20	0.30		520	
Excessive levels in soils	100	250	100	100	200	5	1,500	30	

(*Source:* Angelone and Bini, 1992).

Land application of municipal sewage sludge resulted in a marked increase in metal content in soils. Most heavy metals tend to accumulate in topsoil with little evidence of downward movement. However, under conditions of high leaching and with heavy loading of metals, due to the application of high rates of sewage sludge, enrichment of metals in subsurface has been detected.

In some cases, heavy metal pollution of agricultural soils was due to misuse of routine agricultural production practices. A case in point is Cu contamination of sandy soils in the citrus production region of Florida. In the early 1900s, soil and foliar application (as fungicidal spray) of Cu accounted for 34 and 10kg Cu/ha/year, respectively. Repeated applications of the above Cu levels resulted in accumulation of Cu up to 600 kg/ha in the top 15 cm soil (Alva and Graham, 1991).

8.3 Industrial Sources of Soil Pollutants

Improper management of industrial wastes can have profound implications on soil pollution (Corey, 1986). The US 'Superfund' is a legislative approach to correct the past mis-takes through improper waste management. The US Resource Conservation Recovery Act of 1976 was designed to prevent the improper management of industrial waste in the future. Indiscriminate practices of land application of hazardous wastes could result in pollution of soils which would make that soil unsuitable for any productive purposes. Soil plays an important role when an industrial waste is deposited in a landfill or placed on the soil to promote degradation or break-down. This will decrease the quantity of material available for transport into groundwater.

8.4 Petroleum Products as Soil Pollutants

Disposal of petroleum constituents, i.e., bcnzo[a]pyrene (B[a]P), benzene, toluene, and xylene, can cause soil pollution and, in turn, depending on their bioavailability, can result in human health risk. Benzene and B[a]P are carcinogens and are responsible for a significant fraction of potential health risks associated with petroleum-contaminated soils (US EPA, 1991).

Soil and groundwater contamination by refined petroleum products may also occur due to leaking underground storage tanks. Analyses of benzene, toluene, xylene, ethylbenzene, and total petroleum hydrocarbons are used to detect such contamination.

SUGGESTED FURTHER READINGS

Buckman, H.O., and Brady, N.C., 1969. *The Nature and Properties of Solids.* New York: Macmillan.

Correy, J.C. 1986. Management of soil systems for industrial wastes. In D.W. Nelson *et al.* (eds.), *Chemical Mobility and Reactivity in Soil Systems.* Madison, Wisc.: American Society of Agronomy.

STUDY QUESTIONS

1. What is soil pollution? Describe major routes of contamination of land.
2. Write in details on major types of soil pollution.

Chapter 9

ENVIRONMENT AND COMMUNITY HEALTH

Ecology is a key word in present-day health philosophy. It comes from the Greek "Oikos" meaning a house. Ecology is defined as the science of mutual relationship between living organisms and their environments. Human ecology is a subset of more general science of ecology.

A full understanding of health requires that humanity be seen as part of an ecosystem. The human ecosystem includes in addition to the - natural environment, all the dimensions of the man-made environment - physical, chemical, biological, psychological: in short, our culture and all its products. Disease is embedded in the ecosystem of man. Health, according to ecological concepts, is visualized as a state of dynamic equilibrium between man and his environment.

By constantly altering his environment or ecosystem by such activities as urbanization, industrialization, deforestation, land degradation, construction of irrigation canals and dams, man has created for himself new health problems. For example, the greatest threat to human health in India today is the ever-increasing, unplanned urbanization, growth of slums and deterioration of environment. As a result, diseases at one time thought to be primarily "rural" (e.g., filariasis, leprosy) have acquired serious urban dimensions. The agents of a number of diseases, for example, malaria and kala-azar, which were effectively controlled have shown a recurrence. The reasons for this must be sought in changes in the human ecology. Man's intrusion into ecological cycles of disease has resulted in zoonotic diseases such as kyasanur forest disease, rabies, yellow fever, monkeypox, lassa fever, etc. The Bhopal gas tragedy in 1984 highlights the danger of locating industries in urban area. The nuclear disaster in Soviet Russia in April 1986 is another grim reminder of environmental pollution. The construction of dams, irrigation systems and artificial lakes has created ecological niches favouring the breeding of mosquitos, snails and spread of filariasis, schistosomiasis and Japanese encephalitis. In fact, ecological factors are at the root of the geographic distribution of disease. Therefore it. has been said that good public health is basically good ecology.

Some have equated ecology with epidemiology. The main distinction between epidemiology and ecology is that while epidemiology is the study of the relationship between variations in man's environment and his state of health (or disease), ecology embraces the interrelationship of all living things. In this regard, epidemiology constitutes a special application of human ecology or that part of ecology relating to the state of human health.

It is now being increasingly recognized that environmental factors and ecological considerations must be built into the total planning process to prevent degradat!on of ecosystems. Prevention of disease through ecological or environmental manipulations or interventions is much safer, cheaper and a more effective rational approach than all the other means of control. It is through environmental manipulations that diseases such as cholera, typhoid, malaria and hookworm disease could be brought under control or eliminated. The greatest improvement in human health thus may be expected from an understanding and modification of the factors that favour disease occurrence in the human ecosystem. Professor Rene Dubos believes that man's capacity to adapt himself to ecological changes is not unlimited. Man can adapt himself only in so far as the mechanisms of adaptations are potentially present in his genetic code.

It was Hippocrates who first related disease to environment e.g., air, water, climate etc. Centuries later, Pettenkofer in Germany revived the concept of disease – environment association.

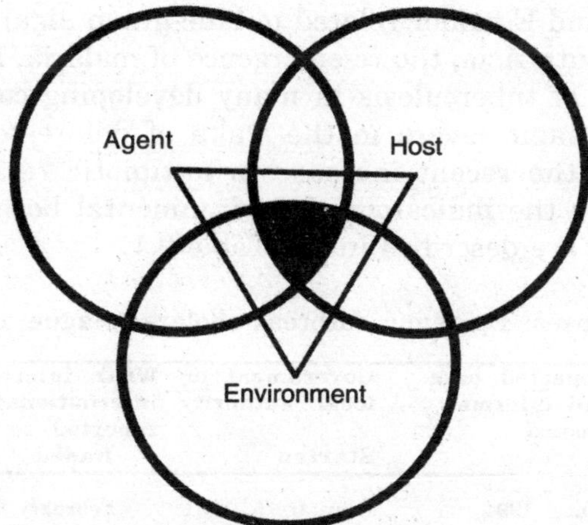

Fig. 9.1 Epidemiologic concepts of disease.

Environment is classified as "internal" and "external". The internal environment of man pertains to each and every component, part, every tissue, organ and organ system and their harmonius functioning within the system. The external or macro-environment consists of all those components to which man is exposed after conception. It is defined as "all that which is external to the individual human host", living or non-living, and with which he is in constant interaction. For descriptive purposes, the environment of man has been divided into three components

    i.   Physical environment,
    ii. Biological environment, and
    iii. Psycho-social environment

All of these can affect the health of man and his susceptibility to a disease.

Biological environment includes all living things viz: viruses and other microbial agents, insects, rodents, animals and plants. These are constantly struggling for their survival, and in this process some of them act as disease producing agents, reservoirs of infection, intermediate hosts and vectors of disease. For the most part, the partners manage to affect a harmonius inter-relationship, to achieve a peaceful co-existence. However, if for any reason this harmonius relationship is disturbed, ill health results. Biological agents of disease include viruses, rickettsiae, fungi, bacteria, protozoa and metazoa. These agents exhibit following biological properties.

i. *Infectivity:* This is the ability of an infectious agent to produce an infection in the host.
ii. *Pathogenicity:* This is the ability to induce clinically apparent illness.
iii. *Virulence:* This is defined as the proportion of clinical cases resulting in severe clinical manifestations.

9.1 Infectious Diseases, Environment and Ecology

An everchangig scenario of infectious disease persists in rich and poor countries, particularly the latter. Emerging and re-emerging diseases are an index of large scale environmental change. The rise of HIV/AIDS throughout the world, especially in third world, cholera in Peru and Equador related to blue-green algal blooms, the emergence of V-cholerae 0139 in South Asia, the re-emergence of malaria, kala-azar and plague in India, the accentuation of tuberculosis in many developing countries and the recent spread of viral hemorrhagic fevers in the wake of deforestation and extension of agricultural irrigation, the recent increase in antibiotic resistance of a variety of microbial pathogens, are the indicators of environmental borne diseases. Emergency trends of these diseases are described in the Table 9.1.

Table 9.1 Emerging diseases: Epidemic outbreak-cholera, plague and ebola

Epidemic outbreak	First suspected case found and informal report issued	Government or local authority Started	WHO intervention international team reported to WHO Ended	Current status
Cholera Peru and later in other Latin American countries	January, 1991	February 5, 1991	February 7, 1991	Low level of endemic
Cholera, Rwandan refugee camps in Zaire report of endemic (Goma)	July 20, 1994	July 20, 1994	July 25, 1994	No
Plague in India (Maharashtra)	August 26, 1994	September 28, 1994	October 13, 1994	No report of endemic
Plague in India (Surat)	September 22, 1994	October 13, 1994	October 26, 1994	No report of endemic
Ebola haemorrhagic fever in Zaire (Kikwit)	January 1995	May 6, 1995	May 9, 1995-	Low level transmission continues

Many health problems today reflect population pressure, human mobility, local climatic change, invasion of Nature's fringes all affecting the prevalence, distribution and spread of infectious diseases. The resurgence of plague in India in 1994 after three decades of absence from the country is probably related to environmental disturbances

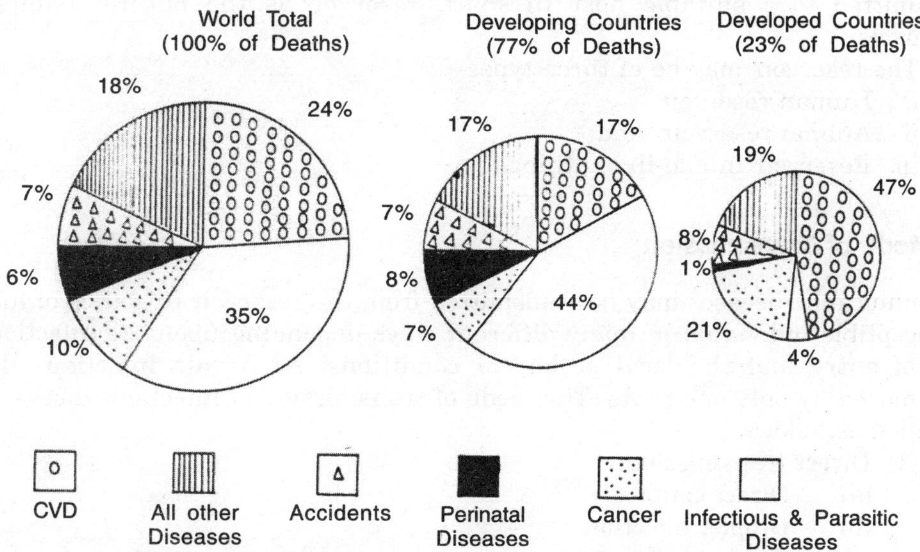

Fig. 9.2 : Estimated distribution of causes of death, 1990 *(Source:* WHO, 1992)

caused by a major earthquake in one area and heavy monsoon rains in the other. Disruption and destruction of the world's natural life support system constitute the greatest threats to human health. Ecological infringement, human mobility and human social change are potent force for new infectious diseases.

9.2 Infection

The entry and development or multiplication of an infectious agent in the body of man or animals is known as infection. A clinically manifested disease of man or animals resulting from an infection is known as *infectious disease.* However, an illness due to a specific infectious agent or its toxic products capable of being directly or indirectly transmitted from man to man, animal to animal or from environment to man or animal is called as *communicable disease.* A communicable disease is transmitted from the source of infection to the host. Basically there are three links in the chain of transmission viz. the reservoir, modes of transmission and the susceptible host (Fig. 9.3).

A reservoir is defined as any person, animal, arthropod, plant, soil or substance (or a combination of these) in which an infectious agent lives, multiplies and on which

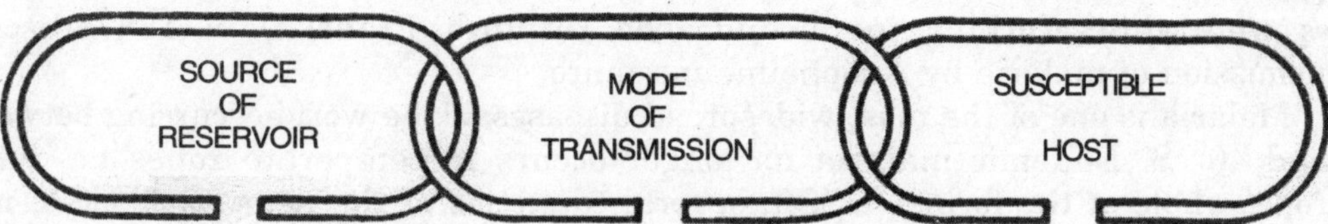

Fig. 9.3 : Chain of infection

it depends for survival and where it reproduces itself in such a manner that it can be transmitted to a suitable host In short, reservoir is the natural habitat of the organisms.

The reservoir may be of three types--
i.   Human reservoir
ii. Animal reservoir, and
iii. Reservoir in non-living things.

9.3 Mode of Transmission

Communicable diseases may be transmitted from the reservoir or source of infection to a susceptible individual in many different ways depending upon the infectious agent, port of entry and the local ecological conditions. As a rule infectious disease is transmitted by only one route. The mode of transmission of infectious diseases may be classified as follows.

(1) Direct transmission
    (a)   Direct contact
    (b)   Droplet infection
    (c)   Contact with soil
    (d)   Inoculation into skin or mucosa
    (e)   Transplacental

(2) Indirect transmission
    (a)   Vehicle born
    (b)   Vector borne
         (i)   Mechanical
         (ii)   Biological
    (c)   Air borne
    (d)   Fomite borne
    (e)   Unclean hands and fingers

A few of infectious diseases viz: malaria, tuberculosis, cholera, viral hepatitis, leishmaniasis are discussed in this chapter.

9.4 Malaria

Malaria is a protozoan disease caused by infection with parasites of the genus *Plasmodium* and transmitted to man by certain species of infected female Anopheline mosquito. Malaria is one of the oldest recorded diseases in the world. In 18th century, people in Italy associated malaria with "bad air". In 1880, Laveran, a French army surgeon discovered malaria parasite in Algiers. The main credit goes to *Sir Ronald Ross* who while working in Secundrabed (Andhra Pradesh) in 1887 discovered transmission of malaria by Anopheline mosquito.

Malaria is one of the most widespread diseases of the world occurring between 60^0 N and 40^0 S. Endemic malaria no longer occurs in temperate zones i.e. whole of Europe, whole of the former USSR, several countries of the near East, USA, most of the Carribbean, large areas of the northern and southern parts of South America, Australia and large parts of China (Hempel, 1982).

An estimate of the problem of malaria in India made in 1953 furnishes following information.

1. An annual incidence of 75 million cases
2. An annual incidence of 8 lakh deaths directly due to malaria.
3. 10.8% of cases of diagnosed malaria
4. A spleen rate of 15.7%
5. A child parasite rate of 3.8%
6. An infant parasite rate of 1.6%

According to World Health Organization, malaria afflicts about 500 million people annually the world over; 90 percent of the cases occur in Africa alone, and the rest mainly in India, Brazil, Sri Lanka, Afghanistan, Thailand, Vietnam, and Colombia. The disease claims about 2.7 million lives annually, including an estimated one million African children under five.

The magnitude of the problem was great. As a result National Malaria Control Programme (NMCP) and National Malaria Eradication Programme (NMEP) were initiated in 1953 and 1958 respectively. In 1961, there were only 40,000 cases of mal aria in the country. However, massive resurgence of malaria occurred in India in mid 1970s. In 1976 the incidence of malaria rose to a peak of 6.4 million cases with 59 deaths. A "Modified plan of Operation" was launched in 1977. As a result, the number of malaria cases dropped to 2.1 million by 1984. However, epidemiological situation has not shown any improvement since then. The entire population of India (95.9%) is now deemed to be under malaria risk (SEARO, 1985). The state wise condition of malaria cases is shown by Fig 9.4.

Malaria is usually accompanied by anaemia and enlargement of the spleen.

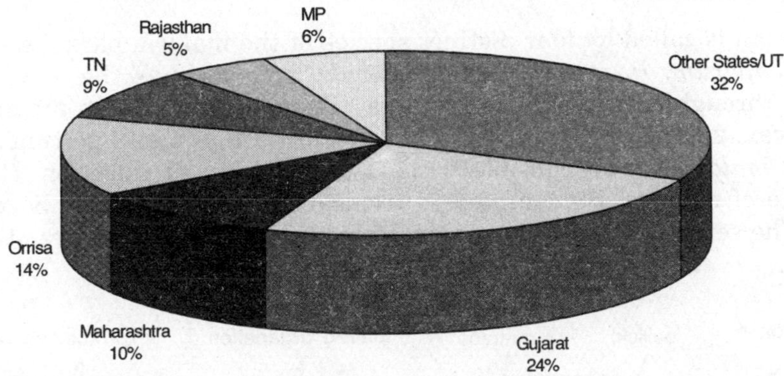

Fig. 9.4 : The state – wise contribution of malaria cases during 1992 (Prov.)

Hyperpyrexia (persistent high temperature), can cause febrile fits and brain damage. Occasionally, destruction of RBC can lead to hypoxia (reduced oxygen supply to the tissues). Malaria can also seriously jeopardize the outcome of pregnancy, the chances of abortion, low birth weight, still births and complication during delivery are enhanced.

Several factors including environmental factors, agent factors and host factors are responsible for the spread of malaria.

9.4.1 Environmental Factors

India's geographic position and climatic conditions had been, for long favourable to the transmission of malaria.

(a) *Season:* Malaria is a seasonal disease. In most parts of India, the maximum prevalence is from July to November.

(b) *Temperature:* Temperature affects the life cycle of malarial parasite. The optimum temperature for the development of parasite in insect vector is 20° C to 30° C. The parasite ceases to develop if the temperature is below 16° C. Temperature higher than 30° C is lethal to the parasite.

(c) *Humidity*: A relative humidity of 60% is considered necessary for mosquitoes to live their normal span of life. If the humidity is low, mosquitoes do not live a long.

(d) *Rainfall:* Rain increases the atmospheric humidity which is necessary for the survival of mosquitoes. However, heavy rain may have an adverse effect in flushing out the breeding places. The relationship between rainfall and mosquito breeding is of fundamental importance.

(e) *Altitude:* As a rule Anopehlines are not found at altitudes above 2,000-2,500 metres, due to unfavourable climatic conditions.

(f) *Man made Malaria:* Burrow pits, garden pools, irrigation channels and engineering projects have led to the breeding of mosquitoes and an increase in malaria. Malaria consequent on such undertakings is called man made malaria.

9.4.2 Agent Factors

Malaria in man is called by four distinct species of the malaria parasite - *Plasmodium vivax, P. falciparum. P. malariae and P. ovale. P. vivax* has the widest geographic distribution through out the world. In India, 70 percent infections are reported to be due to *P.* vivax, 25 to 30 percent due to *P. falciparum* and 4 to 8 percent due to mixed infection. *P. malariac* is responsible for less than 1 percent infection. *P. ovale* is the rare parasite of man mostly confined to tropical Africa. It has also been reported in Viet nam. The severity of malaria is related to the species of the parasite.

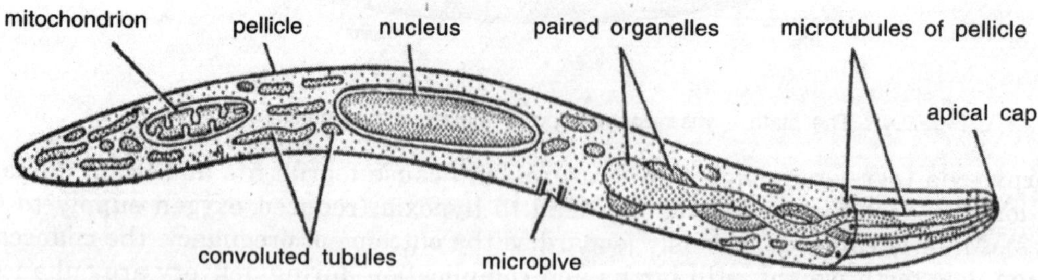

Fig. 9.5 : Plasmodium: Structure of a sporozoite as revealed by electron microscope.

9.4.3 Host Factors

The main variables of the human element that have an influence on malaria epidemiology include the following:

(a) *Age* : Malaria affect all ages. Newborn infants have considerable resistance to infection with *P. falciparum*. This has been attributed to high concentration of fetal haemoglobin.

(b) *Sex:* Males are more susceptible to malaria than females. Females in India are better clothed than males.

(c) *Race:* Individuals with AS haemoglobin (sickle cell trait) have a milder illness with falciparum infection than do those with normal *(AA)* haemoglobin.

(d) *Pregnancy:* Pregnancy increases the risk of malaria in women. Malaria during pregnancy may cause the intrauterine death of the foetus.

(e) *Socio-economic development:* It is generally accepted that malaria has disappeared from most developed countries as a result of socioeconomic development.

(f) *Housing:* Ill-ventilated and ill-lighted houses provide ideal indoor resting places for mosquitoes. Malaria is acquired in most cases by mosquito bites within the houses.

(g) *Population mobility:* Labourers, nomads and wandering tribes are outstanding examples of internal migration. Some of them may import malarial parasites in their blood and reintroduce malaria into areas where malaria has been controlled or eliminated. Imported malaria has become a public health problem in Europe, North America and other temperate parts of the world owing to increased air travel, tourism and migration.

(h) *Occupation:* Malaria is predominantly a rural disease and is closely related to agricultural practices.

(i) *Human habits:* Habits such as sleeping outdoors, nomadism, not using measures of personal protection influence man-vector contact and obviously the choice of control measures.

(j) *Immunity:* The epidemic of malaria is influenced by the immune status of the population. People living in endemic areas exposed continually to malaria develop considerable degree of resistance. Immunity to malaria is acquired only after repeated exposures after several years.

9.4.4 Vector of Malaria

Out of about 45 species of anopheline mosquitoes in India only a few are regarded as vectors of primary importance. These are *Anopheles culcifacies, A. fluviatilis, A. stephensi, A. minimus, A. philippinensis, A. sundiacus and A. maculatus.* The vectors of major importance *are A. culcifacies* in rural areas and *A. stephensi* in urban areas.

9.4.5 Mode of Transmission

1) *Vector transmission:* Malaria is transmitted by the bite of certain species of infected, female, anopheline mosquitoes. A single infected vector during her life time may infect several persons. The mosquito is not infective unless the

sporozoites are present in the salivary glands.

2) *Direct transmission:* Malaria may be induced accidentally by hypodermic intramuscular and intravenous injections of blood or plasma e.g. blood transfusion, malaria in drug addicts. Blood transfusion poses a problem because the parasites keep their infective activity during at least 14 days in blood bottles stored at -4°C.

3) *Congenital malaria:* Congenital infection of the new born from an infected mother may also occur but it is comparatively rare.

9.4.6 Different Faces of Malaria

Malaria varies with the type of parasite or vector involved, the value of resistance and eco-epidemiological considerations:

Tribal Malaria: Generally prevalent in the tribal areas of deep forests, forest-fringes (subtype I) and surrounding ecologically disturbed areas (subtype II), it poses a management problem due to poor health infrastructure and inadequate drugs. *P falciparum* is predominant with multiple drug resistance; deaths are common. Subtype II regions are prone to epidemics due to population migrations. According to NMEP data, tribal areas, with about eight percent of the total Indian population, account for 39 percent of all malaria cases and 68 percent of *P falciparum* cases in the country.

Rural Malaria: Moderately endemic in the irrigated areas of arid and semi-arid plains (subtype I), rural malaria has a predominance of *P vivax*. Widespread vector resistance to multiple insecticides and localized parasite resistance to chloroquine pose problems for the moderately developed health infrastructure. Poor irrigation, water-logging, seepages from canals and poor drainage systems are responsible for the increase of malaria. The rural areas around the Upper Krishna irrigation project in Karnataka, Sardar Sarover Dam in Gujarat and the Indira Gandhi Canal in Rajasthan have witnessed outbreaks recently.

Urban Malaria: Towns and cities (subtype II) are moderately endemic with *P vivax* predominance and focal incidence of *P falciparum*. Sporadic epidemics occur, especially around construction projects. The main vectors are *A stephansi* and *A culicifacies*. Sub-urban and per-urban areas (subtype II), with unplanned settlements, slums and poor sanitary conditions are prone to epidemics and deaths due to *P faciparum* infection.

Industrial Malaria: Development projects are highly prone to malaria epidemics resulting from drug resistant parasites, one or more vectors refractory to transmission control and limited health infrastructure. NMEP data clearly shows that the construction of the Mirzapur thermal power project, Uttar Pradesh, saw a five-fold increase in malaria incidence in the district in 1980. The Mathura oil refinery and the National Thermal Power Corporation unit in Dadri also contributed to the rise of malaria incidence in UP.

Border Malaria: Mainly caused by migrations across International borders, this type cuts across all epidemiological boundaries and often contributes new and drug resistant parasite strains. Migration even within the country complicates the epidemiological classification. Kondrashir, a WHO malaria expert, estimated in 1991 that about one-sixth of India's population moves annually during the transmission season from non-malarious areas to malarious areas and vice versa.

9.4.7 Malaria Control Strategies

A country-wide review of malaria control strategies was undertaken in 1995, following which a Malaria Action Programme (MAP) was launched, as a commitment to WHO Global Malaria Control Strategy (1992). The new strategy emphasizes decentralization, epidemiological approaches in malaria control, community participation, management information systems and health education among other things. Under the proposed new strategy, the financial, technical and administrative responsibilities will be shared by major establishments in the private and public sectors. Health impact assessment has been made an integral part of the environmental impact assessment of all developmental projects.

Malaria research has several aspects: developing new drugs, vaccines, simple diagnostic methods, insecticides and larvicides, methods of sterilizing mosquitoes, epidemiological and entomological studies, host parasite interactions, immunology of the disease and finally ecology.

9.4.8 Biological Control

Efforts have been recently made to control mosquito breeding by spraying larvicidal microbes. The fish *Gambusia affinis* popularly known as mosquito fish has been used throughout the tropical world to control malaria spreading mosquitoes. Four other species of the fish *Daneo rario*, *Aplocheilus panchax*, *Oryzias melastigma* and *Tilapia mossambica* are also known to feed on mosquito larvae (Fig 9.6).

Fig. 9.6: Mosquito fish *Gambusia affinis.*

Impregnated bednets are being increasingly recommended for malaria control following encouraging results in Orissa and the north-east. These bednets are impregnated twice a year with synthetic pyrethroids offering effective protection for six-eight months from mosquitoes) bed bugs and headlice.

Among the new drugs waiting to be introduced in the market is a Chinese herb derivative *Artemisia accua*. Artemisin (the raw material) has been identified as a potent antimalarial drug. It has cured multi-drug resistant malaria cases in Viet nam, Thailand and Combodia.

In India clinical trials of Arteether, a derivative developed by Central Drug Research Institute, Lucknow are currently underway. Indigenous production and sale of mafloquin was cleared by Govt. of India in March, 1996.

9.4.9 Vaccine Development

There are three major stages in the life cycle of the parasite which are targets of vaccine development.

- A vaccine based on sporozoites is designed to prevent infection.
- Vaccines based on the asexual blood stages of the parasite (mesozoites) will not prevent infection) but can reduce or eliminate parasites in the blood.
- Vaccines directed at the sexual stages of the parasite (gametocytes) aim to interfere with the ability of the parasite to infect mosquitoes and thereby prevent transmission of the disease. A vaccine developed by a group of Australian researchers is undergoing trials since 1994 in USA and Africa.

There are no magic bullets to beat malaria as yet. Setting aside eradication, better public health management practices, strong primary health care infrastructure, emphasis on preventive measures and community participation have the potential to reduce malaria incidence and make the WHO's dream of a one-fifth reduction in malaria incidence by 2000 A.D. in at least 75 percent of the affected countries- a reality.

9.5 Tuberculosis

Tuberculosis is a specific infectious disease caused by *Myobacterium tuberculosis*. The pathogen primarily affects lungs and causes pulmonary tuberculosis. It can also infect intestine, meninges, bones, ioints, lymph gland, skin and other tissues of the body. The disease also affects animals and is called as bovine tuberculosis. Tuberculosis remains a world wide public health problem. There are 15-20 million cases of infectious tuberculosis in the world. Eight million new victims and 2.9 million deaths are known to occur every year. The disease is credited with killing over one million women and 1,70,000 children every year. WHO has declared TB as a global emergency.

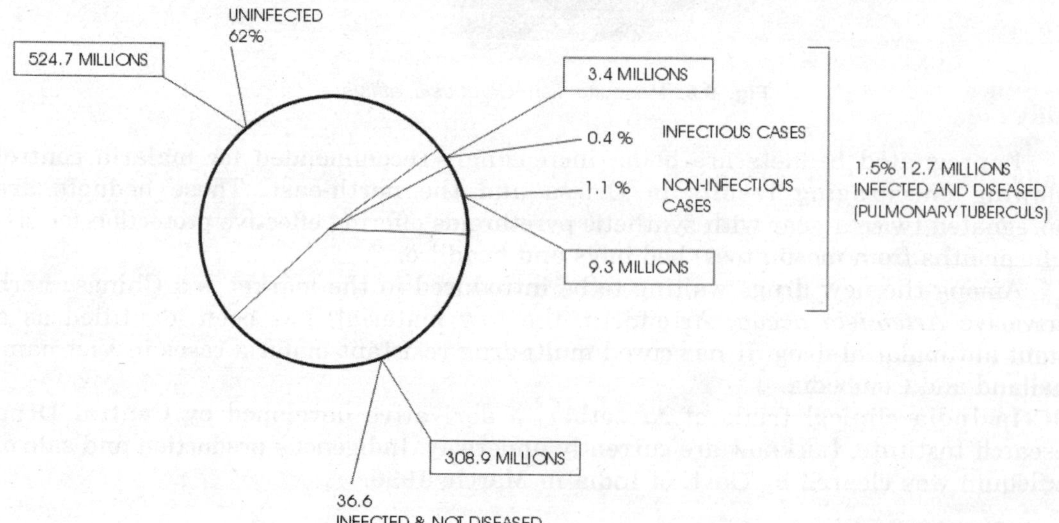

Fig. 9.7: Tuberculosis problem in India [Source: National Tuberculosis Institute Bangalore]

The incidence of TB rose by 12 percent in the US between 1986-1991; Italy reported a 28 percent jump between 1988 and 1990 and Switzerland a 33 percent increase from 1986 to 1990. Experts have attributed this upward trend to changes in the social structure of the cities, the HIV epidemic and a failure in certain regions to improve public health treatment programmes. The problem has assumed a far more fearsome aspect in developing countries. Asian countries are the home to majority of the world's TB infected population. In India, 30 percent male and 35 percent female population suffer from TB. According to National Tuberculosis Institute, Bangalore 30 percent of population is infected in contrast to 2-3 percent in developed countries (Table 9.2).

Table 9.2 Prevalence of Infection

Age group (years)	Infected (%)
0-4	1.0
5-10	6.4
10-14	15.4
15-24	31.9
25-34	47.3
35-44	54.8
45-54	60.7
55+	62.1
Total	30.4

9.5.1 Natural History of Tuberculosis

A few factors viz: agent factors, host factors and social factors can be discussed to understand the natural history of TB.

9.5.2 Agent Factors

It is caused by a facultative intracellular parasite *Myobacterium tuberculosis*. It is readily ingested by pahagocytes and is resistant to intracellular killing.

The most common source of infection is human. The bovine source of infection is usually the infected milk. Patients are infective as long as they remain untreated. Effective antimicrobial treatment reduces infectivity by 90 percent within 48 hours.

9.5.3 Host Factors

Tuberculosis affects all ages. It is more prevalent in males than females. It is not a hereditary disease but inherited susceptibility is an important risk factor. Man has no inherited immunity against tuberculosis. It is acquired as a result of natural infection or BCG vaccination.

9.5.4 Social Factors

Tuberculosis has been described as a barometer of social welfare. Social factors include many non-medical factors such as poor quality of life, poor housing and overcrowding, population explosion, malnutrition, lack of education, large families, early marriages, lack of awareness of causes of illness etc. In most cases, bacteria attacks the lungs. Pulmonary TB destroys the lung tissue, rupturing blood vessels in the process. Victims are virtually consumed by the disease.

9.5.5 Mode of Transmission

Tuberculosis is transmitted mainly by droplet infection and droplet nuclei generated by sputum positive patients with pulmonary tuberculosis. The frequency and vigour of cough and the ventilation of the environment influence transmission of infection.

9.5.6 The Control of Tuberculosis

According to WHO, "tuberculosis control" is said to be achieved when the prevalence of natural infection in the age group 0-14 years is of the order of 1 percent. At present, this is about 40 percent in India.

The control measures should consists of

   i.   Curative component,

   ii. Preventive component

Curative component includes case finding and chemotherapy whereas preventive component includes vaccination. The first step in tuberculosis control is early detection of sputum positive cases, sociological studies on pulmonary tuberculosis and mass miniature radiographic examination. Case finding should follow chemotherapy.

9.5.7 Chemotherapy

Thirteen drugs are now available, active against *M tuberculosis* of which six are considered to be essential. They include rifampicin (RMP), 1 NH streptomycin, pyrazinamide, ethambutol and thioacetazone.

9.5.8 BCG Vaccination

Ever since Koch discovered *M. tuberculosis,* attempts have been made to prepare a prophylactic vaccine against tuberculosis. After a continuous series of experiments two fresh scientists calmette and Guerin were able to raise a strain known as Bacille Calmette Guverin or BCG that was avirulent for man. BCG was accepted as a safe preventive measure in 1948.

Today BCG vaccination is a fundamental component of a national tuberculosis programme. It is a part of WHO Expanded Programme on Immunization. It is badly needed by the developing countries where tuberculosis is still a major health problem.

9.5.9 National Tuberculosis Programme (NTP)

The National Tuberculosis Programme was initiated in 1962. The long term goal of NTP is "to reduce the problem of tuberculosis in the community sufficiently quickly to the level where it ceases to be a public health problem". The backbone of National Tuberculosis programme is the District Tuberculosis programme (DTP). It was evolved by the National Tuberculosis Institute, Bangalore and was accepted by Govt. of India for implementation in 1962.

The District Tuberculosis Programme consists of one District Tuberculosis Centre (DTC) and on an average 50 peripheral health centers. To implement the programme, the Govt. has posted a specially trained team into function. The team includes-

1. District Tuberculosis officer
2. Laboratory technician
3. Treatment Organizer
4. X-ray technician
5. Non-medical team leader
6. Statistical assistant

The programme works under the ambit of the District Health Organization.

Despite effective case finding and therapeutic tools and declined mortality and morbidity in some countries, tuberculosis still remains to be a serious communicable disease worldwide. The chronic nature of the disease, the ability of the tubercle bacilli to remain alive in the human body for years, the concentration of the disease in the older age groups, the increased expectation of life, high prevalence infection rates in some countries, the relatively high reactivation rate, the emergence of drug resistant strains, association of tuberculosis and HIV infection and above all, the perpetuation of the "non-specific determinations" of the disease in the third world countries impede the rapid conquest of the disease.

9.6 AIDS (HIV)

AIDS, the acquired immuno-deficiency syndrome (sometimes called "Slim disease") is a newly described usually fatal illness caused by a retrovirus known as the human immuno-deficiency virus (HIV) which breaks down the body's immune system, leaving the victim vulnerable to a host of life threatening opportunistic infections, neurological disorders or unusual malignancies (WHO, 1986). Strictly speaking, the term AIDS refers only to the last stage of HIV infection. It can be called as modern pandemic, affecting both industrialized and developing countries.

HIV has infected millions of women, men and children worldwide. According to an estimate of WHO, there were 2.5 million adult full blown AID cases recorded throughout the world in 1993. AIDS was first recognized in USA in 1981, earlier cases were found by retrospective analysis to have occurred in 1978 in USA and in late 1970s in equatorial Africa (WHO, 1986). An explosion of HIV has recently occurred in South-East Asia, particularly in Thailand, Burma and India. HIV/AIDS is now known to occur in countries such a Paraguay, Greenland and Pacific Island nations of Fiji, Papua New Guinea and Samoa. The cumulative AIDS cases year-wise and country-wise are shown in the Table 9.3.

9.6.1 Epidemiological Features

i. *Agent Factors:* In May 1986, the International Committee on the Taxonomy named the AIDS causing virus as human immunodeficiency virus. It is 1/10,000[th] of mm in diameter. It is a protein capsule containing two short strands of genetic material (RNA) and a few enzymes. The virus uses human cells to perpetuate itself (Fig. 9.8).

Table 9.3 Reporting of HIV and AIDS in some countries

Country	HIV First infection reported	No. of AID cases reported 1991	1992	Cumulative no. of AIDS cases reported by 1 July 1993
Bangladesh	1989	1	0	1
India	1986	45	140	312
Nepal	1988	4	3	18
Sri Lanka	1986	3	0	24
Thailand	1985	398	730	1569
Pakistan	1987	4	8	26
Saudi Arabia	1986	3	8	50
Sudan	1986	182	184	777
Austria	1985	176	170	943
France	1982	4248	4151	24226
Italy	1982	3607	3812	16860
Russian Federation	1987	25	26	127
Spain	1981	3469	3469	18347
Switzerland	1985	432	445	3028
United States of America	1981	45524	49566	289320
Brazil	1980	8746	7640	36481
United Republic of Tanzania	1983	12059	4579	38719
Kenya	1980	9202	6762	31185
Uganda	1983	8706	4421	34611
Zaire	1986	3120	1181	21008
Australia	1985	670	432	3697
China	1985	3	3	11
Japan	1985	82	90	543
Singapore	1985	12	18	58

The virus replicates in actively dividing T4 lymphocytes and like other retroviruses can remain in a latent state in lymphoid cells that can be activated. The virus has the unique ability to destroy human T4 helper cells, a subset of the human

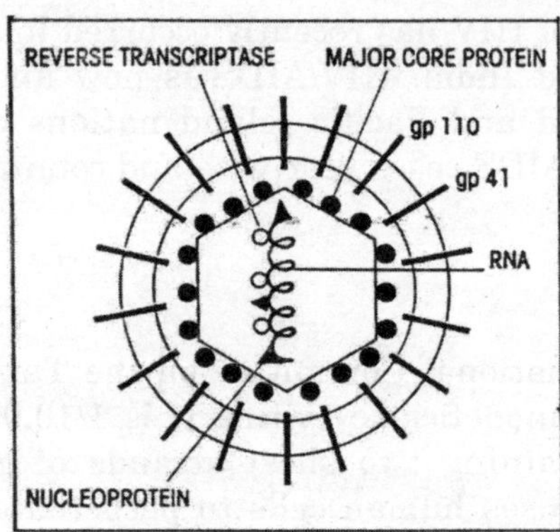

Fig. 9.8a: HIV Structure

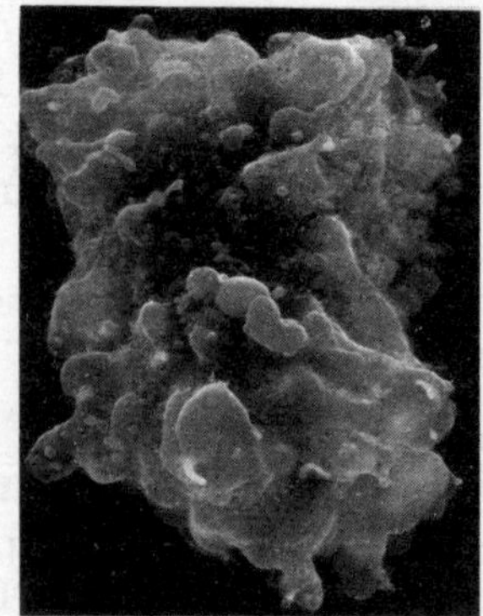

Fig. 9.8b: A helper T cell under the attack of AIDS virus.

T4 lymphocytes. The virus can spread throughout the body. It can pass through the blood-bran barrier and can destroy some brain cells. HIV mutates rapidly. There are two types of HIV-the most common HIV-I and a more recently recognized HIV-2.

The virus has been found in greatest concentration in blood, semen and CSF. However, concentrations have been detected in tears, saliva, breast-milk, urine, cervical and vaginal secretions. To date only blood and semen have been conclusively shown to transmit the virus.

ii. *Host Factors*: Age, sex and immunological status of the host are important host factors. Most cases of infection occur in sexually active persons between the age group of 20-49 years. Children under 15 make up less than 3 percent of cases.

In North America, Europe and Australia, about 70 percent of cases are homosexual or bisexual man. Multiple sexual partners, anal intercourse and male homosexuality increase the risk of infection. Higher rate of HIV infection is found in prostitutes. The immune system disorders associated with HIV infection are considered to occur primarily from the gradual depletion of a specialized group of white blood cells (lymphocytes) called T-helper and T-4 cells.

9.6.2 Mode of Transmission

The causative virus is transmitted from person to person, most frequently through sexual activity or blood transfusion.

i. *Sexual Transmission*: AIDS is first and foremost a sexually transmitted disease. Any vaginal, anal or oral sex can spread AIDS. Every single act of unprotected intercourse with an HIV infected person exposes the uninfected partner to the risk of infection. Anal intercourse carries a higher risk of transmission than vaginal intercourse.

ii. *Blood Contact*: AIDS is also transmitted by contaminated blood-transfusion of whole blood cells, platelets and factors VIII and IX derived from human transfusion of infected blood is estimated to be 95%. Needle sharing by drug users is a major cause of AIDS transmission.

iii. *Maternal-foetal Transmission*: HIV may pass from the infected mother to her foetus, through placenta or by breast feeding. Risks are higher if the mother is newly infected or if she has already developed AIDS.

There is no evidence that HIV is transmitted through mosquitoes or any other insects, casual social contact with infected persons even within households or by food or water. There is no evidence of spread to health case workers in their professional contact with people with AIDS.

9.6.3 Control of AIDS

There are three basic approaches to the control of AIDS

i. *Education*: Until a vaccine or cure for AIDS is found, the only means at present available is health education on avoiding indiscriminate sex, and using condoms. One should also avoid the use of shared razors and toothbrushes) needles and syringes. Infected women should avoid becoming pregnant. All mass media channels should be involved in educating the people on AIDS, its nature, transmission and prevention.

ii. ***Prevention of Bloodborne HIV Transmission:*** People in high risk groups should be asked to refrain from donating blood, body organs, sperm or other tissues. Blood should be screened for HIV_1 and HIV_2 before transfusion. Strict sterilization practices should be ensured in hospitals and clinics. One should avoid injections unless they are absolutely necessary.

iii. ***Specific Prophylaxis:*** At present there is no vaccine or cure for treatment of HIV infection/AIDS. Antiviral chemotherapy with the chemical compound Zidovudine (AZT) while not a cure, has proved to be useful in prolonging the life of severely ill patients. The AZT, however, neither restores the immune system, nor does in destroy the HIV virus already installed in cells.

iv. ***Primary Health Care:*** AIDS touches all aspects of primary health case, including mother child health, family planning and education, it is important therefore, that AIDS control programmes are not developed in isolation. Integration into countries primary health care system is essential.

The WHO has launched a "Global Programme on AIDS" on Feb. 1, 1987 to provide global leadership and to support the development of National AIDS programme.

9.7 Cholera

Cholera is both an epidemic and endemic disease. The 7th pandemic began in 1961 and today it involves more than 92 countries in Asia, Africa and Europe. Global situation shows definite downward trend of reported cases.

Table 9.4 Number of cholera cases and percent deaths notified to WHO 1961-1993

Year	Cases (000)	Deaths, %
1961	62	49.3
1971	176	14.8
1981	51	4.8
1991	595	3.2
1992	46.2	1.7
1993	297	1.7

Source: Weekly Epidemiological Record No.3, 29 Jan. 1994.

Cholera has been present in India since antiquity. In the history of cholera, four phases have been described.

a) First phase (prior to 1817) - the disease was confined to the east.
b) Second period (1819 - 1923) - six pandemics were described, out of which five originated from India.
c) Third period (1923-1960) - the disease retreated from Europe and came to be known as "Asiatic Cholera'.
d) Fourth period (1961-todate) - it started off as the famous 7th pandemic. After spreading to Hong-Kong and Philippines, it moved steadily north and west ward and reached India in 1964. Then it spread to the middle East, and North, East and West Africa, Israel, Italy and Portugal. Cases have also

occurred in Australia, New Zealand. USA and other developed countries but the disease could not establish itself in these countries. The 7th pandemic is still continuing.

The geographical distribution of cholera in India, keeps on changing. West Bengal is no more a "home of cholera", now. Andaman and Nicobar Islands were infected for the first time in 1966, and the desert areas of Rajasthan in 1969. Currently the large endemic foci of cholera are found in Maharashtra, Tamil Nadu, Karnataka, Delhi and Kerala These states account for about 80 percent of reported incidences in the counlry.

9.7.1 Epidemiological Features

i. **Agent Factors:** Cholera is caused by *Vibrio cholerae* 0 group 1, two biotypes - classical and El Tor vibrios are further divided into three serological types namely Inaba, Ogawa and Hikojima.

ii. **Host Factors:** Cholera affects all ages and both sexes. In endemic areas, attack rate is highest in children. Mass population movements viz. pilgrimages, fairs, and festivals result in increased risk of exposure to infection. The incidence of cholera tends to be the highest in the lower socio-economic groups and this is attributable to poor hygiene.

iii. **Environmental Factors:** Vibrio transmission is readily possible in a community with poor environmental sanitation. Important environmental factors are contaminated water and food. Flies may carry *V. cholerae*. Numerous social factors have also been found responsible for endemicity of cholera in India. These comprise certain human habits favouring water and soil pollution, low standards of personal hygiene, lack of education and poor quality of life.

9.7.2 Mode of Transmission and Pathogenicity

Transmission occurs from man to man via faecally contaminated water, contaminated food and drinks and the direct contact.

The main symptom of cholera is diarrhoea. According to current concepts, the cholera vibrio gets through the mucus which overlies the intestinal epithelium. It probably secretes mucinase which helps it to move rapidly through the mucus. Then it gets adhered to the intestinal epithelial cells with the help of a probable adhesence factor. After adhering to mucosa, the pathogen produces the enterotoxin. The enteriotoxin consists of two parts - the light or L-toxin and the heavy or H-toxin. The L-toxin combines with substances in the epithelial cell membrane called gangliosides. The binding is irreversible. The mode of action of H-toxin is not fully clear. What we know is that H-toxin activates adenyl cyclase in the intestinal epithelial cells. The activated adenyl cyclase causes a rise in another substance called 3,5-adenosine monophosphate better known as cyclic AMP or C-AMP provides energy which divides fluid and ions into the lumen of the intestine. The increase in the fluid is the cause of diarrhoea and not the increased peristalsis. There is no evidence that *V. cholerae* invades any tissue, nor the entero-toxin to have any direct effect on any organ other than the small intestine.

9.7.3 The Control of Cholera

WHO (1980) has proposed "guidelines for cholera control". These include- verification of the diagnosis, notification, early case finding, establishment of treatment centers, rehydration therapy (oral rehydration and intravenous rehydration), adjuncts to therapy, epidemiological investigations, sanitation measures, chemoprophylaxis, vaccination and health education programme.

9.8 Poliomyelitis

Poliomyelitis is an acute viral infection caused by a RNA virus. It is primarily an infection of the human alimentary tract but the virus may infect the central nervous system in about 1 percent of cases resulting in varying degrees of paralysis and possibly death. The extensive use of polio vaccines since 1954 has virtually eliminated the disease in developed countries. However, polio continues to be a serious health problem in India. The first epidemic of polio was reported in and around Bombay City in 1949. Sincethen, more epidemics have occurred at various times in Andhra Pradesh, Uttar Praesh, Gujarat, Maharashtra, Rajasthan, Tamil Nadu and Delhi. More recently widespread epidemics have occurred in 1981 and 1987, despite the fact that a national immunization programme (EPI) has been in operation since 1978.

Table 9.5 Reported cases of Poliomyelitis in India

Year	No. of cases
1980	19051
1981	38090
1982	26302
1983	24727
1984	23250
1985	32584
1986	20169
1987	28264
1988	24257
1989	13866
1990	10408
1991	6028
1992	9440

Source: Govt. of India, Annual Report 1993-94, DGHS

9.8.1 Epidemiological factors

Poliomyelitis can occur sporadically, endemically or epidemically.

i. *Agent Factors:* The causative agent is the poliovirus which has three serotypes 1,2 and 3. Most outbreaks of paralytic polio are due to type-1 virus. It can survive for long periods in the external environment. In a cold environment, it can live in water for 4 months and in faeces for six months.

ii. *Host Factors:* In India, polio is essentially a disease of infancy and childhood. About 50 percent of cases are reported in infancy. The most vulnerably age is between 6 months and 3 years. Sex differences have been noted in the ratio of 3 males to 1 female.

iii. *Environmental Factors:* Polio is more likely to occur during the rainy season. Approximately 60 percent of cases recorded in India were June to September. The environmental sources of infection are contaminated water, food and flies. Polio virus survives for a long time in cold environment. Overcrowding and poor sanitation provide opportunities for exposure to infection.

9.8.2 Mode of Transmission

Main route of infection is faecal-oral route. The infection may spread directly through contaminated fingers where hygiene is poor, or indirectly through contaminated water, milk, foods etc. Droplet infection may occur in the acute phase of disease when the virus occurs in the throat. Incubation period is usually 7 to 14 days.

9.8.3 Prevention

Immunization is the only effective means of preventing poliomyelitis. It is essential to immunize all infants by 6 months of age to protect them against polio. Two types of vaccines are used throughout the world.
    (i) Inactivated (SALK) Polio Vaccine (IPV)
    (ii) Oral (Sabin) Polio Vaccine (OPV)

9.8.4 Inactivated (SALK) Polio Vaccine

This vaccine contains all three types of poliovirus, inactivated by formalin. The initial course of immunization consists of four inoculations. IPV induces humoral antibodies (1gM, IgG and IgA) but does not induce intestinal or local immunity. Thus it gives protection from paralysis but offers nothing because the wild viruses can still multiply in the gut and be a source of infection for others. This is a major drawback of IPV. However, it is safe to administer as it does not contain live virus.

9.8.5 Oral (SABIN) Polio Vaccine (OPV)

OPV was described by Sabin in 1957. It contains live attenuated cell cultures. Ideally each virus type should be given separately as monovalent vaccine, but for administrative convenience, it is given as trivalent vaccine.

9.8.6 National Immunization Programme

The WHO programme on Immunization (EPI) and the National Immunization Programme in India recommend a primary course of 3 doses of OPV at one month intervals, commencing the first dose when infant is 6 weeks old, OPV is given currently with DPT. BCG can be given with the first dose of OPV. It is very important to complete vaccination of all infants before 6 months of age. This is because most polio cases occur between the ages of 6 months and 3 years. One booster dose of OPV is recommended 12 to 18 months later.

9.9 Viral Hepatitis

Viral hepatitis may be defined as the infection of liver caused by half a dozen viruses viz: hepatitis A virus (HAV), hepatitis B virus (HBV), hepatitis D virus (HDV or the delta agent), epidemic non-A hepatitis virus and by atleast two non-A and non-B viruses. Other viruses may be implicated in hepatitis such as cytomegalo virus, Epstein-Barr virus, yellow fever virus and rubella virus. Viruses of herpes-simplex, varicalla and adenoviruses can also cause severe hepatitis in immuno-compromised individuals but are rare.

HAV causes a relatively benign disease and dose not cause a major public health problem. However, HBV is responsible for severe liver damage and is associated with chronic liver diseases and hepatocellular carcinoma. It has been estimated that about 4 million people suffer every year from one or the other form of acute viral hepatitis.

9.9.1 Hepatitis A

It was formerly known as "infectious" hepatitis or epidemic jaundice. It is endemic in most of the developing countries. Epidemics of hepatitis A often evolve slowly; involve wide geographic areas and last many months but faecal contamination of drinking water may cause the epidemic to evolve explosively.

9.9.2 Epidemiological Factors

   i. *Agent Factors:* The causative agent is an enterovirus (type 72) of the Picornqviridae family. It multiplies only in hepatocytes. Virus is fairly resistant to heat and chemicals. Man is the only reservoir of infection.

   ii. *Host Factors:* HAV is more frequent among children than in adults. Both sexes are equally susceptible.

   iii. *Environmental Factors:* In India, the disease tends to be associated with heavy rainfall. Poor sanitation and overcrowding favour the spread of infection, giving rise to water borne and food borne epidemics.

9.9.3 Mode of Transmission

Faecal-oral route is the major route of transmission. It may occur by direct contact or indirectly by way of contaminated water, food or milk

9.9.4 Prevention and Containment

Hepatitis A may be controlled through control of reservoir, control of transmission, control of susceptible population, and vaccination, The best method to curtail infection is by promoting simple measures of personal and community hygiene. Community water supplies should be purified by flocculation, filtration and adequate chlorination. Vaccine is under the process of development.

9.9.5 Hepatitis B

Hepatitis B is the major health problem in India. It is estimated that there will be 400 million HBV patients in the world if HBV vaccine is not used widely.

9.9.6 Epidemiological Factors

i. ***Agent Factors:*** Hepatitis B virus was discovered by Blumberg in 1963. HBV is complex, 42 run, double shelled DNA virus, originally known as the "Dane particle". It replicates in the liver cells. Contaminated blood is the main source of infection. Man is the only reservoir of infection.

ii. ***Host Factors:*** Highest infection is found in 20-40 year age group. Risk groups comprise surgeons; health care and laboratory personnel, recipients of blood transfusion, homosexuals, prostitutes and precutaneous drug abusers. Serological screening and vaccination of high risk groups is highly recommended.

9.9.7 Prevention and Containment

Different measures are available to contain the disease. Hepatitis B vaccine, RDNA-- yeast derived vaccine, hepatitis B immunoglobin (HB16), simultaneous administration of HBIG and the hepatitis vaccine are presently being practiced.

9.10 Lymphatic Filariasis

The disease "lymphatic filariasis" covers infection with three closely related microfilariae nematode worms – *Wuchereria bancrofti, Brugia malayi.* All three infections are transmitted to man by the bite of infective mosquitoes. The disease manifestations range from none to both acute and chronic manifestations, such as lymphangitis, lymphadenitis, elephantiasis of genital organs legs and arms or as a hypersensitivity state such as tropical pulmonary eosinophilia or a typical filarial arthritis. Though not fatal, the disease is responsible for considerable suffering deformity and disability.

Filariasis is a global problem. A WHO expert committee has estimated that 750 million are actually infected. Lymphatic filariasis is a major public health problem in India. 396 million people were found to be infected in 1993. Heavily infected areas are found in Uttar Pradesh, Bihar, Andhra Pradesh, Orissa, Tamil Nadu, Kerala and Gujarat.

9.10.1 Epidemiological Factors

i. ***Agent Factors:*** These are at least 8 species of filarial parasites that are specific to man.

The first three are responsible for lymphatic filarias is and the rest for non-lymphatic filariasis. Man is the definitive host and mosquito the intermediate host. The adult worms are usually found in the lymphatic system of man. The males are about 40 mm long and the females 50 to 100 mm long. The females are viviparous. The adult worms may survive for 15 years or more (Fig. 9.9).

ii. ***Host Factors:*** Man is the natural host. All ages are susceptible to infection. Rate of infection is higher in man. Lymphatic filariasis is often associated with urbanization, industrialization, migration of people, illiteracy, poverty and poor sanitation.

Table 9.6 Human filarial infections

Organism	Vector	Disease
1. *Wuchereria bancrofti*	Culex Mosquitoes	Lymphatic filariasis
2. *Brugia malaya*	Mansonia Mosquitoes	Lymphatic filariasis
3. *Brugia timori*	Anopheles Mosquitoes	Lymphatic filariasis
4. *Onchocerca volvulus*	Mansonia Mosquitoes	Lymphatic filariasis
5. *Loa loa*	Chrysops flies	Subcutaneous nodules River blindness
6. *T. perstans*	Culicoides	Probably rarely any clinical illness
7. *T. streptocerca*	Culicoides	Probably rarely any clinical illness
8. *Mansonella azzardi*	Culicoides	Probably rarely any clinical illness

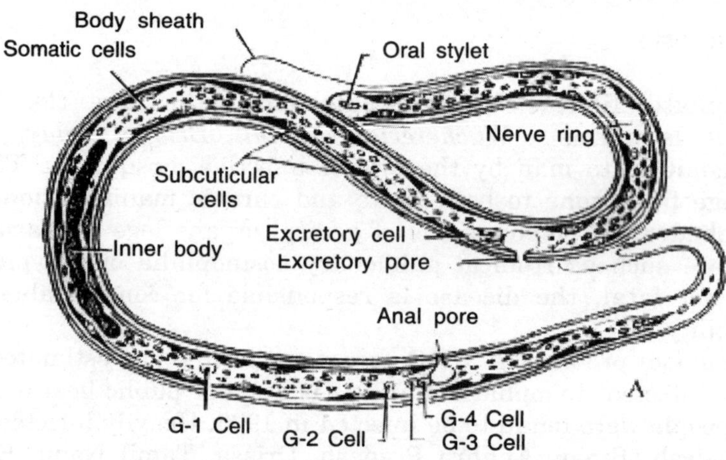

Fig. 9.9: *Wuchereria bancrofti*. (After Brown, 1950.)

iii. ***Environmental Factors:*** Climate is an important factor in the epidemiology of filariasis. It influences breeding of mosquitoes. Filariasis is also associated with bad drainage. Inadequate sewage disposal and lack of town planning have aggravated the problem of filariasis in India.

9.10.2 Mode of Transmission

Filariasis is transmitted by the bite of infected vector mosquitoes. The parasite is deposited near the site of puncture. It finally reaches the lymphatic system. The dynamics of transmission depend upon man-mosquito contact.

Fig. 9.10: Filaria map of India

9.10.3 Control Measures

The current strategy of filariasis control is based on:

    a. Chemotherapy

    b. Vector control

Diethycarbamazine (DEC) is the only drug available at present against lymphatic filariasis. Use of DEC medicated salt is a special form of mass treatment using very low doses of the drug over a long period of time.

Vector control may also be beneficial when used in conjunction with mass treatment with DEC. The most important element in vector control is the reduction of the target mosquito population. The mosquito population can be controlled by antilarval measures. Antilarval activities comprise chemical control, removal of *Pistia* plant and minor environmental measures.

9.10.4 National Filarial Control Programme

National filarial control programme (NFCP) has been in operation since 1955. According to recent estimates about 396 million people are exposed to the risk of infection; 19 million manifest the disease and 25 million have filarial parasites in their blood. Under the NFCP, the following activities are being undertaken and supervised by the Director, National Institute of Communicable Diseases, Delhi.

   i)   required anti larval measures

   ii) antiparasitic measures

So far, 238 out of 300 districts situated in endemic areas have been surveyed, and 175 have been found to be endemic for filariasis. At present there are about 206 filaria control units, 27 survey units and 195 filaria control clinics functioning in the endemic areas. The population protected so far is 43.43 million out of 304 million at risk.

9.11 Leishmaniasis

Leishmaniasis is a collective term used for a group of protozoan diseases caused by parasites of the genus *Leishmania* and transmitted to man by the bite of female phelbotomine sandfly. They are responsible for various syndromes viz: Kala azar or visceral leishmaniasis (VL), cutaneous leishmaniasis (CL), muco-cutaneous

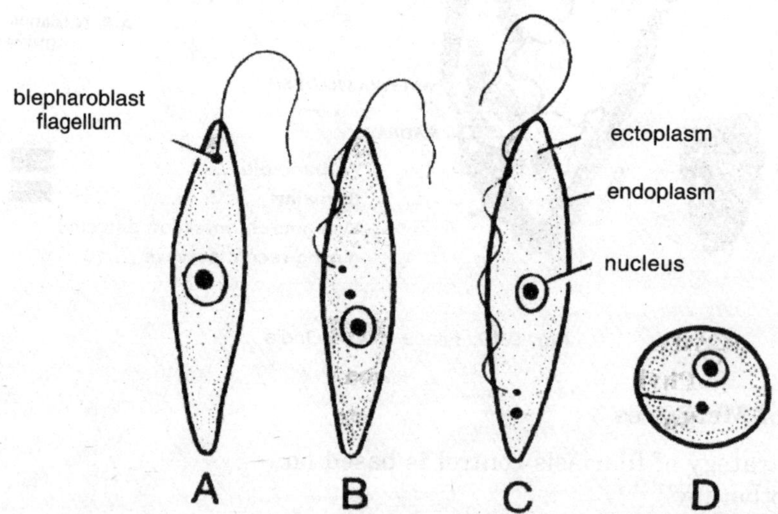

Fig. 9.11: *Leshmania* Stages A-*Leptomonas*. B-*Crithida*, C-*Trypanosoma*, D-*Leishmania*

leishmaniasis (MCL), anthrophonotic cutaneous leishmaniasis (ACL), zoonotic cutaneous leishmaniasis (ZCL) and post kala azar dermal leishmaniasis (PKDL) etc.

However, visceral leishmaniasis or kala azar is by far the most important diseases of India.

Kala azar occurs widely throughout the world viz: South America, South Africa, the Mediterranean countries, India and China. It is endemic in 82 countries. 350 million people are at the risk. New agro-industrial projects, large scale migration of populations, unplanned urbanization and man made environmental changes are responsible for its epidemics.

In India, kala azar has its home in the plains of Ganges and Brahamputra. It is known to occur in Assam, West Bengal, Bihar Eastern districts of Uttar Pradesh, foothills of Sikkkim and to a lesser extent in Tamil Nadu and Orrissa. The disease is on increasing trend and the incidence for a few years is shown in the Table 9.7.

Table 9.7 Kala-Azar incidence in India

Year	Bihar		West Bengal		Total Country	
	Cases	Deaths	Cases	Deaths	Cases	Deaths
1986	14079	47	3718	25	17806	72
1987	19179	77 (+19 suspected)	4447	10	23685	94
1988	19639	123 (+19 suspected)	3068	2	22739	137
1989	30903	447	3573	20	34489	497
1990	54650	589	30347	16	57742	606
1991	59614	834	2030	3	61670	838
1992	66959	1266	1212	1	68175	1267

9.11.1 Epidemiology of Leishmaniasis

i. ***Agent Factor***: *Leishmania donovani* is the causative agent of kala azar, (VL). *L. tropica* is the causative agent of cutaneous leishmaniasis (CL). *L. braziliensis* is the causative agent of muco-cutaneous leishmaniasis. The life cycle is completed in two different hosts- a vertebrate and an insect. There are a variety of animal reservoirs, e.g. dogs, jackals, foxes, rodents and other mammals. Indian kala azar is considered to be a non-zoonotic infection with man as the sole reservoir

ii. ***Host Factors***: Kala azar can occur in all age groups. In India the peak-age is 5-9 years. Males are affected twice as often as females. Kala azar usually strikes the poorest of the poor. People who work as farm labourers, in forestry, mining and fishing are at a great risk of bitten by sandflies.

iii. ***Environmental Factors:*** Kala azar is mostly confined to the plains. It does not occur in altitudes over 2,000 feet. The disease is generally confined to rural areas. *P. argentipes* is the proven vector of kala-azar. Forest clearing, cultivation projects, large water resource schemes and colonization and resettlement programmes are responsible for the infection.

9.11.2 Mode of Transmission

In India, kala azar is transmitted from person to person by the bite of the female phlebotomine sandfly. *P. argentipes* which is a highly anthrophilic species. After an infective blood meal, the sandfly becomes infective in 6 to 9 days. Transmission of Kala azar has also been recorded by blood transfusion.

9.11.3 Control

In absence of an effective vaccine, the control measures comprise the following:
1. Control of reservoir
2. Sandfly control

Since man is the only reservoir of kala azar in India, active and passive case detection and treatment of those found to be infected may be sufficient to abolish the human reservoir and control the disease. The application of residual insecticides has proved defective in the control of sandflies. For long lasting results, insecticideal (DDT) spraying should be combined with sanitation of cattlesheds and poultry at a far distance from human dwellings and improvement of housing and general sanitation.

The risk of infection can be reduced through health education and by the use of individual protective measures such as avoiding sleeping on floor, using fine mesh nets around the bed. Insect repellants may also be used for temporary protection.

SUGGESTED FURTHER READINGS

Hempel, J.H.G. 1982. World Health. April.

Park, K. 1994. Park's Text Book of Preventive and Social Medicine. Banarsidas Bhanot Publishers, Jabalpur.

Searo 1985. Annual Report. R.D. p 191.

Toman, K. 1979. Tuberculosis: Case finding and Chemotherapy. Questions and Answers. WHO, Geneva.

WHO 1986. Techn. Repr. Ser. 736.

WHO 1980. Guidelines for Cholera Control. WHO/CDD/SER/804. Geneva.

Youmans, G.P. *et. al.* 1980. The Biological and Clinical Basis of Infectious Diseases. 2nd ed., Saunder Philadelphia.

STUDY QUESTIONS

1. Define infection? How is it related with ecology? Give a brief description of ecology and disease.
2. Give a detailed description of any two of the following environment-borne diseases.
   i.   Malaria
   ii. AIDS
   iii. Cholera
   iv. Filariasis

3. Write short notes on the following:
  i.   Malaria action progrmmne (MAP)
  ii.   National tuberculosis progrmmne (NTP)
  iii.   Control of AIDS
  iv.   Viral hepatitis
  v.   Kala azar

Chapter 10

ECOTOXICOLOGY

Ecotoxicology can be defined as "the study of harmful effects of chemicals upon ecosystems". This implies that ecotoxicology is concerned not only with the detections of chemicals per se, but with biological effects of toxic chemicals that contaminate or have contaminated the environment. These biological effects may be anything from a molecular effect in a species to the effects on the biosphere as a whole.

Public concern about global and principal aspects of environmental pollution has shed new light on the role of ecotoxicology in the society. It is interdisciplinary in nature and includes experimental and theoretical methods from related sciences like biology, chemistry, toxicology, geology, geography and even physics. Ecotoxicology perhaps started with the green wave in 1960s so as to reveal the effects of toxic substances in the environment. The first ecotoxicological model emerged in late 1970s and the first conference on the development of ecological models was held in 1980 in Copenhagen.

Subsequently, application of ecotoxicological models in environmental management of toxic substances quickly became a powerful tool. The range of available ecotoxicological models is very wide with respect to chemical compounds, environmental problems and the ecological component and processes involved. Toxicological processes are a concern relevant for all structural and functional levels within and around ecosystems. Some toxicological processes occurring naturally are described below

A. At the Species Level (All Biotic)

Plants may use toxic substances in competing with each other. Plants can use toxic substances against some animals. Animals can use toxic substances to hunt their prey. Fish and amphibians use poisonous secretions for defense. Some algal blooms in oceans and wetlands can produce poisonous chemicals.

B. At the Ecosystem Level

- Some forest systems can alter the soil surface in such a manner that many other species can not germinate.
- Eutrophication of water bodies can kill many species by oxygen depletion.
- Salinization of lowlands in semideserts can render such places uninhabitable for plants and animals.

10.1 Ecotoxicological Episodes

10.1.1 DDT

Perhaps most widely discussed case of biological damage as a result of anthropogenic chemicals is the relationship between application of p, p' -DDT (dichloro diphenyl trichloroethane; systematic name: 1,1, I-trichloro- 2, 2-di -(4-chlorophenyl)-ethane) as insecticide and a dramatic population decline of the peregrine -falcon *(Falco peregrinus)* in the 1950s and 1960s. In USA DDT was first synthesized in 1874 by the German Chemist Othmar Zeidler, and its insecticidal activity was detected in 1939 by Paul Muller, who was honoured for this finding with the Nobel prize for medicine in 1946.

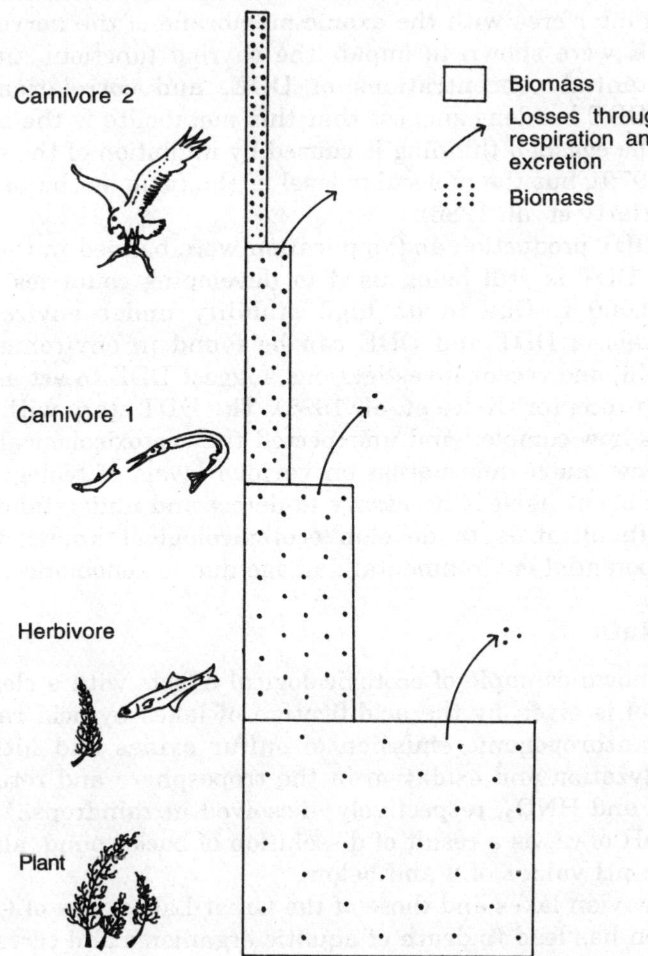

Fig. 10.1 Concentration of DDT residues being passed along a single food chain is indicated schematically in this diagram. As "biomass" is transferred from one trophic level to another, usually more than half is consumed in respiration or excreted (arrows): the remainder forms the new biomass. The losses of DDT residues along the chain, on the other hand, are small in proportion to the amount that is transferred from one link to the next. For this reason high concentration of DDT occur in carnivore.

From 1940 to 1970, DDT was used in high volumes throughout the world and proved its merits in the control of malaria and typhus. It had reached annual levels of around 100,000 t in 1963 (Metcalf 1973). One major cause for the observed population decline of the falcons was their reduced breeding success, which could be traced back to a significant eggshell thinning caused by DDT and its metabolite DDE in the bird. Uptake of DDT occurred with contaminated prey species, illustrating that persistent insecticides may well reach and affect non-target animals at higher trophic levels through bioaccumulation along the food chain.

The relatively high persistence of DDT is associated with its lipophilicity, which in turn enables the compound to penetrate insect cuticles much better than animal skins and thus makes it appear safe and target-specific for infield applications, DDT is neurotoxic and interferes with the axonic membrane of the nervous system, and DDT as well as DDE were shown to impair the thyroid function. Analysis of falcon eggs revealed substantial concentrations of DDE, and correlations between eggshell thickness and DDE burdens suggest that this metabolite is the active agent. There is evidence that the eggshell thinning is caused by inhibition of the enzyme Ca^{2+}- ATPase (Cooke 1973, 1979), but the molecular level of the toxic mechanism has not been fully unraveled (Moriarty et. al. 1986).

In 1972, DDT production and application were banned in the USA and Germany, and nowadays DDT is still being used in developing countries in estimated annual volumes of 40,000 t. Due to its high stability under environmental conditions, substantial levels of DDT and DDE can be found in environmental compartments allover the world, and recent investigations suggest DDE to act as a potent antagonist of the androgen receptor (Kelce et. al. 1995). The DDT story with the peregrine falcon thus illustrates how complex and unexpected the ecotoxicological effect of xenobiotics may be, and how much information on various levels of biological organization and about the toxic agent itself is necessary to detect and understand such events. It also shows how difficult it is to develop ecotoxicological knowledge for a predictive assessment of potential environmental damage due to xenobiotics.

10.1.2 Acid Rain

Another well-known example of ecotoxicological effects with a clear chemical cause at the system level is given by the acidification of lakes by acid rain. Major sources of acid rain are anthropogenic emission of sulfur oxides and nitrogen oxides, which undergo hydrolyzation and oxidation in the troposphere and return to the ground as H_2SO_3, H_2SO_4, and HNO_3, respectively, dissolved in raindrops. Uncontaminated rain already has a pH of ca. as a result of dissolution of background, airborne CO_2, and acid rain may reach pH values of 4 and below.

In Scandinavian lakes and those of the Great Lakes area of Canada and the USA, acidic deposition has lead to death of aquatic organisms and corresponding reductions in species diversity, disruption of normal food-chain relations, and shifts to greater abundance of acidophilic species. A common geogenic feature of these lakes is their relatively low buffer capacity, which makes them particularly susceptible to a drop in pH due to the uptake of airborne acids. The resultant ecotoxicological effects could be traced back to (at least) two components: acidic water is directly toxic for biota"

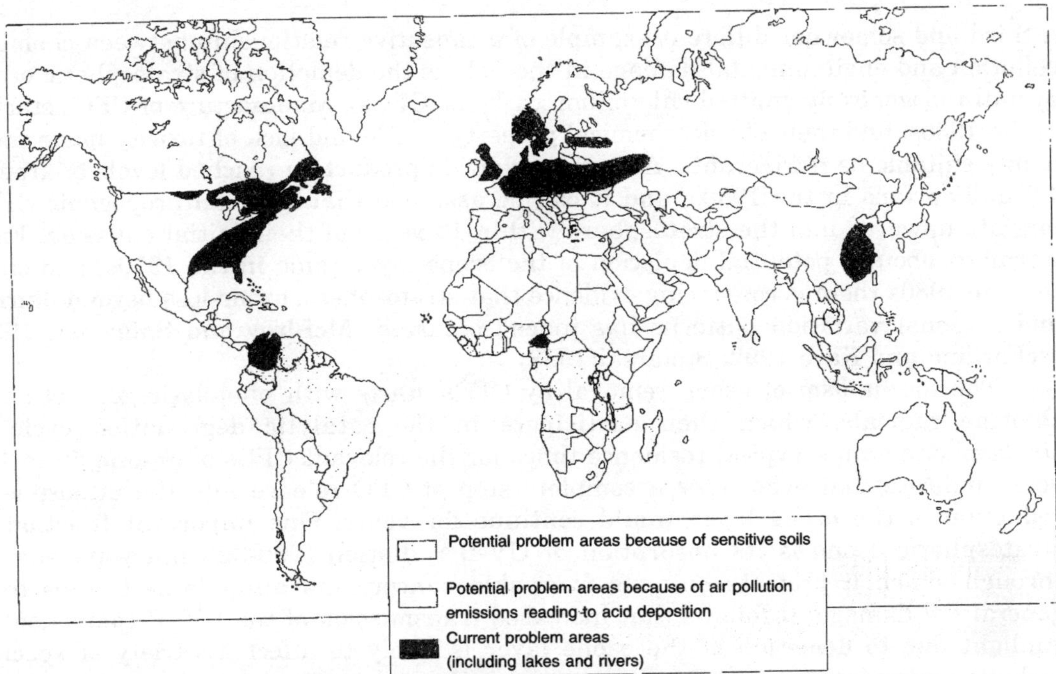

Fig. 10.2 Regions showing the areas facing the problems of acid rain.

perturbating the osmoregulation caused by disruption of the transepithelial electrochemical gradient, and acidification of lake water leads to mobilization of heavy metals with subsequent uptake and more specific toxic effects in aquatic organisms. A typical example is the generation of toxic Al^{3+} by acidic hydrolysis of $Al(OH)_3$, a natural component of sediments and soils. In contrast to the DDT example outlined above, the primary action of the chemical contaminent (the acid) is the alteration of an important physicochemical characteristic (the pH) of the abiotic environmental compartment, and the associated ecotoxicological effect results from a superposition of chemical stress by the primary agent as well as by other mobilized toxicants.

Much work has been done to identify ecological consequences in the acidified lakes. In the northwestern of Ontario in Canada, the effect of pH on aquatic communities has been studied in long-term ecosystem-level experiments with poorly buffered small lakes (Schindler 1987; Schindler et al. 1985). Artificial acidification was achieved by gradual addition of sulfuric acid to decrease the lake pH to 5.0 over a period of eight years. Small species with high reproduction rates and wide dispersal powers, such as phytoplankton, turned out to be most sensitive at the early stage of acidification, and other early indicators of acidic stress were provided by morphological abnormalities in benthic invertebrates. On the other hand, global ecosystem functions, such as primary production, nutrient cycling and respiration, remained essentially unaffected by acidification of the lake water.

10.1.3 Chlorofluorocarbons

A third and somewhat different example of a causative relationship between chemical pollution and environmental damage in the field is the depletion of stratospheric ozone by anthropogenically emitted chlorofluorocarbons (CFCs). Manufacture of CFCs started in the 1930s, and their physicochemical property profile and lack of toxicity made them highly suitable as refrigerants. Annual worldwide production reached levels of around 1.2 million tons in the 1980s, and it can be assumed that most anthropogenic CFCs migrate upwards into the stratosphere within 10 years of their initial emission. First warnings about a potential depletion of the ozone layer came in the 1970s, and since the late 1980s there is convincing evidence that stratospheric ozone loss beyond diurnal and seasonal variation patterns has indeed occurred (McElroy and Salawitch 1989; McFarland and Kaye 1992; Stolarski 1988).

The mechanism of ozone removal by CFCs starts with photolytic generation of chlorine radicals, which then participate in the catalytic degradation cycle of stratospheric ozone. Typical residence times for the relevant CFCs of around 50 to 100 years indicate that even after a complete stop of CFC release into the atmosphere, reduction of the ozone layer would continue for years. One important function of stratospheric ozone is its absorption of UV-B radiation (290-320 nm wavelength), through which it protects humans from skin cancer, and animals and crops from general UV damage. It follows that increased transmission of the UV -B component of sunlight due to depletion of the ozone layer is likely to affect a variety of species, including microorganisms.

Current discussion focus on potential deleterious effects on krill in the Antarctic with subsequent implications for associated food chains. Another matter of concern is the potential sensitivity of crop plants to increased UV-B radiation, which may also

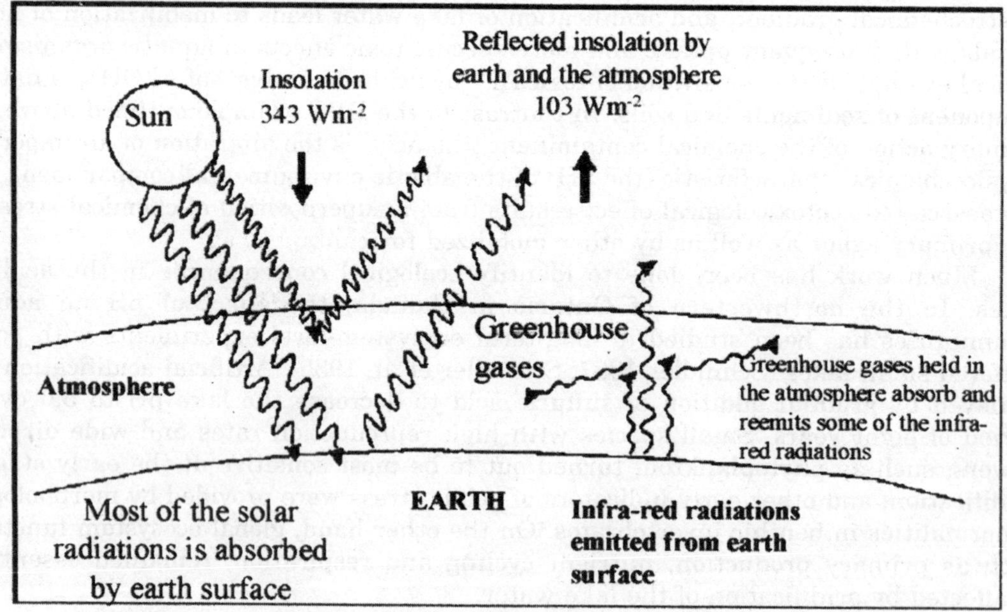

Fig. 10.3 Global warming by greenhouse gases – a schematic representation

make the plant more susceptible to other natural and anthropogenic stressors. Finally, depletion of stratospheric ozone might lead to climatic changes with a corresponding impact on the global ecosystem. These scenarios demonstrate that although CFCs do not act directly on biological systems, their interference with the ozone layer may well have dramatic indirect effects on the population, community and ecosystem level. Because the initial stressor would be electromagnetic radiation and thus non-chemical, the CFC problem would not fall within the focus of classical ecotoxicology. However, the potentially hazardous consequences for biological systems could still be traced back to interferences with natural ecosystems.

10.1.4 Minamata Disease

During the 1960s and 70s, the Minamata Bay mercury pollution disaster received global media attention, opening the world's eyes to the negative health effects of methyl mercury. Between 1932 and 1968, the Japanese Chisso Corporation discharged about 27 tonnes of methyl mercury with its wastewater into the bay. The pollution caused severe damage to the central nervous system of the people who ate large quantities of contaminated fish and shellfish from the bay. In addition, congenital Minamata disease occurred as many infants were born with a condition resem-bling cerebral palsy caused by methyl mercury poisoning of the foetus during pregnancy. The disease, which was officially recognized on 1 May 1956, caused many people to lose their lives or suffer from physical deformities.

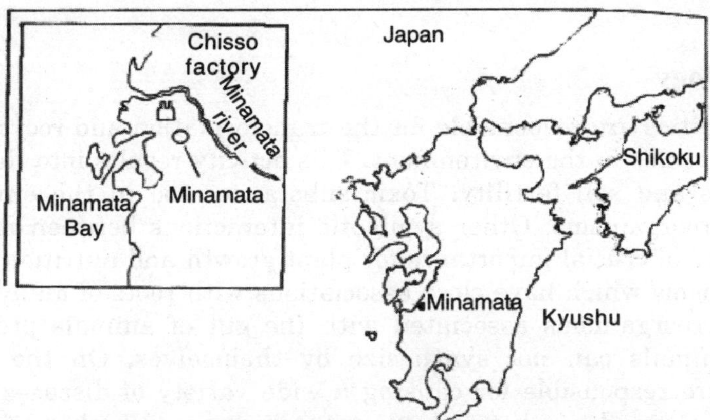

Fig. 10.4 Sites associated with Minamata disease.

After the cause of the disease was finally confirmed, a number of measures were implemented, ranging from regulation of the factory effluent, voluntary restrictions on harvesting of fish and shellfish from the bay, installation of dividing nets to enclose the mouth of the bay and prevent the spread of contaminated fish, and dredging of mercury-containing sediments. It was only in October 1997 that the dividing nets that had closed off the bay for 23 years were removed. After several studies confirming that mercury levels in fish were below regulatory levels and had remained so for three years, Minamata Bay was reopened as a general fishing zone.

Till 1992, 2,252 people were diagnosed with" Minamatc Disease", with 1,043 deaths reported.

10.2 Scientific Approach to Ecotoxicology

Ecotoxicology is shared by several disciplines of science including chemistry, biochemistry, microbiology, biotechnology, ecology, toxicology, anthropology, wildlife and forestry etc. Currently, there is growing concern internationally about human health. Scope of ecotoxicology in relation to a few scientific disciplines is discussed below-

10.2.1 Chemistry

Chemical approach to ecotoxicology is often analytical. Quick screening of a toxicological effect is essential. For example Ames Test for testing mutagenicity and related carcinogenicity in selected bacteria can be used for polyaromatic hydrocarbon, benzo (a) pyrene. Such mutagenicity due to DNA damage may manifest as cancer in mammalian organs. Ames Test is widely used in biological monitoring practices today.

10.2.2 Biochemistry

The main principle of biochemistry that can be applied to ecotoxicology is the modern concept of enzyme regulation by metabolites of xenobiotics. Many xenobiotics can interfere with the finely balanced biochemical reactions of living cells by disturbing the complex web of molecular interactions necessary for life.

Additional problems are DNA damage and disruption of cell membrane by reactive oxygen species generated by reactions between chemicals and atmospheric molecular oxygen.

10.2.3 Microbiology

Microbial communities are responsible for the transformation and recycling of organic and inorganic molecules in the environment. This activity results into the maintenance of nutrient cycles and soil fertility. Toxic substances too in the environment are detoxified by microorganisms. Other symbiotic interactions between microorganisms and plants that are of crucial importance for plant growth and nutrition are dinitrogen fixing microorganisms which have close associations with roots of many plant species. Furthermore, microorganisms associated with the gut of animals provide essential vitamins that animals can not synthesize by themselves. On the negative side microorganisms are responsible for causing a wide variety of diseases in plants and animals. The interdependence between microorganisms and higher life forms means that toxins that effects either group of organisms can have important consequences for the other.

10.2.4 Biotechnology

Biotechnology is an applied science that aims to harmers different life forms for the benefit of man. Agriculture, antibiotic production, and bioremediation are examples of such applications. With the advance of biological sciences comes the responsibility for biotechnologists to ensure that the integrity of the environment is maintained by ensuring that the biological functions that allow life on this planet to thrive are not impaired.

Interestingly, recombinant technology has also opened the way to construct organisms that can be used to monitor the environmental impact of toxicants. The creation of super bug by Indian born scientist Anand Chakrobroty that selectively feeds on hydrocarbons has been used to clean oil spills in the sea. The development of Bt cotton is another step taken to protect cotton plants from pests. Several other examples of bioremediation support a relationship between biotechnology and ecotoxicology.

10.2.5 Toxicology

Important principles of toxicology that relate to ecotoxi-cology include the concepts of extent of exposure, per-sistence, and distribution of chemicals in the environ-ment. Subsequently predictions can be made on the toxicity of such chemicals to individual organisms or populations. The starting point of such analysis is often the chemical structure of each toxin, which will allow some prediction of the behavior of the chemical in the environment to be made. One of the best examples of this kind is the environmental impact assessment of the insecticide DDT. DDT is almost completely insoluble in water but readily soluble in fat. Furthermore, DDT and its degradation product DDE are highly persistent. Its low solubility means that it is easily dispersed in aquatic environments (in the case of DDT it is found at low concentrations all over the globe) and only accumulates in places with a high fat content (i.e., living organisms). Once it has entered living organisms, it persists and accumulates in the fat tissues of organisms that are higher up the food chain resulting in toxicity. For this reason, populations of both fish eating birds and bird eating raptors such as peregrines and hawks were badly affected by DDT, even in cases where they inhabited pristine environments.

Another aspect of toxicology is risk analysis. This quantitative topic is most easily studied when death rates within a population can be quoted. Whereas the rates can be determined experimentally for animals, plants, and microorganisms, the determination of the effects on the human population is inevitably more difficult to determine, and one of the few situations where effects are well known is the mutagenic effects of radionucleotides, resulting from contamination caused by accidents at nuclear power installations or nuclear warfare.

10.3 Entry, Movement and Fate of Pollutants in Ecosystems

Organic and metallic (including radionuclide) toxi-cants that enter the environment are regarded as pollut-ants of air, soil, and water. Their movement and trans-port is through air and water depending on their volatility and solubility, respectively. Soil and the detri-tus of sediments provide important solid supports for adsorption to regulate the movement and flow of the pollutants through terrestrial ecosystems, as well as influencing the localization and persistence of the pol-lutants in the environment. In this respect clay and humus particles that are electrostatically charged have an important role to play. Some organisms will take up the pollutant and concentrate them in their cells, a phenomenon known as bioaccumulation or biomagni-fication. This is particularly serious where the toxicant enters the food chain leading to progressive accumula-tion of toxic molecules higher up the chain, including humans. This is often the case with molecules that are relatively biologically inert such as heavy metals, PCBs, and organochlorine insecticides. Accumulation of such molecules in different species can lead to toxicity ex-pressed as reduced growth, reduced fecundity, changes in behavior, susceptibility to diseases, or increased mor-tality. However, on the positive side, where there is storage of the pollutant in the cell, as is normally the situation with (heavy) metals, it is possible that the pollutant might be harvested and therefore removed from the environment. This especially applies to plants that hyper accumulate metals in their tissues. The appli-cation of this process for the cleanup of contaminated land is known of phytoremediation. Also, many pollut-ants are biologically degradable and can therefore be detoxified. Both plants and mammals have a range of enzymes that are involved in detoxification of molecules that are potentially harmful to them (cytochromes P-450, for example), whereas the almost limitless meta-bolic capacity of a wide variety of microorganisms allows degradation of pollutants, especially hydrocar-bons, into nontoxic molecules such as water and carbon dioxide. The stimulation of microbial degradation and metabolism of pollutants is known as bioremediation and is currently used to clean up contaminated land while microbial degradation of pollutants present in sewage and water is used in a variety of water purifica-tion systems. The fate of pollutants in the environment is therefore not only dependent on the molecular char-acteristics of the pollutant but also on a range of biotic and abiotic factors.

10.4 Biological Monitoring

Monitoring literally means "to keep an eye"; therefore, biological monitoring obviously indicates the practice of monitoring the environment in terms of pollution and its effects. There are three sequential steps in biomonitoring viz. survey, surveillence and monitoring. In preventive health care, monitoring can be called as a systematic health related activity designed to lead if necessary to corrective action. CEC, NIOSH and OSHA define biological monitoring (BM) as the "measurement and assessment of work place agents of their metabolites either in tissues, secreta, excreta or any combination of these systems", in occupationally exposed subjects. BM in principle refers to "toxico-kinetic parameters" viz: uptake, distribution, bio-transformation, accumulation and excretion whereas biological effect monitoring (BEM) is the assessment of early biological effects in workers so as to evaluate exposure and/or health risk (Zielhuis and Henderson, 1986).

These definitions clearly conclude that biological monitoring pro-vides occupational health personnel with a tool for assessing worker's exposure to chemicals. Practice of biological monitoring can be based on following concepts.

10.4.1 Biological Exposure Indices (BEI)

BEI are the reference values for the determinants which are indica-tors of the degree of integrated exposure to industrial chemicals. To apply the BEI, the determinant must be measured in the specified biological specimen, collected at the specified time with respect to the occupational exposure. BEIs are intended to be guidelines for the evaluation of potential health hazards in the practice of industrial hygiene. Due to biological variability, it is possible 'for an individual to exceed the BEI without incurring an increased health risk.

The determinants are usually the chemical, itself or its metabolites and occasionally a reversible biochemical change induced by the chemi-cal. The specificity of the determinant as an indicator of exposure is important but not critical. The most important factors are: (1) the quantitative relationship between intensity of exposure and biological level of the determinant and (2) the ability of the test to measure the degree of pre-effect exposure.

In some instances markers can be qualitative or semiquantitative indicators of exposure, unusual values indicate an overexposure to a toxic chemical. A few examples are the measurements of acetylcholinesterase activity, methamoglobin, amino levulinic acid or Zn-protoporphyrine. ACGIH (1984) recommends measurements of reversible biochemical changes for exposure monitoring if no quantitative specific indicator is available. However, measurement of irreversible biochemical changes or functional changes are not recommended.

10.4.2 Biological Monitoring in European Countries and USA

The great occasion that gave rise to further studies on biological monitoring in Europe was the "International Conference on Environ--mental Health Aspects of Lead", organized jointly by the European Communities Commission and Environmental Protection Agency in Amsterdam in 1972. Following Amsterdam Congress, several other symposia, congresses and workshops were held in Paris in (1974), Luxembourg (1977, 1984) and Odense (1987). In Luxembourg work-shop, possible methods were identified which can today be used to evaluate the internal dose, biological effective dose and the early biological effects.

Of primary importance was the adoption of atomic absorption spectrophotometry for the determination of metals and gas chromatogra-phy for the quantification of solvents and their metabolites. The induction of these techniques revolutionized the potentiality of analysis permitting on one hand, the determination of a large number of toxic agents in biological materials and on the other hand greatly enhancing the speci-ficity and sensitivity of the method used.

Although, biological monitoring has been conducted for decades in USA, there has been a recent explosion in its applications. The usage of the term biological monitoring and its relationship to environmental monitoring are consistent with the recent American Conference of Governmental Industrial Hygienists (ACGIH) publication on biological exposure indices. These recommendations are a joint publication from NIOSH, Occupational Safety and Health Administration and the Com-mission of European Communities.

The National Institute for Occupational Safety and Health (NIOSH) registered over 90,000 chemicals with some toxic effect in 1988. Of these over 60,000 chemicals were commonly used by Industry in the U.S. However approximately 6,000 chemicals used in commerce could un-dergo toxicity testing. Finally definitive biological monitoring data are available for less than 100 of these chemicals. In these perspectives, biological monitoring has been used with increasing frequency over recent years.

10.4.3 Biological Monitoring in Policy Development

The mandated responsibilities of Ministry of Industries, Environ-ment and Labour should include the development of criteria that assure every man and woman a safe and healthy working environment. When such criteria are released, most interested parties focus on the limit of exposure in air, all but ignoring other components of the criteria. Measurement of exposure agents in breathing air does not assure that the worker is totally protected from adverse health effects but it is the actual body burdens of the agent resulting from all routes of the exposure directly related with health effects.

The uptake of the workplace agent by inhalation, absorption, through skin or gastrointestinal tract and non-occupational exposure to the agent, all influence the body burden. Moreover, interaction of the agent with the other environmental or workplace agents and factors may stimulate or inhabit its metabolism and elimination and may thus influence the toxicity of the agent in the worker. Biological monitoring methods that have been validated so far can safely be used as a complement to environmental monitoring. In order to fully evaluate the potential health hazard in the workplace, both environmental and biological monitoring data provide independent information that can help in overall assessment of exposure and potential health effects.

Major constraints in incorporating biological monitoring techniques in policy document and recommendations result from the need to validate monitoring method and to account for factors effecting measurements.

While a number of biological monitoring methods for uptake have been validated and evaluated such as those in NIOSH criteria documents and in the ACGIH-BEI,

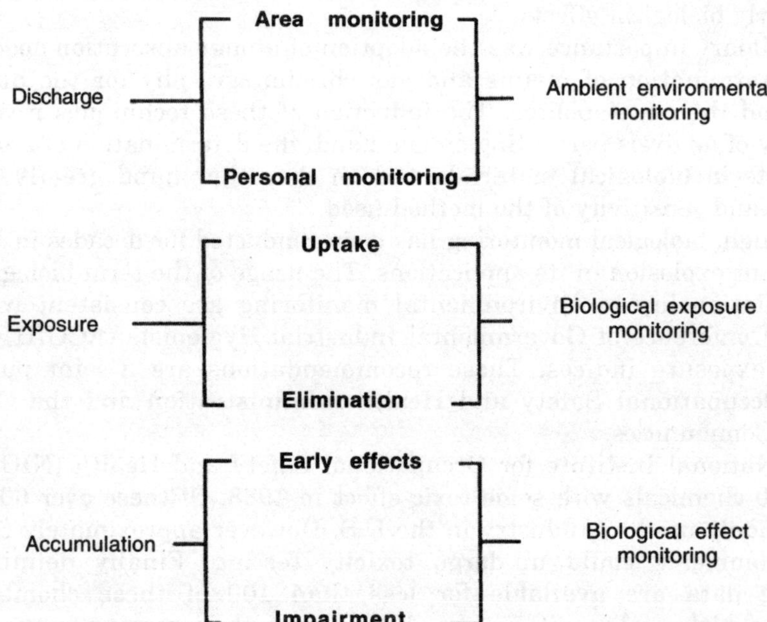

Fig. 10.5 Role of Biological monitoring in health surveillance

insufficient data exist on many agents for which biological monitoring methods exist and could be useful if validated and thoroughly evaluated for influencing factors. Therefore, more research is needed on biological markers before it can be determined whether, it would be appropriate to include biological monitoring based on these biological effects in exposure monitoring recommendations.

10.5 Challenges for the Prevention of Occupatinal Diseases

A major challenge for occupational-environmental health profes-sionals is how to determine individual's true exposure status in the absence of objective definitive evidence. Another major challenge involves ethical use of data and action to remedy hazards. Concern is often raised that workers with elevated biological monitoring measure-ments might be released from employment and replaced with new workers. Such a practice might be rationalized on the basis of identifying susceptible workers, however, some of the issues dealing with differential susceptibility, need more attention.

Yet another challenge to be resolved is whether the legal and regulatory context is sufficiently developed for the prevention of occu-pational diseases. Additional legislative and regulator clarification is needed on the rights and duties of all parties involved in the control and prevention of occupation hazards.

SUGGESTED FURTHER READINGS

G. Schuurmann and B. Markert (1998), Ecotoxicology. Wiley, New York.

STUDY QUESTIONS

1. Define ecotoxicology. Discuss the scope of ecotoxicology with suitable examples.
2. Discuss the ecotoxicology of -
   a. DDT
   b. Chlorofluorocarbons
   c. Acid rain
   d. Minamata disease
3. Discuss the importance of biological monitoring in occupational health.

Chapter 11

OCCUPATIONAL HEALTH

ILO/WHO Committee on Occupational Health (1950) defined it as "the subject that aims at the promotion and maintenance of the highest degree of physical, mental and social well being of workers in all occupations". Preventive medicine and occupational health have the same objectives. In the past, it was customary to think of occupational health in relation to factories and mines, hence the terms "industrial hygiene" or "industrial health" were in vague. Modern concepts of occupational health now embrace all types of employment including mercantile and commercial enterprises, service places, forestry and agriculture and includes the subjects of industrial hygiene, industrial diseases, industrial accidents and industrial toxicology. Occupational health in agriculture and ergonomics (human engineering) are relatively new concepts.

Occupational environment is defined as "the sum total of all external conditions and influences that prevail at the place of work and leave a bearing on the health of working population". Occupational environment and domestic environment overlaps each other. From an ecological viewpoint, occupational health represents a dynamic equilibrium or adjustment between the industrial workers and his work environment.

According to the United Nations, 180,000 workers die every year throughout the world as a result of accidents and occupational diseases. Another 110 million suffer non-fatal injuries. In India, the problems of occupational diseases are further magnified due to rampant malnutrition and endemic diseases. Tropical diseases further add considerably to the plight of low paid workers. Other contributions include hot and dry climate, migration of working population, socio-economic conditions and illiteracy. The problems of occupational health has considerably grown along with industrialization.

11.1 Occupational Hazards

An industrial worker may be exposed to five types of hazards, depending upon his occupation:
    (a) Physical hazards
    (b) Chemical hazards
    (c) Biological hazards
    (d) Mechanical hazards
    (e) Psychosocial hazards

11.1.1 Physical Hazards

(1) **Heat and Cold**: The common physical hazard in most industries is heat. The direct effects of heat exposure are burns, heat exhaustion, heat stroke and heat cramps; the indirect effects are decreased efficiency, increased fatigue and enhanced accident rates. Many industries have local "hot spots"- ovens and furnaces, which radiate heat. Radiant heat is the main problem in foundry, glass and steel industries, while *heat stagnation* is the principal problem in jute and cotton textile industry. High temperatures are also found in mines for instance in the Kolar Gold mines of Mysore which is the second deepest mine of the world (11,000 feet), temperatures as high as 150^0 F are recorded. Physical work under such conditions is very stressful and impairs the health and efficiency of the workers. For, gainful work involving sustained and repeated effort, a reasonable temperature must be maintained in each work room. The Indian Factories Act has not laid down any specific temperature standard. However, the work of Rao (1952, 1953) and Mookerjee *et al.* (1953) indicate that a corrected effective temperature of 69 to 80^0 F is the comfort zone in this country and temperatures above 80^0 F cause discomfort.

Important hazards associated with cold work are chilburns, erythrocyanosis, immersion foot, and frostbite as a result of cutaneous vasoconstriction. General hypothermia is not unusual.

(2) **Light**: The workers may be exposed to the risk of poor illumination or excessive brightness. The *acute* effects of poor illumination are eyestrain, headache, eye pain, lacrymation, congestion around the cornea and eye fatigue. The *chronic* effects on health include *"miner's nystagmus"*. Exposure to excessive brightness or "glare" is associated with discombrt, annoyance and visual fatigue. Intense direct glare may also result in blurring of vision and lead to accidents. There should be sufficient and suitable lighting, natural or artificial, wherever persons are working.

(3) **Noise**: Noise is a health hazard in many industries. The effects of noise are of two types: (i) *Auditory effects* which consist of temporary or permanent hearing loss (ii) *Non-auditory effects* which consist of nervousness, fatigue, interference with communication by speech, decreased efficiency and annoyance. The degree of injury from exposure to noise depends upon a number of factors such as intensity and frequency range, duration of exposure and individual susceptibility.

(4) **Vibration**: Vibration, especially in the frequency range 10to 500 Hz, may be encountered in work with pneumatic tools such as drills and hammers. Vibration usually affects the hands and arms. After some months or years of exposure, the fine blood vessels of the fingers may become increasingly sensitive to spasm (white fingers). Exposure to vibration may also produce injuries of the joints, of the hands, elbows and shoulders.

(5) **Ultraviolet Radiation**: Occupational exposure to ultraviolet radiation occurs mainly in arc welding. Such radiation mainly affects the eyes, causing intense conjunctivitis and keratitis (welder's flash). Symptoms are redness of the eyes and pain, these usually disappear in a few days with no permanent effect on the vision or on the deeper structures of the eye.

(6) **Ionizing radiation**: Ionizing radiations are finding increasing application in medicine and industry, e.g. x-rays and radio active isotopes. Important radio-isotopes

are Co^{60} and P^{32}. Certain tissues such as bonemarrow are more sensitive than others and from a genetic standpoint, there are special hazards when the gonads are exposed. The radiation hazards comprise genetic changes, malformation, cancer, leukaemia, depilation, ulceration, sterility and in extreme cases death. The International Commission of Radiological Protection has set the maximum permissible level of occupational exposure at 5 rem per year to the whole body.

11.1.2 Chemical Hazards

There is hardly any industry which does not make use of chemicals. The chemical hazards are on the increase with the introduction of newer and complex chemicals. Chemical agents act in three ways: local action, inhalation and ingestion. The ill-effects produced depend upon the duration of exposure, the quantum of exposure and individual susceptibility.

11.1.2.1 Local Action

Some chemicals cause dermatitis, eczema, ulcers and even cancer by *primary irritant action;* some cause dermatitis by an allergic action. Some chemicals, particularly the aromatic nitro and amino compounds such as TNT and aniline are absorbed through the skin and cause systemic effects. Occupational dermatitis is a big problem in industry. Rao and Banerji (1952) were the first to draw attention in India to the prevalence of occupational dermatitis due to machine oil, rubber, x-rays, caustic alkalies and lime.

11.1.2.2 Inhalation

(i) *Dusts:* Dusts are finely divided solid particles with size ranging from to 150 microns. They are released into the atmosphere during crushing, grinding, abrading, loading and unloading operations. Dusts are produced in a number of industries mines, foundry, quarry, pottery, textile, wood or stone working industries. Dust particles larger than 10 microns settle down from the air rapidly, while the smaller ones remain suspended indefinitely. Particles smaller than 5 microns are directly inhaled into the lungs and are retained there. This fraction of the dust is called "respirable dust", and is mainly responsible for pneumoconiosis. Dusts have been classified into *inorganic* and *organic* dusts; *soluble* and *insoluble dusts*. The inorganic dusts are silica, mica, coal, asbestos dust, etc; the organic dusts are cotton, jute and the like. The soluble dusts dissolve slowly, enter the systemic circulation and are eventually eliminated by body metabolism. The insoluble dusts remain, more or less, permanently in the lungs. They are mainly the cause of pneumoconiosis. The most common dust diseases in this country are silicosis and anthracosis,

(ii) *Gases:* Exposure to gases is a common hazard in industries. Gases are sometimes classified as *simple gases* (e.g. oxygen, hydrogen), *asphyxiating gases* (e.g. carbon monoxide, cyanide gas, sulphur dioxide, chlorine) and *anesthetic gases* (e.g., chloroform, ether, trichlorethylene). Carbon monoxide hazard is frequently reported in coal-gas manufacturing plants and steel industry.

(iii) *Metals and their compounds:* A large number of metals, and their compounds are used throughout industry. The chief mode of entry of some of them is by inhalation as dust fumes. The industrial physician should be aware of the toxic effects of lead, antimony, arsenic, beryllium, cadmium, cobalt, manganese, mercury, phosphorus, chromium, zinc and others. The ill-effects depend upon the duration of exposure and the dose or concentration of exposure. Unlike the pneumoconiosis, most chemical intoxication respond favourably to cessation, exposure and medical treatment.

11.1.2.3 Ingestion

Occupational diseases may also result from ingestion of chemical substances such as lead, mercury, arsenic, zinc, chromium, cadmium, phosphorus etc. Usually these substances are swallowed in minute amounts through contaminated hands, food or cigarettes. Much of the ingested material is excreted through faeces and only a small proportion may reach the general blood circulation.

11.1.3 Biological Hazards

Workers may be exposed to infective and parasitic agents at the place of work. The occupational diseases in this category are brucellosis, leptospirosis, anthrax, hydatidosis, psittacosis, tetanus, encephalitis, fungal infections, schistosomiasis and a host of others. Those working among animal products (e.g., hair, wool, hides) and agricultural workers are specially exposed to biological hazards.

11.1.4 Mechanical Hazards

The mechanical hazards in industry centre around machinery, protruding and moving parts and the like. About 10 per cent of accidents in industry are said to be due to mechanical causes.

11.1.5 Psychosocial Hazards

The psychosocial hazards arise from the workers' failure to adapt to an alien psychosocial environment. Frustration, lack of job satisfaction, insecurity, poor human relationships, emotional tension are some of the psychosocial factors which may underline both physical and mental health of the workers. The capacity to adapt to different working environments is influenced by many factors such as education, cultural background, family life, social habits, and what the worker expects from employment.

The health effects can be classified in two main categories: (a) Psychological and behavioural changes: including hostility, aggressiveness, anxiety, depression, tardiness, alcoholism, drug, sickness, absenteeism; (b) Psychosomatic health: including failgue, headache; pain in the shoulders, neck and back; propensity to peptic ulcer, hypertension, heart disease and rapid ageing.

Reports from various parts of the world indicate that physical factors (heat, noise, poor lighting) also play a major role in adding to or precipitating mental disorders among workers. The increasing stress on automation, electronic operations and nuclear energy may introduce newer psychological health problems in industry. Psychological hazards are therefore assuming more importance than physical or chemical hazards.

11.2 Occupational Diseases

There is no internationally accepted definition for the term "occupational disease". However, occupational diseases are usually defined as diseases arising out of or in the course of employment. For convenience, they may be grouped as under:

I. Diseases due to Physical Agents:

(1)	Heat	Heat hyperpyrexia, heat exhaustion, heat syndrome, heat cramps, burns and local effects such as prickly heat.
(2)	Cold	Trench foot, frost bite, chilblains
(3)	Light	Occupational cataract, Miner's nystagmus
(4)	Pressure	Caisson disease, air embolism; blast (explosion)
(5)	Noise	Occupational deafness
(6)	Radiation	Cancer, leukaemia, aplastic anaemia, pancytopenia
(7)	Mechanical Factors	Injuries, accidents
(8)	Electricity	Burns

II. Diseases due to Chemical Agents:

(1) Gases: CO_2, CO, HCN, CS_2, NH_3, N_2, H_2S, HCI, SO_2 - these cause gas poisoning.

(2) Dusts Pneumoconiosis

(i) Inorganic Dusts:

(a) Coal dust	Anthracosis
(b) Silica	Silicosis
(c) Asbestos	Asbestosis, lung cancer
(d) Iron	Siderosis

(ii) Organic (vegetable) Dusts:

(a) Cane fibre	Bagassosis
(b) Cotton dust	Byssinosis
(c) Tobacco	Tobacossis (11)
(d) Hay orgrain dust	Farmers's lung

(3) Metals and their compounds:

Toxic hazards from lead, mercury, cadmium, manganese, beryllium, arsenic, chromium etc.

(4) Chemicals : Acids, alkalies, pesticides

(5) Solvents : Carbon bisulphide, benzene, trichloroethylene, chloroform, etc.

III. Diseases due to Biological Agents:

Brucellosis, leptospirosis, anthrax, actinomycosis, hydatidosis, psittacosis, tetanus, encephalitis, fungal infections, etc.

IV. Occupational Cancers:

Cancer of skin, lungs, bladder

V. Occupational Dermatosis:
Dermatitis, eczema

VI. Diseases of psychological origin:
Industrial neurosis, hypertension, peptic ulcer, etc.

11.3 Pneumoconiosis

Dust within the size range 0.5 to 3 microns is a health hazard, producing after a variable period of exposure, a lung disease known as pneumoconiosis, which may gradually cripple a man by reducing his working capacity due to lung fibrosis and other complications. The hazardous effects of dusts on the lungs depend upon a number of factors such as (a) chemical composition (b), fineness (c) concentration of dust in the air (d) period of exposure and (e) health' status of the person exposed. Therefore, the threshold limit values for different dusts are different. In addition to the toxic effects of the dust on the lung tissues, the super-imposition of infections like tuberculosis may also influence the pattern of pneumoconosis. The important dust diseases are silicosis, anthracosis, byssinosis, bagassosis, asbestosis and farmer's lung. As no cure for the pneumoconiosis in known, it is essential to prevent these diseases from arising. A brief account of these conditions is given in the following pages:

11.3.1 Silicosis

Among the occupational diseases, silicosis is the major cause of permanent disability and mortality. It is caused by inhalation of dust containing free silica or silicon dioxide (SiO_2). It was first reported in India from the Kolar Gold Mines (Mysore) in 1947. Ever since, its occurrence has been uncovered in various other industries, e.g., mining industry (coal, mica, gold, silver, lead, zinc, manganese and other metals), pottery and ceramic industry, sand blasting, metal grinding, building and construction work, rock mining, iron and steel industry and several others.

In the mica mines of Bihar, out of 329 miners examined, 34.1 per cent were found suffering from silicosis. In a ceramic and pottery industry, the incidence of silicosis was found to be 15.7 per cent. The incidence of silicosis depends upon the chemical composition of the dust, size of the particles, duration of exposure and individual susceptibility. The higher the concentration of free silica in the dust, the greater the hazard. Particles between 0.5 to 3 microns are the most dangerous because they reach the interior of the lungs with ease. The longer the duration of exposure, the greater the risk of developing silicosis. It is found that the incubation period may vary from a few months up to 6 years of exposure, depending upon the above factors.

The particles are ingested by the phagocytes which accumulate and block the lymph channels. Pathologically, silicosis is characterized by a dense "nodular" fibrosis, the nodules ranging from 3 to 4 mm in diameter. Clinically the onset of the disease is insidious. Some of the early manifestations are irritant cough, dyspnoea on exertion and pain in the chest. With more advanced disease, impairment of total lung capacity (TLC) is commonly present. An x-ray of the chest shows "snow-storm" appearance in the lung fields. Silicosis is progressive and what is more important is that silicotics are prone to pulmonary tuberculosis, a condition called "silic, tuberculosis." In recent years doubts have been raised, whether silico-tuberculotics are really tubercular or purely silicotics. It is because, sputum in silico-tuberculotics rarely shows" tubercle bacilli; children and women of silico-iuberculotics do not develop tuberculosis; post-mortems on silico-tuberculotics failed to prove the existence of tuberculosis disease, but showed them to be cases of pure silicosis. The radiological evidence in the two conditions is so similar that one is apt to mistake a case of silicotic to be a case of tuberculosis of lungs. The final answer to this question is still awaited.

There is no effective treatment for silicosis. Fibrotic changes that have already taken place cannot be reversed. The only way that silicosis can be controlled (if not altogether eliminated) is by (a) rigorous dust control measures, e.g., substitution, complete enclosure, isolation, hydroblasting, good house-keeping, personal protective measures and (b) regular physical examination of workers.

Silicosis was made a notifiable disease under the Factories Act 1948 and the Mines Act 1952.

11.3.2 Anthracosis

Previously it was thought that pulmonary "anthracosis", was inert. Recent studies indicate that there are two general phases in coal miner penumoconiosis - (1) the first phase is labeled *simple, pneumoconiosis* which is associated with little ventilatory impairment. This phase may require about 12 years of work exposure for its development (2) the second phase is characterised by *progressive massive fibrosis* (PMF); this causes severe respiratory disability and frequently results in premature death. Once a background of simple penumoconiosis has been attained in the coal worker, a progressive massive fibrosis may develop out of it without further exposure to it. From the point of view of epidemiology, the risk of death among coal miners has been nearly twice that of the general population. Coal-miners pneumoconiosis has been declared a notifiable disease in the Indian Mines Act of 1952 and also compensable in the Workmen's Compensation (Amendment) Act of 1959.

11.3.3 Byssinosis

Byssinosis is due to inhalation of cotton fibre dust over long periods of time. The symptoms are chronic cough and progressive dyspnoea, ending in chronic bronchitis and emphysema. India has a large textile industry employing nearly 35 per cent of the factory workers. Incidence of byssinosis is reported to be 7 to 8 percent in three independent surveys carried out in Bombay, Ahmedabad and Delhi (12).

11.3.4 Bagassosis

Bagassosis is the name given to an occupational disease of the lung caused by inhalation of bagasse or sugar-cane dust. It was first reported in India by Ganguli and Pal in 1955 in a cardboard manufacturing firm near Calcutta. India has a large sector of cane-sugar industry. The cane-sugar fibre which until recently went to waste is now utilized in the manufacture of paper. cardboard and rayon.

Bagassosis has been shown to be due to a thermophilic actinomycete for which the name *Thermoactinomyces sacchari* was suggested. The symptoms consist of breathlessness, cough, haemoptysis and slight fever. Initially there is acute diffuse bronchiolitis. Skiagram may show mottling in lungs or shadow. There is impairment of pulmonary function. If treated early, there is resolution of the acute inflammatory condition of the lung. If left untreated, there 1s diffuse fibrosis, emphysema and bronchiolitis.

11.3.4.1 Preventive Measures

(1) *Dust control*: Measures for the prevention and suppression of dust such as, wet process, enclosed apparatus, exhaust ventilation etc., should be used.

(2) *Personal Protection*: Personal protective equipment (masks or respirators with mechanical filters or with oxygen or air supply) may be necessary.

(3) *Medical Control*: Initial medical examination and periodical medical check-ups of workers are indicated.

(4) *Bagasse Control*: By keeping the moisture content below 20 per cent and spraying the bagasse with 2 per cent propionic acid, a widely used fungicide, bagasse can be rendered safe for manufacturing use.

11.3.5 Asbestosis

Asbestos is the commercial name given to certain types of fibrous materials. They are silicates of varying composition; the silica is combined with such bases as magnesium, iron, calcium, sodium and aluminium. Asbestos is of two types - serpentine or chrysolite variety and the amphibole type. Ninety per cent of the world's production of asbestos is of the serpentine variety, which is hydrated magnesium silicate, the amphibole type contains little magnesium. The amphibole type occurs in different varieties, e.g., crocidolite (blue), amosite (brown), and anthrophyllite (white). Asbestos fibres are usually from 20 to 500 μ in length and 0.5 to 50 μ in diameter. Asbestos is used in the manufacture of asbestos cement, fire-proof textiles, roof tiling, brake lining, gaskets and several other items. Asbestos is mined in Andhra Pradesh (Cudappah), Bihar, Karnataka, and Rajasthan - but most of it is imported from USSR. Canada, US and South Africa.

Asbestos enters the body by inhalation, and fine dust may be deposited in the alveoli. The fibres are insoluble. The dust deposited in the lungs causes pulmonary fibrosis leading to respiratory insufficiency and death; carcinoma of the bronchus; mesothelioma of the pleura or peritoneum; and cancer of the gastro-intestinal tract. In Great Britain, an association was reported between mesothelioma and living within a half-mile (1 km) of an asbestos factory. The risk of bronchial cancer is reported to be high if occupational exposure to asbestos is combined with cigarette smoking. Mesothelioma, a rare form of cancer of the pleura and peritoneum, has been shown to have a strong association with the crocidolite variety of asbestos. The disease does not usually appear until after 5 to 10 years of exposure. The fibrosis in asbestosis is due to mechanical irritation, and is peri-bronchial, diffuse in character, and basal in location in *contrast* to silicosis in which the fibrosis is nodular to character and present in the upper part of the lungs. Clinically the disease is characterised by dyspnoea which is frequently out of proportion to the clinical signs in the lungs. In advanced cases, there may be clubbing of fingers, cardiac distress and cyanosis. The sputum shows "asbestos bodies" which are asbestos fibres coated with fibrin. An x-ray of the chest shows a ground-glass appearance in the lower two thirds of the lung fields. Once established, the disease is progressive even after removal of the worker from contact.

The *preventive measures* consist of : (1) use of safer types of asbestos (chrysolite and amosite) : (2) substitution of other insulants: glass fibre, mineral wool, calcium silicate, plastic foams, etc: (3) rigorous dust control; (4) periodic examination of workers; biological monitoring (clinical, x-ray, lung function), and (5) continuing research.

11.3.6 Farmers Lung

Farmers lung, is due to the inhalation of mouldy hay or grain dust. In grain dust or hay with a moisture content of over 30 per cent bacteria and fungi grow rapidly, causing a rise of temperature to 40 to 50^0 C. This heat encourages the growth of thermophilic atinomycetes, of which *Micropolyspora faeni* is the main cause of farmer's lung. The acute illness is characterised by general and respiratory symptoms and physical signs. Repeated attacks cause pulmonary fibrosis and inevitable pulmonary damage and corpulmonale. It is quite possible that this condition might be widespread in India considering the bulk of the population engaged in agricultural work.

11.4 Lead Poisoning (Plumbism)

More industrial workers are exposed to lead than to any other toxic metal. Lead is used widely in a variety of industries because of its properties: (1) low boiling point (2) mixes with other metals easily to form alloys (3) easily oxidised and (4) anticorrosive. All lead compounds are toxic - lead arsenate, lead oxide and lead carbonate are the most dangerous; lead sulphide is the least toxic.

Industrial Uses: Over 200 industries are counted where lead is used - manufacture of storage batteries; glass manufacture; ship building; printing and potteries; rubber industry and several others.

Non-occupational Sources: The greatest source of environmental (non-occupational) lead is gasoline. Thousands of tons of lead every year is exhausted from automobiles. Lead is one of the few trace metals that is abundantly present in the environment. Lead exposure may also occur through drinking water from lead pipes chewing lead paint on window sills or toys in case of children.

Mode of Absorption: Lead poisoning may occur in three ways: (1) *Inhalation*: Most cases of Industrial lead poisoning is due to inhalation of fumes and dust of lead or its compounds. (2) *Ingestion*: Poisoning by ingestion is of less common occurrence. Small quantities of lead trapped in the upper respiratory tract may be ingested. Lead may also be ingested in food or drink through contaminated hands. (3) *Skin*: Absorption through skin occurs only in respect of the organic compounds of lead, especially tetraethyl lead. Inorganic compounds are not absorbed through the skin.

Body Stores: The body store of lead in the average adult population is about 150 to 400 mg and blood levels average about 25 µg/100 ml. An increase to 70 µg/100 ml blood is generally associated with clinical symptoms. Normal adults ingest about 0.2 to 0.3 mg of lead per day largely from food and beverages.

Distribution in the Body: Ninety per cent of the ingested lead is excreted in the faeces. Lead absorbed from the gut enters the circulation, and 95 per cent enters

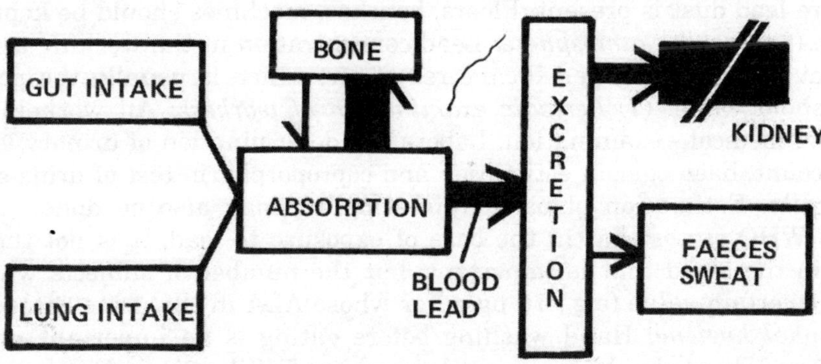

Fig. 11.1 A partial block in renal excretion (indicated by diagonal line) results in increased blood lead with lowered urine lead

the erythrbcytes. It is then transported to the liver and kidneys and finally transported to the bones where it is laid down with other minerals. Although bone lead is thought to be 'metabolically inactive', it may be released to the soft tissues again under conditions of bone resumption. Lead probably exerts its toxic action by combining with essential SH-groups of certain enzymes, for example some of those involved in porphyrin synthesis and carbohydrate metabolism. Lead has an effect on membrane permeability and potassium leakage, has been demonstrated from erythrocytes exposed to lead.

Clinical Picture: The clinical picture of lead poisoning or plumbism is different in the inorganic and organic lead exposures. The toxic effects of *inorganic* lead exposure are abdominal colic, obstinate constipation, loss of appetite, blue-line on the gums, stippling of red cells, anaemia, wrist drop and foot drop. The toxic effects of *organic* lead compounds are mostly on the central nervous system - insomnia, headache; mental confusion, delirium, etc. -

Diagnosis : Diagnosis of lead poisoning is based on : (1) *History:* a history of lead exposure (2) *Clinical features* : such as loss of appetite, intestinal colic, persistent headache, weakness, abdominal cramps and constipation, joint and muscular pains, blue line on gums, anaemia, etc. (3) *Laboratory tests*: .(a) *Coproporphyrin in urine* (CPU) : Measurement of CPU is a useful screening test. In non-exposed persons, it is less than 150 micrograms/litre. (b) *Amino levulinic acid in urine (ALAU)* : If it exceeds 5 mg/litre, it indicates clearly lead absorption. (c) *Lead in blood and urine:* Measurement of lead in blood or urine requires refined laboratory techniques. They provide quantitative indicators of exposure. Lead in urine of over 0.8 mg/litre (normal is 0.2 to 0.8 mg) indicates lead exposure and lead absorption. A blood level 70 µg/1 00 ml is associated with clinical symptoms. (d) *Basophilic stipling of RBC* : It is a sensitive parameter of the heamatological response.

Preventive Measures: (1) *Substitution:* That is, where possible lead compounds should by substituted by less toxic materials. (2) *Isolation:* All processes which' give rise to harmful concentration of lead dust or fumes should be enclosed and segregated. (3) *Local exhaust ventilation:* There should be adequate local exhaust ventilation system to remove fumes and dust promptly (4) *Personal protection:* Workers should be protected by approved respirators. (5) *Good housekeeping:* Good housekeeping is essential where lead dust is present. Floors, benches, machines should be kept clean by wet sweeping. (6) *Working atmosphere:* Lead concentration in the working atmosphere should be kept below 2.0 mg per 10c.metres of air, which is usually the permissible limit or threshold value. (7) *Periodic examination of workers:* All workers must be given periodical medical - examination. Laboratory determination of urinary lead, blood lead, red cell count, haemoglobin estimation and coproporphyrin test of urine should be done periodically. Estimation of basophilic stippling may also be done. An Expert Committee of WHO states that in the case of exposure to lead, it is not the average level of lead in the blood that is important, but the number of subjects whose blood level exceeds a certain value (e.g., 70 µg/ml or whose ALA in the urine exceeds 10 mg/litre, (8) *Personal hygiene:* Hand washing before eating is an important measure of personal hygiene. There should be adequate washing facilities in industry. prohibition on taking food in work places is essential. (9) *Health education:* Workers should be educated on the risks involved and personal protection measures.

Management: The major objectives in management of lead poisoning are the prevention of further absorption, the removal of lead from soft tissues and prevention of recurrence. Early recognition of cases will help in removing them from further exposure. A saline purge will remove unabsorbed lead from the gut. The use of d-penicillamine has been reported to be effective. Like Ca-EDTA, it is a chelating agent and works by promoting lead excretion in urine, Lead poisoning is a notifiable and compensable disease in India since 1924.

11.5 Occupational Cancer

Occupational cancer is an important problem in Industry. The sites of the body most commonly affected are skin, lungs, bladder and blood-forming organs.

11.5.1 Skin Cancer

Percival Pott was first to draw attention to cancer of scrotum in chimney sweeps in 1775. It was subsequently found that cancer of the scrotum and of the skin in other parts of the body was caused by coal tar, x-rays, certain oils and dyes. Statistics now show that nearly 75 per cent of occupational cancers are skin cancer. Skin cancers are an occupational hazard among gas workers, coke oven workers, tar distillers, oil refiners, dye-stuff makers, road makers and in industries associated with the use of mineral oil, pitch, tar and related compounds.

11.5.2 Lung Cancer

Lung cancer is a hazard in gas industry, asbestos industry, nickel and chromium work, arsenic roasting plants and in the mining of radio-active substances (e.g., uranium). Nickel, chromates, asbestos, coal tar (presumably 3.4 benzpyrene), radioactive substances and cigarette-smoking are proved carcinogens for the lung. Arsenic, beryllium and isopropyl oil are suspected carcinogens. More than nine-tenths of lung cancer are attributed to tobacco smoking, air pollution and occupational exposure.

11.5.3 Bladder Cancer

Cancer bladder was first noted in man in aniline industry in 1895. In more recent years, it was noted in the rubber industry. It is now known that bladder cancer is caused by aromatic amines, which are metabolised in the body and excreted in the urine. The industries associated with bladder cancer are the dye-stuffs and dyeing industry, rubber, gas and the electric cable industries. The following have been mentioned as possible bladder carcinogens: beta-naphthylamines, benzidine, paraamino-diphenyl, auramine and magenta.

11.5.4 Leukaemia

Exposure to benzol, roentgen rays and radio-active substances give rise to leukaemia. Benzol is a dangerous chemical and is used as a solvent in many industries. Leukaemia may appear long after exposure has ceased.

The characteristics of occupational cancer are: (1) they appear after prolonged exposure, (2) the period between exposure and development of the disease may be as long as 10 to 25.years, (3) the disease may develop even after the cessation of exposure, (4) the average age incidence is earlier than that for cancer in general, (5) the localisation of the tumours is remarkably constant in anyone occupation. Personal hygiene is very important in -the prevention of occupational cancer.

11.5.5 Control of Industrial Cancer

The control measures comprise the following: (1) elimination or control of industrial carcinogens. Technical measures like exclusion of the carcinogen from the industry, well-designed building or machinery, closed system of production, etc., (2) medical examinations, (3) inspection of factories, (4) notification (5) licensing of establishments, (6) personal hygiene measures, (7) education of workers and management, and (8) research.

11.6 Occupational Dermaititis

Occupational dermatitis is a big health problem in many industries. The causes may be, *Physical* - heat, cold, moisture, friction, pressure, x - rays and other rays; *Chemical-* acids, alkalies, dyes, solvents, grease, tar, pitch, chlorinated phenols etc. *Biologic*-living agents such as viruses, bacteria, fungi and other parasites; *Plant products* - leaves, vegetables, fruits, flowers, vegetable dust, etc.

The dermatitis - producing agents are further classified into: (1) primary irritants and (2) sensitizing substances. Primary irritants (e.g. acids, alkalies, dyes, solvents, etc.) cause dermatitis in workers exposed in sufficient concentration and for a long period of time. On the other hand, allergic dermatitis occurs only in small percentage of cases, due to sensitization of the skin.

11.6.1 Prevention

Occupational dermatitis is largely preventable if proper control measures are adopted: (1) *Pre-selection:* The workers should be medically examined before employment, and those with an established or suspected dermatitis or who have a known pre-disposition to skin disease should be kept away from jobs involving a skin hazard. (2) *Protection:* The worker should be given adequate protection against direct contact by protective clothing, long leather gloves, aprons and boots. The protective clothing should be frequently washed and kept in good order. There are also, what are known as *barrier creams* which must be used regularly and correctly. There is no barrier cream so far invented which will prevent dermatitis in all occupations. (3) *Personal hygiene:* There should be available a plentiful supply of warm water, soap and towels. The worker should be encouraged and educated to make frequent use of these facilities. Adequate washing facilities in industry are a statutory obligation under the Factories Act. (4) *Periodic inspection:* There should be a periodic medical check-up of all workers for early detection and treatment of occupational dermatitis. If necessary, the affected worker may have to be transferred to a job not exposing him to risk. The worker should be educated to report any skin irritation, no matter how mild or insignificant.

11.7 Radiation Hazards

A number of industries use radium and other radio-active substances, e.g., painting of luminous dials for watches; and other instruments, manufacture of radio-active paints. Exposure to radium also occurs in mining of radio-active ores, monozite sand workers and handling of their products. X-rays are used both in medicine and industry. Exposure to ultraviolet rays occurs in arc and other electric welding processes. Infrared rays are produced in welding, glass blowing, foundry work and other processes where metal and glass are heated to the molten state and in heating and drying of painted and lacquered objects.

11.7.1 Effects of Radiation

Occupational hazards due to ionizing radiation may be acute burns, dermatitis and blood dyscrasias; chronic exposure may cause malignancies and genetic effects. Lung cancer may develop in miners working in uranium mines due to inhalation of radio-active dust.

11.7.2 Preventive Measures

(1) Inhalation, swallowing or direct contact with the skin should be avoided, (2) In case of .X-rays, shielding should be used of such thickness and of such material as to reduce the exposure below allowable exposures, (3) The employees should be monitored at intervals not exceeding 6 months by use of the film badge or pocket electrometer devices, (4) Suitable protective clothing to prevent contact with harmful material should be used, (5) Adequate ventilation of work-place is necessary to prevent inhalation of harmful gases and dusts, (6) Replacement and periodic examination of workers should be done every 2 months. If harmful effects are found, the employees should be transferred to work not involving exposure to radiation, (7) pregnant women should not be allowed to work in places where there is continuous exposure.

11.8 Occupational Hazards of Agricultural Workers

Occupational health in agriculture is a new concept. From the standpoint of capital investment and number of persons employed, agriculture may be termed as "big buziness". Agricultural workers have a multitude of health problems - a fact which is often forgotten because of the widespread misconception that occupational health *is* mainly concerned with industry and industrialized countries. The health problems of workers in agriculture may be enumerated as below.

(1) **Zoonotic Diseases**: The close contact of the agricultural worker with animals or their products increases the likelihood of his contracting certain zoonotic diseases such as brucellosis, anthrax, leptospirosis, tetanus, tuberculosis (bovine) and Q fever. The extent of the occupational occurrence of these diseases in most parts of the world is not known.

(2) **Accidents**: Agricultural accidents are becoming more frequent, even in developing countries, as a result of the increasing use of agricultural machinery. Insect and snake bites are an additional health problem in India.

(3) **Toxic Hazards**: Chemicals are being used increasingly in agriculture either as fertilizers, insecticides or pesticides. Agricultural workers are exposed to toxic hazards from these chemicals. Associated factors such as malnutrition and parasitic infestation may increase susceptibility to poisoning at relatively low levels of exposure.

(4) **Physical Hazards**: The agricultural worker may be exposed to extremes of climatic conditions such as temperature, humidity, solar radiation, which may impose additional stresses upon him. He may also have to tolerate excessive noise and vibrations, inadequate ventilation and the necessity of working in uncomfortable positions for long periods of time.

(5) **Respiratory Diseases**: Exposure to dusts of grains, rice husks, coconut fibres, tea, tobacco, cotton, hay and wood are common where these products are grown. The resulting diseases -e.g., byssinosis, bagassosis, farmer's lung and occupational asthma, appear to be widespread.

11.9 Health Problems due to Industrialization

Industrialization implies the transformation of a peasant society into a community dependent upon the industries. It involves individual and collective technical skills for the manufacture of particular goods through highly specialized processes. There is division of work under the same roof with emphasis on mass production and community profit. In short, industrialization means a social and economic revolution in the culture of a nation. Any such revolution is bound to carry with it hazards.

The community health problems arising out of industrialization may be enumerated as follows.

11.9.1 Environmental Sanitation Problems

(a) *Housing:* A rise in the number of slums and insantary dwellings is one of the chief problems in all industrial areas due to migration of people from the country-side for employment. The effect of substandard housing on the health of the population is discussed elsewhere in detail. (b) *Water pollution*: Water pollution is one of the tragic aftermaths of rapid industrialization due to discharge of industrial wastes without treatment, into water courses. *Industrial wastes* may contain acids, alkalies, oils and other organic and inorganic chemicals, some of which may be toxic; synthetic detergents and radioactive substances. It requires legal, administrative and technical measures to deal with the situation. Pollution control measures should be instituted in the planning stage itself in the process of industrialization to which the country is committed. (c) *Air pollution*: This is an important problem in industrial areas which may have an adverse effect on the health of the population. Air pollution is due to the discharge of toxic fumes, gases, smoke and dusts into the atmosphere. It requires proper town planning and zoning to, eliminate this hazard. (d) *Sewage disposal*: There is bound to be pressure on the existing sanitation services if proper planning is not undertaken' before locating industries. Lack of facilities for the disposal of sewage leads- to pollution of water supply, contamination of soil with parasites and their ova.

11.9.2 Communicable Diseases

The main problems in industrial areas are tuberculosis, venereal diseases, and food and water borne infections. These are in addition to the specific diseases associated with specific industries. Industrial areas without proper sewage disposal have become hot-beds for filariasis owing to the breeding of the mosquito vectors in contaminated water.

11.9.3 Food Sanitation

The standards of food sanitation are bound to be lowered due to industrialization, if proper precautions are not taken. Food-borne infections such as typhoid fever and viral hepatitis are all too common in India.

11.9.4 Mental Health

Mental health problems are due to altered living conditions. People are removed from the warmth of village community life and are transplanted in an alien environment which calls for certain adjustments. Failure of adjustment leads to mental illness, psychoneurosis, behaviour disorders, delinquency etc.

11.9.5 Accidents

Accidents are a public health problem in industrial areas due to congestion, vehicular traffic and the increased tempo of life. These accidents are in addition to those that occur in the factories.

11.9.6 Social Problems

Alcoholism, drug addiction, gambling, prostitution, increased divorces, breaking up of home, juvenile delinquency, increased incidence of crime are some of the social problems due to industrialization.

11.9.7 Morbidity and Mortality

Vital statistical rates indicate that industrial areas are characterized by high morbidity and mortality from certain diseases. For example the incidence of chronic bronchitis and lung cancer is higher in industrialized areas than in rural areas. The crude death rate, the infant mortality rate tend to be high in industrial areas. India has special problems of her own due to industrialization. It is because the levels of public health are generally low the average expectation of life is less than that in industrially advanced countries. The resulting health problems must become particularly acute.

SUGGESTIVE FURTHER READINGS

Inoue, T. et al. (1983). A nation wide survey of organic solvents components in various solvent products. *Industrial Health*. 21, 175-183.

STUDY QUESTIONS

1. Give a detailed description of occupational hazards with suitable examples.
2. Write an essay on occupational diseases.
3. Write briefly on the following.
   i.   Pneumoconiosis
   ii.   Asbestosis
   iii. Plumbism
   iv.   Occupational dermatitis
   v.   Occupational cancer
4. Discuss diseases caused by industries in man.

Chapter 12

FUNDAMENTALS OF TOXICOLOGY

Toxicology can be defined quite simply as the branch of science dealing with poisons. Broadly speaking, a poison is any substance causing harmful effects in an organism to which it is administered, either deliberately or by accident. Clearly, this effect is closely related since any substance, at a low enough dose, is without effect, while many, if not most, substances have deleterious effects at some higher dose. Much of toxicology deals with compounds exogenous to the normal metabolism of the organism. Such compounds are referred to as foreign compounds or more recently, as xenobiotics. However, many compounds endogenous to the organism e.g., metabolic intermediates such as glutamate, and hormones such as thyroxine, are toxic when administered in unnaturally high doses. Similarly, trace elements such as selenium, which are essential in the diet in low concentrations, are frequently toxic at higher levels. Whether the harmful effects of physical phenomena such as irradiation, sound, temperature, and humidity are included in toxicology, appear to be largely dependent on the preference of the writer, it is convenient, however, to include them under the broad definition of toxicology.

The method of assessing toxic effects is another parameter of considerable complexity. Acute toxicity, usually measured as mortality and expressed as the LD_{50}- the dose required to kill 50% of a population of the organism in question under specified conditions- is probably the simplest measure of toxicity. Even so, reproducibility of LD_{50} values is highly dependent upon the extent to which many variables are controlled. These include the age, sex, and physiological condition of the animals, their diet, the environmental temperature and humidity, and the method of administering the toxicant.

Chronic toxicity may be manifested in a variety of ways - carcinomas, cataracts, peptic ulcers, and reproductive effects, to name only a few. Furthermore, compounds may have different effects at different doses. Vinyl chloride, a potent hepatotoxin at high doses, is a carcinogen with a very long latent period at low doses. Most drugs have therapeutic effects at low doses but are toxic at higher levels. The relatively nontoxic acetylsalicyclic acid (aspirin) is a useful analgesic at low doses, *is* toxic at high doses, and may cause peptic ulcers with chronic use.

Considerable variation also exists in the toxic effects of the same compound administered to different animals, or even to the same animal by different routes. The

150

insecticide malathion has a low toxicity to mammals, whereas it is toxic enough to insects to be a widely used commercial insecticide. The route of entry of toxicants into the animal body is frequently oral, in the food or drinking water in the case of many chronic environmental contaminants such as lead or insecticide residues, or directly as in the case of accidental or deliberate acute poisoning. Other routes for non experimental poisoning include dermal absorption and pulmonary absorption. The above routes of administration are all used experimentally, and in addition, several types of injection are also common -intravenous, intraperitoneal, intramuscular, and subcutaneous. The toxicity of many compounds varies tenfold or greater depending upon the method of administration.

12.1 Relationship of Toxicology to other Sciences

Toxicology is frequently said to be a branch of pharmacology, a science that deals primarily with the therapeutic effects of exogenous substances and with all the chemical and biochemical ramifications involved in those effects. Since the therapeutic dose range of pharmacological compounds is usually quite small, and most of these compounds are toxic at higher doses, it may be more appropriate to consider pharmacology a branch of toxicology.

Toxicology is clearly related to the two applied biologies-medicine and agriculture. In the former, clinical diagnosis and treatment of poisoning as well as the management of toxic side effects are areas of significance, while in the latter the development of agricultural biocides such as insecticides, herbicides, nematicides, and fungicides is of great importance. The detection and management of the off-target effects of such compounds is also an area of increasing importance that is essential to their continued use. Toxicology may also be considered an area of fundamental biology since the adaptation of organisms to toxic environments has important implications for ecology and evolution.

The tools of chemistry and chemical biology since the adaptation of organisms to and progress in toxicology are closely related to the development of new methodology. Those of chemistry provide analytical methods for toxic compounds, particularly for forensic toxicology and residue analysis, and those of biochemistry provide the techniques to investigate the metabolism and mode of action of toxic compounds. On the other hand, studies of the chemistry of toxic compounds have contributed to fundamental organic chemistry, and studies of the enzymes involved in detoxication and toxic action have contributed to our basic knowledge of biochemistry.

12.2 Scope of Toxicology

Toxicology in the most general sense may be one of the oldest practised sciences. From his earliest beginnings, man must have been aware of numerous toxins such as snake venoms and those of poisonous plants. From the earliest written records it is clear that the ancients had considerable knowledge of poisons. The Greeks made use of hemlock as a method of execution, more particularly, the Romans made much use of poisons for political and other assassinations. Indeed, it was *Dioscorides*, a Greek at the court of Nero, who made the earliest known attempt to classify poisons. Although poisoning has

PARACELSUS
(The father of Toxicology)
(1493 – 1541)

enjoyed a considerable vogue at many times and places, the scientific study of toxicology can probably be dated from *Paracelsus,* who in the sixteenth century, put forward the necessity for experimentation and included much in his range of interests that would today be classified as toxicology.

The modem study of toxicology is usually dated from the Spaniard, Orfila (1787-1853), who, at the University of Paris, identified toxicology as a separate science. Among his many contributions, he devised chemical methods for the detection of poisons and stressed the value of chemical analysis to provide legal evidence. He was also the author, in 1815, of the first book devoted entirely to the toxic effects of chemicals. Toxicology can be subdivided in a variety of ways. *Loomis* refers to the three "basic" subdivisions as *environmental, economic,* and *forensic.* Environmental toxicology is further divided into such areas as pollution, residues, and industrial hygiene; economic toxicology is said to be devoted to the development of drugs, food additives, and pesticides; and forensic toxicology is concerned with diagnosis, therapy, and medicolegal considerations. Clearly, these categories are not mutually exclusive; for example, the off target effects of pesticides are considered to be environmental, while the development of pesticides is economic.

12.2.1 Environmental Toxicology

Environmental toxicology is the most rapidly growing branch of science. Public concern over environmental pollutants and their possible chronic effects, particularly carcinogenicity, has given rise, in the United States, to new research and regulatory agencies and recently to the Toxic Substances Control Act. Similar developments are also taking place in many other countries. The range of environmental-pollutants is enormous, including industrial and domestic effluents, combustion products of fossil fuels, agricultural chemicals, and many other compounds that may be found in food, air, and water. Such compounds as food additives and cosmetics are also being subjected to the same scrutiny.

Other sub specialties are frequently mentioned that do not fit into the above divisions. Behavioral toxicology, an area of increasing importance, could be involved in any of these and is usually treated as a separate subspeciality. Analytical toxicology provides the methods used in essentially every branch of the subject, while biochemical toxicology, provides the fundamental basis for all branches of toxicology.

12.3 Language of Toxicology

Like any other specialized field, environmental health has its own language. Some of the terms used in this book may need a few words of introduction.

Toxic, a central concept simply means capable of causing illness. The types of illnesses caused by environmental toxins are conventionally divided into acute and chronic. *Acute* illness are those which appear soon after exposure to a toxic compound, last for a relatively short time, and then resolve themselves, even if the resolution is in death. The term *subacute* is also occasionally used to describe disorders with subtle symptoms that are not immediately obvious without special tests. Lead workers, for example, often appear to be much healthier than a thorough medical examination reveals them to be **Chronic** illnesses, by contrast, that may appear years or even decades after exposure, and which may "remain", unresolved, for the victim's lifetime.

There are three special kinds of toxic hazards that have special relevance to environmental health: ***carcinogens, mutagens*** and ***teratogens***. As most of us know, a carcinogenic substance is one that causes cancer. A mutagenic substance is one that causes changes in the genetic material of a cell. Spontaneous, natural mutations occur in our body cells all the time; the vast majority of them cause no damage, and even when they do it is usually limited to the lifetime of the cell they occur in. But in rare cases the cell may continue to grow and divide after a mutagen has altered its basic genetic structure, and if this mutation is passed on to succeeding generations via egg or sperm, it may cause birth defects, inherited diseases, mental deficiency, increased susceptibility to disease, and a host of other abnormalities and disorders. If the mutated genes are recessive, it may take more than one generation for these effects to show up.

Mutagenicity and carcinogenicity are related in some way, but were not yet sure just how. Radiation is probably the best-known example of a mutagenic environmental hazard.

Contamination is measured in terms of the *concentration* of a substance in the environment, and there are a number of different conventions governing the measurement of concentration. The most common system makes use of metric units, particularly the milligram (one-thousandth of a gram, abbreviated mg) and the microgram (one million of a gram, abbreviated ìg). Occ.asionally, in very refined, ultrasensitive measurements, nanograms (one-billionth of a gram), picograms (one-trillionth of a gram), and even smaller units may be used.

Obviously, a contaminant concentration of one milligram per kilogram is the same as one part of contaminant per million parts of noncontaminant, or 1 ppm, and this alternative method of indicating relative concentrations is also widely used, particularly for air pollutants, food additives, and pesticides residues*. The terms *ppm* and *ppb* (parts per billion) are common: *ppt* (parts per trillion) appears only rarely.

Exposure is a way of saying that the contamination in the environment has passed into an organism; a human being is exposed to a toxic compound if some amount of it has entered his or her body. Exposure does *not* mean that a person has merely been in the proximity of a toxic substance. For example, if you walk past a drum bearing a warning label and containing a toxin, you are not necessarily exposed to whatever it contains. But if the drum leaks its contents into the air or soil, and pollutes the air you breathe or the water you drink, you probably will be exposed to its contents.

Like illnesses, exposures may also be subdivided into **acute** and **chronic** types. Acute exposures are those that occur over short periods of time, often to high concentrations of a hazardous substance. Chronic exposures, which are much more common among the general public, involve longer periods of time and for the most part, lower concentrations. *Dose* is the term for measuring exposures. Basically, the dose a person exposed to a toxic substance receives is dependent on its concentration in the immediate environment and the duration of the exposure. However, because the interaction of human beings and the environment is a complex, constantly changing process, numerous other factors may also play a part in determining dose. Dose can be a function of weather conditions, the persistence and solubility of the toxic substance in the biosphere, the size of its molecules or particulates, the presence of other compounds in the environment that it may react with, the age and overall health of the exposed individual, whether the substance is inhaled, swallowed, or absorbed through the skin, and the effectiveness of the body's natural defenses in detoxifying the substance and eliminating it from the body. Some substances, like asbestos, become virtually permanent contaminants in the body once they have penetrated far enough into the lungs or other organs. Others, such as methanol, are metabolized and excreted from the body in a matter of hours at most.

The tendency for some substances to collect in the tissues of a living organism and stay there is known as **bioaccumulation**. Their tendency to move up the food chain as one species consumes another, becoming ever more concentrated as they go, is called **biomagnification**.

Traditionally a *threshold* was a measurable level of exposure to a toxic substance below which there would probably be no adverse health effect and above which there probably would be. The setting of safety standards for the work site as well as the general environment often involves the assumption that approximate thresholds can be determined, monitored and enforced for the toxin in question. But this assumption has been subjected to various criticisms in the past few decades. First, it is often pointed out that, whether we can measure it or not, it is entirely possible that *every molecule* of every substance we take into our bodies has some effect on us. It may not be a detectable effect and it may not be harmful or long lasting, but it is an effect. Thus, this argument goes, the concept of a specific cutoff point below which a substance is treated as though it didn't exist and above which it is considered harmful is misleading. Far more appropriate, proponents of this view argue, is the assumption that these substances have a *range* of effects, beginning at once end with those that are imperceptibly molecular and extending to the catastrophically toxic, ultimately fatal effect at the other end of the spectrum.

An LD_{50} is customarily expressed in terms of milligrams per kilogram of body weight. Thus, a substance whose LD_{50} is 2 mg/kg is five times as toxic as one whose LD_{50} is 10 mg/kg. In general, substances with LD_{50} values below 50 mg/kg are considered highly toxic. Those with values between 50 and 500 mg/kg are considered moderately toxic, and those with values above 500 mg/kg are regarded as less toxic.

12.4 Portals of Entry of a Toxin

The principal portals of entry are the skin, the gastrointestinal tract, and the lungs. The fact is stressed that in all cases the toxicant must pass through a number of biological membranes before it can be distributed throughout the body and that uptake depends on the nature of the membranes as well as the physical properties of the toxicant. The structure of cell membranes, basically bimolecular lipid leaflets with

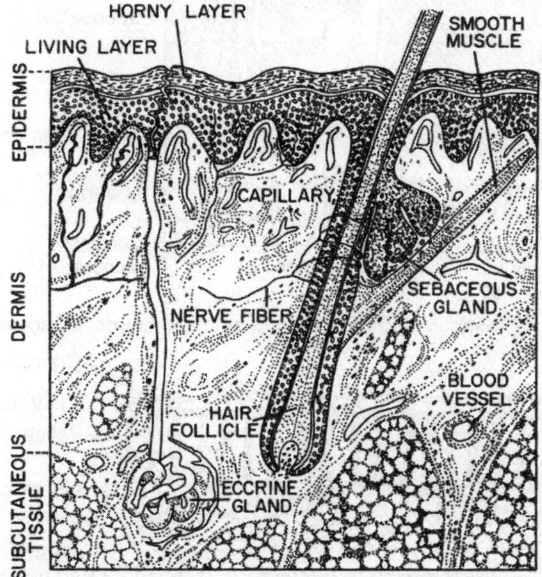

Fig. 12.1 Zones of skin in idealized section. Dermis supported by fat-rich, subcutaneous zone intermingled with other tissues.

associated proteins, and their various modifications are presented in some detail since their nature is responsible for the fact that lipophilicity is the most important determinant of the rate of uptake of exogenous molecules. Active transport, pinocytosis etc. are much less common than diffusion across the lipid membranes.

The insect, with its waxy epicuticle and plants, with a waxy cuticle and a stomatal system represent important special examples of entry of a toxin.

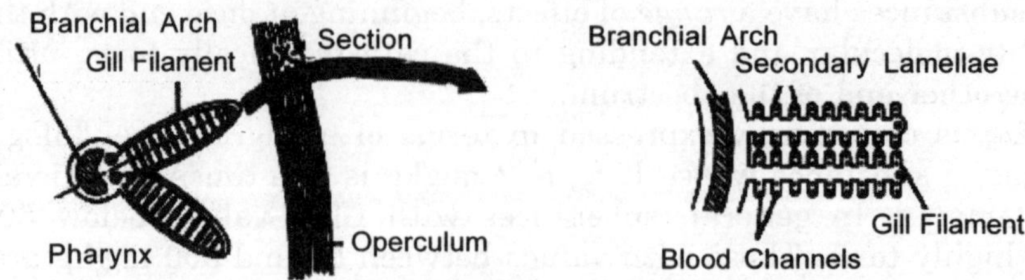

Fig. 12.2: Structure of fish gill. Magnified section along dashed line of filament.

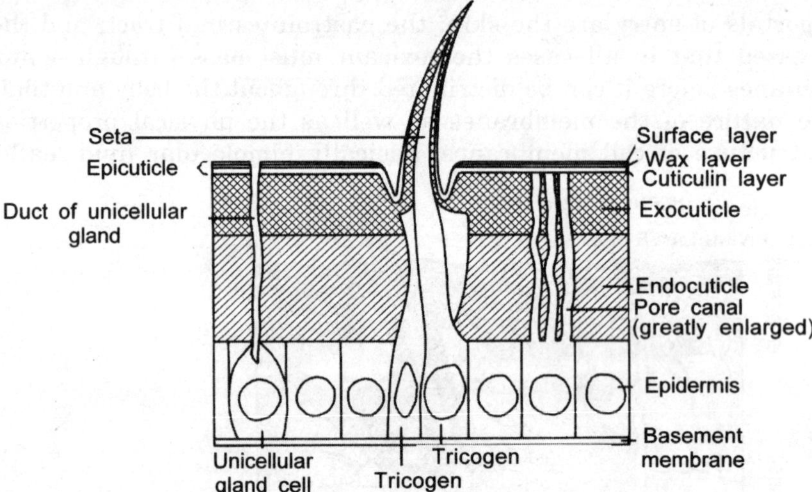

Fig. 12.3: Diagrammatic section of insect culticle illustrating primary layers of accessory components.

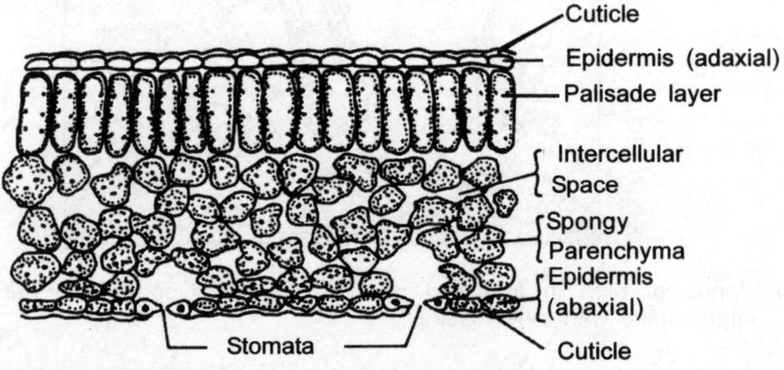

Fig. 12.4: Schematic diagram of cross section of a leaf

12.5 Distribution of a Toxin

Various factors responsible for the distribution of toxicants throughout the body need discussion. This is primarily concerned with the binding of toxicants to blood proteins, particularly lipoproteins. Lipoproteins are an important class of protein, particularly in the vascular fluids. They vary in molecular weight from 200,000 to 10,000,00 and the lipid content varies from 4% to 95%, being composed of triglycerides, phospholipids, and free and esterified cholesterol. Although they are classified into groups based on their flotation constants, each group is, in fact, a mixture of many similar lipoproteins.

The nature and importance of various types of ligand-protein interactions are assessed, including covalent binding, ionic binding, hydrogen bonding, vander Waals forces, and hydrophobic interactions. Many of the same binding forces are also important in toxicant receptor interactions.

It deals with the mathematical approach to the distribution of toxicants, or toxicokinetics. It provides a simplified, but still mathematically rigorous, treatment of distribution data, including both analysis and the formulation of mathematical models.

12.6 Metabolism of a Toxin

The majority of xenobiotics that enter the body do so because they are lipophilic. The metabolism of xenobiotics, which is carried out by a wide range of relatively nonspecific enzymes, serves to increase their water solubility and make possible their elimination from the body. This process consists of two phases. In phase I, a reactive polar group is

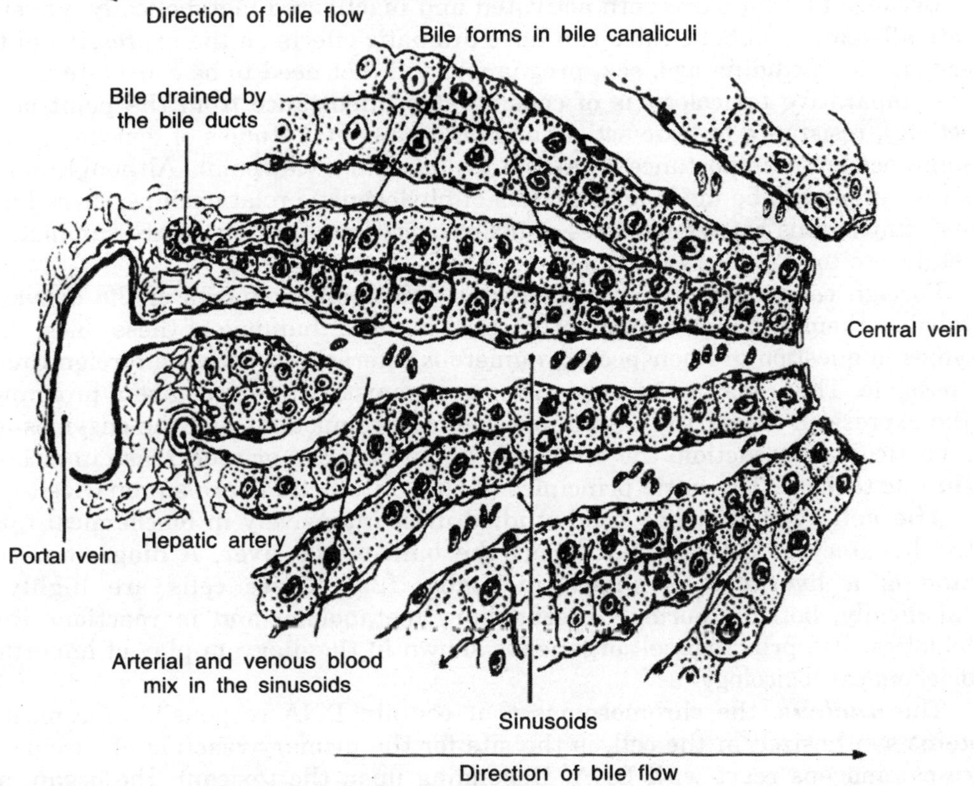

Fig. 12.5 Drawing to show the relationship between the sinusoids and the bile canaliculi in the liver parenchyma.

introduced into the molecule, rendering it a suitable substrate for phase II reactions. Phase I reactions include the well-known cytochrome P_{450} -dependent mixed-function oxidations as well as reductions, hydrolyses, etc. Phase II reactions include all the conjugation reactions in which a polar group on the toxicant is combined with an endogenous compound such as glucuronic acid, glutathione etc. to form a highly water-soluble conjugate that can be eliminated from the body.

It should be pointed out at this early stage that these metabolic reactions are not all detoxications since many foreign compounds are metabolized to highly reactive products that are responsible for their toxic effect. These include the activation of carcinogens and hepatotoxi cants.

Although the liver is the most studied organ with regard to xenobiotic metabolism, several other organs are known to be active in this respect, although neither the specific activity nor the range of substrates metabolized is as large as in the liver. These organs include the lungs and the gastrointestinal tract, as one might expect of organs that are important site for the entry of xenobiotic into the body, and to a lesser extent, the other important portal of entry, the skin. Other organs, such as the kidney, may also be important sites for xenobiotics metabolism. The distribution of xenobiotic-metabolizing enzymes between different organs is touched upon in following chapters.

Because toxicants are both activated and inactivated metabolically, physiological factors affecting metabolic rates can have dramatic effects on the expression of toxicity. These effects, including age, sex, pregnancy, and diet need to be considered.

Comparative toxicology is of considerable importance from the point of view of selectivity, resistance to toxic action, and environmental studies of toxicants, as well as of some academic importance from the evolutionary viewpoint. Although only a few generalizations can be made on the basis of phylogenetic relationships, there have been many comparisons between species of toxicological interest. These are summarized and resistance to toxic action is outlined.

Foreign compounds can be substrates, inhibitors, inducers of the enzymes that metabolize them and, not infrequently, serve in more than one of these roles. Since the enzymes in question are nonspecific, numerous interactions between foreign compounds are possible. These may be synergistic or antagonistic and may have a profound effect on the expression of toxicity. Depending upon the compounds and the enzymes involved in a particular interaction, the effect can be an increase or a decrease in either acute or chronic toxicity. The basic principles of such interactions are summarized.

The cell type that has been studied most intensively in biochemical toxicology is the hepatocyte, the cell that forms the bulk of the liver. A diagrammatic cross section of a liver cell is shown in figure 12.5. These cells are highly active metabolically, both in normal intermediary metabolism and in reactions involving xenobiotics. The principal cell organelles shown in the diagram play of important sole in biochemical toxicology.

The ***nucleus,*** the chromosomes that contain DNA responsible for most of the proteins synthesized in the cell, is the site for the primary reaction of carcinogenesis, since carcinogens react with DNA. Depending upon the toxicant, the organ, and the cell type involved, similar reactions are involved in mutagenesis and teratogenesis. The nuclear envelope has recently been shown to have an active aryl hydrocarbon hydroxylase system.

Mitochondria are the site of electron transport and oxidative phosphorylation pathways that provide sites for the action of many acute toxicants.

The ***endoplasmic reticulum*** exists in two forms: rough, which is associated with protein synthesis, and while both rough and smooth are active in the oxidation of xenobiotics, the latter usually has the highest specific activity. After, disruption of the cells, followed by differential centrifugation, the two types are isolated as rough and smooth microsomes.

12.7 Sites of Action of a Toxin

Compounds of intrinsic toxicity and active metabolites produced in the body ultimately arrive either at a site of action or an excretory organ. Although almost any organ can show the effects of toxicity, some are more easily affected than others by particular classes of toxicants, and some have been studied in greater detail than others. In all cases, however, toxicant-receptor interactions are important. The fundamentals of such interactions are discussed as follows.

Acute toxicants tend to affect either oxidative metabolism, as discussed, the synapses of the nervous system, or the neuromuscular junction. Toxic effects on the central nervous have been widely studied. The cholinesterase inhibitors, are discussed in the chapter dealing with toxicity of pesticides.

The commonest modes of chronic toxicity involve interaction with nucleic acids, causing carcinogenesis or reproductive effects. These effects are discussed in the chapter of environmental carcinogenesis. Although specific organ damage is known for several toxicants, the central role of the liver in studies of toxic action is acknowledged.

While toxicants can be classified in many ways, based either on natural distribution, commercial use patterns, or chemistry, only two such groups viz: metals and pesticides are accorded separate discussion in this book.

Many metabolic pathways are affected by toxicants, but space limitations preclude their detailed description in this introductory chapter. They include glycolysis, the tricarboxylic acid cycle, the pentose cycle, the electron transport system and oxidative phosphorylation, nucleic acid synthesis, protein synthesis, and many others, as well as such specialized systems as photosynthesis in plants. Some of these pathways are discussed and the remainder can be found in any adequate general textbook of biochemistry.

In vivo testing for chronic toxicity in animals and short-term mutagenicity tests are both somewhat remote from a strictly biochemical treatment of the mechanisms involved in toxicology.

12.8 Excretion of Toxin

Either the unmetabolized toxicants or their metabolic products are ultimately excreted, the latter usually as conjugated products resulting from phase II reactions. The two primary routes of excretion (the urinary system and the biliary system), minor routes also (such as the lungs, sweat glands, sebaceous glands, hair, feathers, and nails) and sex related routes (such as milk, eggs, and fetus) constitute the routes of excretion.

12.9 Nature of Toxic Effects

The nature and magnitude of a toxic effect depend on many factors, amongst which are the physicochemical properties of the substance, its bioconversion, the conditions of exposure, and the presence of bioprotective mechanisms. The last factor includes physiological mechanisms such as adaptive enzyme - induction, DNA repair mechanisms and phagocytosis. Some of the frequently encountered types of morphological and biochemical injury constituting a toxic response are listed below. They may take the form of tissue pathology, aberrant growth processes, altered or aberrant biochemical pathways or extreme physiological responses.

Inflammation is a frequent local response to irritant chemicals or may be a component of systemic tissue injury. The inflammatory response may be acute with irritant or tissue damaging materials, or chronic with repetitive exposure to irritants or the presence of insoluble particulate material. Fibrosis may occur as a consequence of the inflammatory process.

Necrosis, used to describe circumscribed death of tissues or cells, may result from a variety of pathological processes induced by chemical injury, e.g. corrosion, severe hypoxia, membrane damage, reactive metabolite binding, inhibition of protein synthesis and chromosome injury. With certain substances, differing patterns of zonal necrosis may be seen. In the liver, for example, galactosamine produces diffuse necrosis of the lobules (Mehendale, 1987), acetominophen (paracetamol) mainly centrilobular necrosis (Goldfrank et al. 1990) and certain organic arsenicals peripheral lobular necrosis (Ballantyne, 1978).

Enzyme Inhibition by chemicals may inhibit biologically vital pathways, producing impairment of normal function. The induction of toxicity may be due to accumulation of substrate or to deficiency of product or function. For example, organophosphate anticholinesterases produce toxicity by - accumulation of acetylcholine at cholinergic synapses and neuromuscular junctions (Ellenhorn and Barceloux, 1988). Cyanide inhibits cytochrome oxidase and interferes with mitochondial oxygen transport, producing cytotoxic hypoxia (Ballantyne, 1987).

Biochemical uncoupling agents interfere with the synthesis of high-enephosphate molecules, but electron transport continues resulting in excess liberation of energy a heat. Thus, uncoupling produces increased oxygen consumption and hyperthermia. Examples of uncoupling agents are dinitrophenol and pentachlorophenol (Williams, 1982; Kurt et at 1988).

Lethal synthesis occurs when foreign substances of close structural similarity to normal biological substrates become incorporated into biochemical pathways and are then metabolized to a toxic product. A classical example is fluoroacetate, which becomes incorporated in the Kreb's cycle as fluoroacetyl coenzyme A, which combines with oxaloacetate to form fluorocitrate. The latter inhibits aconitase, blocking the tricarboxylic acid various system toxicity (Albert, 1979).

Lipid peroxidation in biological membranes by free radicals starts a chain of events causing cellular dysfunction and death. The complex series of events includes oxidation of fatty acids to lipid hydroperoxides which undergo degradation to various products, including toxic aldehydes. The generation of organic radicals during peroxidation results in a self-propagating reaction (Horton and Fairhurst, 1987).

Carbontetrachloride, for example, is activated by a hepatic cytochrome P_{450}-dependent mono-oxygenase system to the trichloromethyl and trichloromethyl peroxy radicals; that covalently bind with macromolcules and the latter initiates the process of lipid peroxidation leading to hepatic centrilobular necrosis. The zonal necrosis is possibly related to high cytochrome P_{450} activity in centrilobular hepatocytes (Albano at al. 1982).

Covalent binding of electrophilic reactive metabolites to nucleophilic macromolecules may have a role in certain genotoxic, carcinogenic, teratogenic and immunosuppressive events. Important cellular defence mechanisms exist to moderate these reactions, and toxicity may not be initiated.

Receptor interaction at a cellular or macromolecular level, with specific chemical structures may modify the normal biological effect mediated by the receptor, these may be excitatory or inhibitory. An important example is effects on Ca channels (Braunwald, 1982).

Immune-mediated hypersensitivity reactions by antigenic materials are particularly important considerations for skin and lung resulting in allergic contact dermatitis and asthma, respectively (Cronin, 1980; Brooks, 1983; Bardana et al. 1992).

Immunosuppression by xenobiotics may have important repercussions in increased susceptibility to infective agents and certain aspects of tumorigenesis.

Neoplasia, resulting from aberration of tissue growth and control mechanisms of cell division, and resulting in abnormal proliferation and growth, is a major consideration in repeated exposure to xenobiotics. The terms tumorigenesis and oncogenesis are general words used to describe the development of neoplasms; the word carcinogenesis should be restricted specifically to malignant neoplasms. In experimental and epidemiological situations, oncogenesis may be exhibited as an increase in specific types of neoplasm, the occurrence of rare, or unique neoplasms or a decreased latency to detection of neoplasm.

The preceding description of the nature and scope of biochemical toxicology should make it clear that the biochemistry of toxic action is a many-faceted subject, covering, all aspects from the initial environmental contact with a toxicant to its ultimate excretion back into the environment. A considerable amount of information is discussed in the following chapters and, in general, it is assumed that the reader has a competent knowledge of basic biochemistry.

SUGGESTIVE FURTHER READINGS

Ariens, E. J., Simonis, A. M., Offermeier, J. Introduction to General Toxicology. New York: Academic Press, 1976.

Casarett, L. J. Origin and scope of toxicology. In Casarett, L. J., Doull, J. (Eds.). Toxicology, The Basic Science of Poisons. New York: Macmillan, 1975.

Hayes, W. J. Toxicology of Pesticides. Baltimore: Williams & Wilkins, 197

Loomis, T. A. Essentials of Toxicology. Philadelphia: Lea & Febiger, 1974.

STUDY QUESTIONS

1. Define toxicology. Discuss scope and relationship of toxicology with other branches of science.
2. Define the following terms -
   a. Acute toxicity
   b. Chronic toxicity
   c. Carcinogen
   d. Teratogen
   e. Mutagen
   f. Bioaccumulation
   g. Biomagnification
3. Describe different portals of entry of a toxin. Write briefly on the distribution of a toxin.
4. Give a detailed account of the metabolism of a toxin. Discuss sites of action of a toxin.
5. Write in details on the nature of effects of toxins.

Chapter 13

TOXICITY OF HEAVY METALS

About three-quarters of all known chemical elements are *metals,* generally characterized by high electrical and thermal conductivity, malleability, ductility and a high reflectivity of light. Aluminium (Al), iron (Fe), calcium (Ca), sodium (Na) and magnesium (Mg) are the most abundant metals in the Earth's crust. Only a few metals, such as copper (Cu), gold (Au), platinum (Pt) and silver (Ag), are found in the free state, while the majority of metals are found in the form of complexes with other elements. Most metals are solids in their pure form, with a relatively simple crystal structure distinguished by a close packing of atoms and a high degree of symmetry. Metal atoms usually contain less than half the full number of valence electrons and combine with non-metals, such as oxygen and sulphur, which generally have more than half the maximum number electrons in their outermost shell. The chemical reactivity of native metals vary, from the highly reactive lithium (Li), Na and K to Au, and Pt, which have very low chemical reactivity.

Metalloids have chemical properties intermediate between those of typical metals and non-metals. The group includes boron (B), silicon (Si), germanium (Ge), arsenic (As), antimony (Sb) and tellurium (Te). Most of these elements have important commercial applications in transistors and other semiconductor devices, ceramics, solar batteries and certain polymers.

Heavy metal is an expression often used for elements, not necessarily 'true' metals, with a specific gravity exceeding 5 mg ml^{-1}, Several toxic elements, such as As, Sb, cadmium (Cd), mercury (Hg), lead (Pb), uranium (U) and bismuth (Bi), are often included in this group.

Alloys are metallic substances, solid components or solutions that are composed of two or more elements, usually themselves metals. The value of certain alloys, such as brass (copper and zinc) and bronze (copper and tin), was discovered in ancient times. Today, the alloy steels are of primary importance. These contain significant amounts of elements other than iron and carbon, such as chromium, nickel, molybdenum, tungsten and vanadium.

Amalgams are alloys of mercury and one or more other metals. An amalgam of Silver and tin, with minor amounts of copper and zinc, is frequently used in dentistry. Amalgamation of precious elements, such as silver and gold, is a common practice in small-scale mining (De Lacerda and Salomons, 1998). Agitating their ores with

163

mercury and allowing the resultant pasty or liquid amalgam to settle can recover fine particles of silver and gold. The mercury is distilled off and the precious metal is isolated as a residue.

Trace elements usually classified as essential to man are iodine, iron, zinc, copper, manganese, selenium, molybdenum and cobalt (WHO, 1996). The uptake and utilization of essential trace elements are under physiological control and a well balanced diet is usually adequate in order to maintain trace element homeostasis. However, deficiencies in the gastro-intestinal uptake of elements or certain inborn errors of metabolism can result in disease due to sustained trace element imbalance (Lentner, 1986). For some essential trace elements, excess and the difference between essential and toxic intake levels may be small (Nord, 1995). Several other trace elements, notably arsenic, cadmium, lead and mercury, have no known function in the human body. Exposure to these elements may cause immediate as well as delayed adverse health effects, eventually developing into permanent impairment of essential bodily functions.

A full coverage of all aspects on metal toxicology is beyond the scope of this chapter. Chronic toxicity of a few non-essential trace elements is discussed below.

13.1 Arsenic (As)

The name arsenic is derived from the Latin word *arsenicum* and the Greek word *arsenikon* (yellow orpiment). It can also be traced to the Arabic word, Az-zernikh, meaning the orpiment from zerni-zar, the Persian word for gold (Hammond, 1995).

Elemental arsenic is a steel grey, very brittle, crystalline, semimetallic solid, Arsenic tarnishes in air and is rapidly oxidized to arsenous oxide (As_2O_3) when health.

13.1.1 Toxicity

Soluble inorganic arsenic salts are highly toxic. The lethal dose of ingested arsenic trioxide is about 1-3 mg (kg body weight)$^{-1}$ in adults while even lower exposure levels, 1-4 mg day^{-1}, have caused serious health effects, including fatalities, in small children (WHO, 1981). The Morinage incident, where arsenic-contaminated powdered milk was fed to children, affected 12000 Japanese children and 130 fatalities were recorded (Yamashita et al. 1972). Symptoms of arsenic intoxication include nausea, headache and severe abdominal pain due to damage to the gastro-intestinal mucous membranes, violent vomiting and diarrhoea caused by paralysis of the capillary control in the intestinal tract. Eventually the gastrointestinal epithelium may be sloughed off, followed by a decrease blood volume, decreased blood pressure, disturbed heart action, failure of vital cardiovascular and brain functions and death. The massive loss of water may lead to renal failure and anuria. Acute arsenic intoxication may also lead to a general paralysis of the capillaries, acute excitability of the brain and death through general paralysis. Non-fatal acute intoxications may result in damage to the peripheral nervous system, manifested as sensory loss, and due to axonal degeneration. Other symptoms include anaemia and leukopenia, fever, anorexia, hepatomegaly and melanosis.

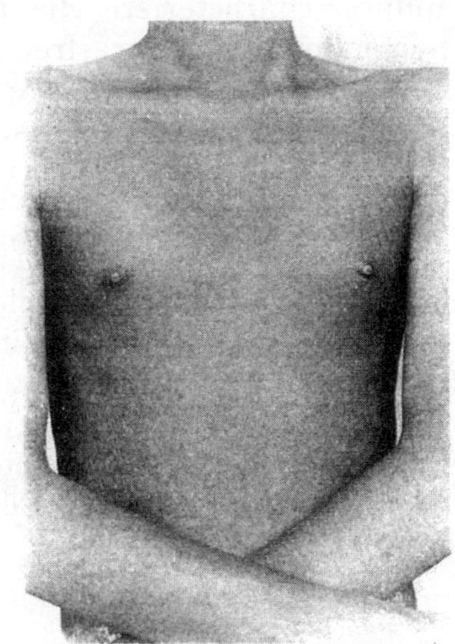

Spotted melanosis (rain drop disease), Village: Bishnupur, Gaighata Block, District: North 24-Parganas.

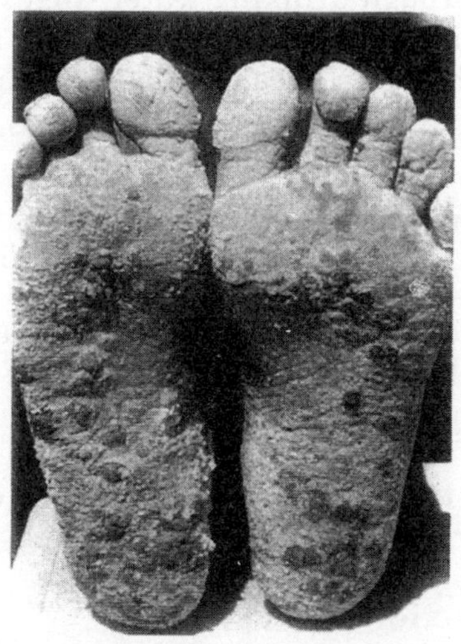

Diffuse with nodular keratosis on sole. Village: Chandpur Rail Line, Basirhat block-II, District: North 24-Parganas.

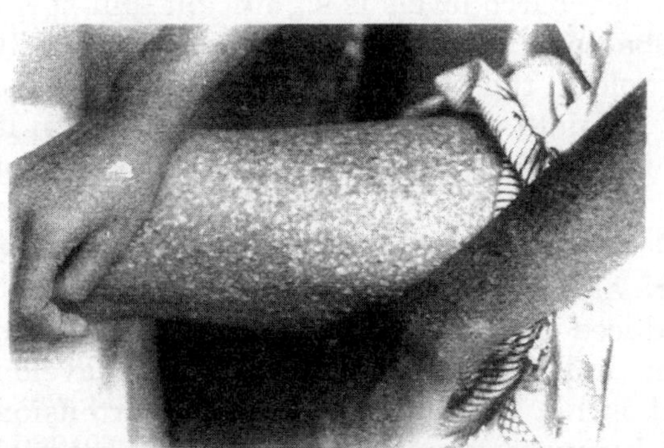

Leucomelanosis, Village: Jampukur, Kaliganj Block, District: Noida.

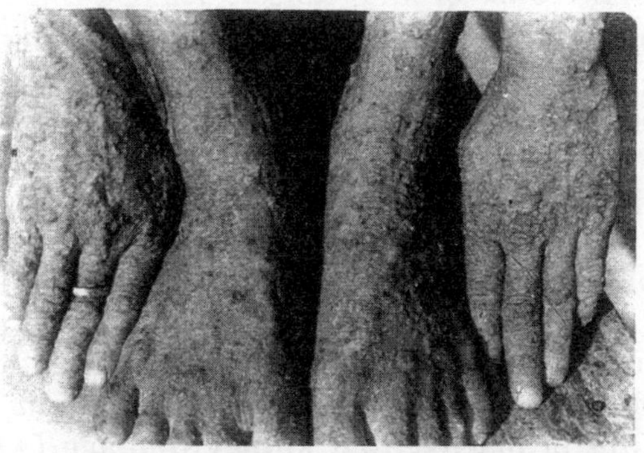

Palpable nodules in dorsum on hands, feet and legs. Village: Mandra, Purbasthali Block I, District: Bardhaman.

Fig. 13.1 Changes in skin pigmentation due to arsenic poisoning. .

Inhalation of arsenic cause respiratory tract symptoms, such as rhinitis, laryngitis and bronchitis (Ishinishi ct al., 1977; Landrigan, 1981; Nemery, 1990). At high exposure levels, hyperplasia and atrophy in the respiratory tract and perforation of the nasal septum have been reported (Pinto .and McGill, 1953; Lundgren, 1954).

Subchronic and chronic exposure to inorganic arsenic has been associated with a reduction in nerve conduction velocity (Blomet al. 1985; Lagerkvist and Zetterlund, 1994) and hepatic injury (Lander et al. 1975). Peripheral vascular damage has been observed among people exposed to arsenic in drinking water in Taiwan and Chile (Tseng, 1977). The disease may develop into gangrene of the lower extremities ('blackfoot disease') and the effect appears to be associated with the cumulative dose of arsenic through drinking water. Ingestion of arsenic induces characteristic changes in skin pigmentation (Fig. 13.1). Areas of hyper- and hypopigmentation are frequently seen on the neck, chest and back. Palmoplantar and popular hyperkeratosis are other dermal manifestations, and these may eventually develop into malignant lesions.

13.2 Cadmium

Although Cd is found in foods, the levels are too low to be of any toxicological significance. Cadmium has many industrial uses, for example in electroplating, in low-melting alloys, in low-friction, fatigue-resistant bearing alloys, in solders, in batteries, in pigments, and as a barrier in atomic fission control. Therefore, it is to be expected that low to moderate Cd content of the environment is widespread. Since chronic exposure to even low levels of trace elements can lead to health problems, Cd is of particular concern to those concerned with environmental quality.

13.2.1 Toxicity

Industrial exposure is the most prevalent cause of chronic and acute Cd toxicity. Chronic toxicity is manifested in humans by anoxia as a result of olfactory nerve damage, kidney dysfunction, and emphysema. Cadmium has also been implicated as a possible cause of lung cancer. The Cd content of tobacco levels is significant, but there is no experimental evidence linking Cd in tobacco to emphysema and lung cancer. It has also been suggested that Cd may play a role in the production of arteriosclerosis, hypertension, and cardiovascular disease, but the data are limited and contradictory. It is worth noting that the body burden of Cd in smokers in 1.5 - 2 times than that of nonsmokers.

Acute Cd toxicity in humans often leads to pneumonitis ranging from severe to fatal. Vomiting, diarrhea, and prostraction are also symptoms of acute Cd poisoning.

In laboratory animals, Cd produces reduced growth, kidney and liver damage, brain hemorrhages, skeletal decalcification, and testicular necrosis. Rats develop hypertension as a result of Cd ingestion. It is not clear whether the hypertension results primarily from kidney damage (which involves lesions in the renal arterial system, glomeruli, and the tubular system) or from the fact that low concentrations of Cd increase pressure response to norepinephrine. The latter result has been demonstrated in isolated arterial strips, and it should be noted that higher concentrations of Cd have the opposite effect.

The main biochemical finding in Cd toxicity is proteinuria as a result of renal damage. It has been postulated that the proteinuria results from Cd transport to the proximal tubules by metallothionein (a low molecular weight 211- and Cd-binding protein). In the tubules Cd acts upon enzymes that are responsible for reabsorption. Although the significance of the finding is unknown, tryptophan is not found in the low molecular weight proteins of serum and urine from the urine of humans with tubular proteinuria.

Cadmium affects the activities of several enzymes. Enhanced activity of δ-aminolevulinic acid dehydratase, pyruvate -dehydrogenase, and pyruvate decarboxylase have been noted, while depressed activity of δ-aminolevulinic acid synthetase, alcohol dehydrogenase, arylsulfatase, and lipoamide dehydrogenase result from Cd intoxication.

13.3 Mercury

There is no known nutritional requirement for Hg and most of the Hg present in foods results from environmental contamination. Because it has many uses, there are numerous opportunities for contamination of food, air, and water with Hg. Elemental Hg is used in thermometers, barometers, diffusion pumps, Hg vapor lamps, electrical switches, dental fillings, paints, batteries, catalysts, and the manufacture of chlorine. Mercury salts are used as medicine, paint pigments, explosive detonators, and in the manufacture of paper. Organic Hg compounds are used as fungicides for seed treatment and in the manufacture of certain types of plastic.

The body's ability to eliminate Hg is limited, and therefore Hg is a cumulative poison. Reduced elimination appears to result from a high affinity of the tissues for Hg rather than poor excretion, since excretion kinetics are first orders. Mercury can be absorbed through the gastrointestinal and respiratory tracts and through the skin. Elemental and organic Hg compounds are volatile, and only small quantities are needed to saturate the atmosphere. Both elemental and organic Hg compounds pass the blood-brain barrier, and thus can induce central nervous system symptomology. Methyl Hg compounds is particularly dangerous since tissue retention is even longer than for other forms of Hg. As a consequence, elemental and organic Hg compounds need to be handled with extreme caution.

13.3.1 Toxicity

Symptoms of Hg intoxication are varied. They range from excessive salivation and diarrhea to tremors, ataxia, irritability, dizziness, moodiness, and depression. Acute exposure to elemental Hg by inhalation results in pulmonary edema and the symptoms closely resemble influenza. If not immediately fatal, recovery is usually complete.

Chronic exposure can lead to symptoms of central nervous system involvement. The character of the Mad Hatter in *Alice in Wonderland* is derived from the fact that hat makers often suffered from neurological disorders resulting from the use of mercuric nitrate to treat felt. Mercury poisoning has also been observed in dental technicians and industrial workers.

Several epidemic-type outbreaks of organic Hg poisoning have been described in the literature. These have occurred in Pakistan, Guatemala, and Iraq as a result of human consumption of seed grains treated with organic Hg. Minamata disease occurred in Japan as a result of local inhabitants consuming fish and shellfish from Minamata Bay, into which a local plastics factory had been dumping methyl Hg.

Mercury compounds are highly reactive and can interact with various chemical groupings of proteins and nucleic acids. The binding of Hg to sulfhydryl groups of membrane proteins causes an inactivation of membrane ATPase and a blockage of glucose transport into the cell. Mercury also reacts with phosphoryl groups - of membranes, sulfhydryl, amino, and carboxyl groups of enzymes, and phosphoryl groups and bases of nucleic acids.

13.4 Lead

Exposure to Pb occurs in many forms, in addition to that of industrial hazards. Although Pb intake from paints, water pipes, tin, cans, and insecticides has decreased, exposure to other forms of Pb such as in motor vehicle exhausts and tobacco smoke has either stabilized or increased. Intake of Pb paint by children is still a problem in poor urban, neighborhoods where Pb-containing painted surfaces still remain. Lead poisoning has been reported in the southern United States as a result of consumption of non-tax-paid, distilled alcoholic beverages commonly known as moonshine whiskey. Old auto radiators, which contain Pb, Cd, and Zn, are often used to distill illegal whiskey. Storage of acidic foods in cans in which solder is exposed or in crockery with Pb-containing glazes has also been reported to result in increased Pb concentration in the food.

13.4.1 Toxicity

Symptoms of Pb poisoning include abdominal pain, anemia, and lesions of the central and peripheral nervous systems. The lesions of the central nervous system cause behavioral problems. The anemia is characterized by a larger than normal number of erythrocytes and is of the hypochronic, microcytic type.

The principal biochemical effect of Pb intoxiciation in humans and animals is defective hemoglobin synthesis. Lead inhibits Fe incorporation into protoporphyrin, which results in lower heme concentrations and higher protoporphyrin concentrations in erythrocytes. Excretion of coproporphyrin is increased, and the Fe content of the blood plasma and bone marrow is elevated. Lead also interferes with an earlier step in

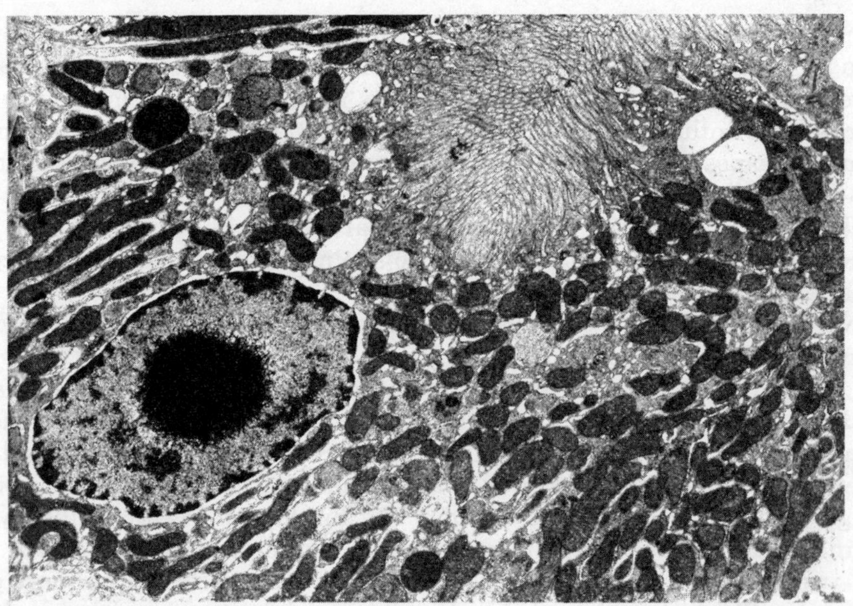

Fig. 13.2 Renal proximal tublar cell from a rat that received 200 µg Pb/ml drinking water reveals a nucleus containing dense staining inclusion body (uranyl acetate and lead citrate, x 10,000).

heme. synthesis by inhibiting δ-aminolevulinic acid dehydratase, which converts δ-aminolevulinic acid to porphobilinogen. The resulting increase in δ-aminolevulinic acid in blood and urine is a sensitive indicator of plumbism. In advanced Pb poisoning, synthesis of the globin moiety of hemoglobin is also inhibited.

The Na, K-APTase of red cell membranes is inhibited by Pb. Serum levels of transaminases and aldolase are increased by Pb exposure, while the serum levels of alkaline phosphatase and cholinesterase are decreased. Lead can also inhibit enzymes with a single, functional sulfhydryl group, but the effect of Pb on sulfhydryl groups is not as marked as that of Hg or Cd.

Lead interferes with tryptophan metabolism, probably by inhibiting monoamine oxidase (MAO). The inhibition of MAO by tetraethyl lead blocks serotonin catabolism in the brain, and the increased serotonin levels may account for some of the psychological and nerve function impairment.

13.5 Aluminium (Al)

Alum was used by the Greeks and Romans as an astringent in medicine and as a mordant in dyeing. The metal was isolated by Wohler in 1827. The element was named aluminium, derived from the Latin *word salurnen* or *alum,* in the early nineteenth century, although the American Chemical Society has used the name aluminium in their publications since 1925. The element is mainly obtained from the mineral bauxite. Aluminium and aluminium alloys are appreciated owing to their strength and lightness and they have found a multitude of applications, including the manufacture of cooking utensils, aircrafts and rockets (WHO, 1997). Natural aluminium minerals are also used for water purification and aluminium salts have found use in antacids and antiperspirants.

13.5.1 Toxicity

The intake of aluminium in the form of antacids, although exceeding the normal dietary intake levels by two or three orders of magnitude, has not been associated with adverse health effects (WHO, 1997). The gastro-intestinal uptake of aluminium is generally very low but can be enhanced in the presence of citrate. Aluminium is a potential neurotoxic agent in humans (WHO, 1997). Cases of encephalopathy have been recorded in patients receiving dialysis treatment for renal failure (Rozas at al. 1978a,b; Platts and Anastassiades, 1981). The dialysis fluids used contained high concentrations of aluminium, usually above 200 igl^{-1}. The pathogenic role of aluminium in the onset and progression of Alzheimer's disease is controversial and not supported by available data (Savory ct al. 1996; WHO, 1997). Occupational exposure to aluminium compounds has been associated with pulmonary fibrosis and irritant-induced asthma (Dimnan, 1988; O'Donnel et al. 1989; Kongerud, 1992; WHO, 1997). However, the exposure situations are usually complex and involve other compounds that may contribute to the observed effects.

13.6 Beryllium (Be)

The name of the element can be derived from the Greek words *beryllos* and *beryl*. Beryllium was discovered as the oxide by Vauquelin in 1798 and the metal was independently isolated in 1828 by Wohler and Bussy. Aquamarine and emerald are precious forms of the mineral beryl.

13.6.1 Toxicity

Exposure to beryllium is primarily an occupational health problem. Massive exposure (>100 ìg m^{-3}) to airborne beryllium causes acute beryllium disease, characterized by chemical pneumonitis, which may be fatal (WHO, 1990b). Long-term exposure to lower concentrations of beryllium may cause chronic beryllium disease, a form of granulomatous interstitial pneumonitis with symptoms including dyspnea, cough and reduced pulmonary function (Rossman, 1996). The disease may have an immunological component. Recently, it was proposed that a genetic modification of the major histocompatibility complex allele HLA-DPB1 may be used as a biomarker of susceptibility to chronic beryllium disease (Richeldi et al. 1993).

13.7 Bismuth (Bi)

The name originates from the German *Weisse Masse* (white mass), later transformed to *Wisuth* and *Bisemutum*. The element was confused with lead and tin in early times but shown to be distinct ftom lead in 1753 by Claude Geofftey the Younger.

Bismuth is found in nature as the pure element and the oxide (Bi_2O_3) and sulphide (Bi_2S_3) and is mainly used in alloys but also a pigment in cosmetic preparations (Fowler and Vouk, 1986; Bradley et al. 1989). Bismuth salts have been used medicinally in the treatment of gastric ulcers (bismuth subcitrate) and diarrhoea (bismuth subsalicylate) (Arduino and DuPont, 1993).

13.7.1 Toxicity

Reports on bismuth toxicity in humans are mainly found in connection with its therapeutic use. Nephropathies, encephalopathy, arthrosis, damage to the oral epithelium (gingivitis, stomatitis) and melanosis have been observed in patients treated with Bi- containing preparations (Bradley at al. 1989; Slikkerveer and de Woltt 1989). The effects depended on the type of preparations used.

During the mid-1970s, reports from France (Buge et al. 1974; Martin-Bouyer et al. 1980) and Australia (Burns et al. 1974; Lowe, 1974; Robertson, 1974) showed a form of bismuth-associated encephalopathy of known aetiology. The clinical symptoms described in both countries (confusion, tremor, motor disturbances) were similar and developed after prolonged therapy (from 4 weeks to 30 years) with high doses (0.7 - 20 g day/l) of mainly bismuth subnitrate (France) and bismuth subgallate (Australia) (Serfontein and Mekel, 1979). The effects were often reversible and disappeared after ceasation of Bi therapy, although fatalities were also reported. In a review of 63 cases of bismuth-associated encephalopathy, blood bismuth levels of 170-2850 µg l^{-1} were reported among the patients (Hillemand et al. 1977). An epidemiological study saw no correlation between encephalopathy, and age, type of bismuth compound administered, dose or length of treatment (Martin-Bouyer et al. 1980). Constipation was more prevalent as a therapeutic indication among the encephalopathy patients, whereas diarrhoea and colon disease were more frequent among the controls (bismuth therapy without encephalopathy). The data suggested that the intestinal microflora may be of importance for the absorption of bismuth.

13.8 Chromium (Cr)

Chromium takes its name from the Greek word for colour, *chroma*. The element was discovered by Vauquelin in 1797.

Major sources of chromium exposure are chromated steel, cement, leather goods, and welding fumes. Chromium is a cause of occupational allergic contact dermatitis and the trivalent form of chromium [Cr(III)] is considered to be the sensitizing agent. Hexavalent chromium may be released from chromium metal by the corrosive action of sweat and penetrate the skin. Cr(VI) is subsequently reduced to Cr(III). Inhalation of corrosive Cr(VI) compounds may lead to ulceration and perforation of the nasal septum.

13.8.1 Toxicity

Acute exposure to chromium cases death preceded by nausea, vomiting, shock and coma. Intravascular hemolysis and acute renal failure have also been reported from ingestion of potassium dichromate. Chronic exposure causes respiratory effects viz: ulcerated performation of nasal septum, irritation of mucous membranes and general bronchiospasm. Effects of chromium on skin are well known. Chromium ulcers or chrome holes are characteristic lesions.

Haxavalent chromium is much more toxic than trivalent. In fact trivalent chromium has such a low order of toxicity that a wide margin of safety exists between the amounts ordinarily ingested and those likely to induce deleterious effects. Cats tolerate 1000 mg/day and rats showed no adverse effects from 100 mg/kg diet. Lifetime exposure to 5 mg/liter of chromium (III) in the drinking water induced no toxic effects in rats and mice, and exposure of mice for three generations to chromium oxide at levels up to 20 ppm of the diet had no measurable effect on mortality, morbidity, growth, or fertility.

Chronic exposure to chromate dust has been correlated with increased incidence of lung cancer, and oral administration of 50 ppm of chromate has been associated with growth depression and liver and kidney damage in experimental animals.

13.9 Nickel (Ni)

The name of the element is derived from the German word *Kupftrnickel* meaning 'Old Nick's (= Satan's) copper'. Nickel was discovered by Cronstedt in 1751.

13.9.1 Toxicity

Exposure to nickel and its salts is regarded as one of the most common causes of human skin sensitization and allergic contact dermatitis (WHO, 1991a). Dermal exposure to nickel may occur through jewellery, wrist watches with metal backs, coins and jeans buttons. The prime cause of nickel sensitization is ear piercing in combination with nickel-containing jewellery (Liden, 1992). The risk of sensitization has led to the adoption of nickel release limits for jewellery and other metal objects in close skin contact. It has been shown that exposure to nickel generates nickel-specific T-lymphocytes (Sinigaglia et al 1985), probably mediated by nickel interaction with the major histocompatibility class II-peptide complex (Sinigaglia, 1994).

13.10 Cobalt (Co)

Cobalt was isolated by the Swedish chemist Georg Brandt in the middle of the eighteenth century, although compounds containing cobalt were used already in ancient Egypt. *Kobold* was a name applied during the sixteenth century to ores eventually found to be toxic, arsenic-bearing cobalt ores.

Cobalt is essential to humans in the form of cyanocobalamin (vitamin B12) (Lentner, 1986). Cobalamins function in rearrangement (adenosylcobalamin) and methyl transfer (methyl cobalamin) reactions. Vitamin B_{12} is essential for the production of red blood cells.

13.10.1 Toxicity

Intoxications (cardiomyopathies) due to its use as foam stabilizer in beer were described by Bonenfant et al. (1967). Inhalation of cobalt-containing dust may cause respiratory irritation and 'hard-metal' pneumoconiosis, developing into interstitial fibrosis.

13.11 Copper (Cu)

The element was discovered in prehistoric times and has its name from the island of Cyprus (Latin: *cuprum)*. Copper is an essential trace element and necessary for the functioning of several enzymes involved in electron transfer (cytochrome oxidase), free radical defence (catalase, superoxide dismutase) and melanin formation (tyrosinase). Copper is also essential for the utilization of iron and formation of haemoglobin.

13.11.1 Toxicity

Intoxication by copper salts results in vomiting, hypertension, coma and death. Excess hepatic copper causes hepatitis leading to cirrhosis, hepatic failure and ultimately death. The liver shows pericentral necrosis. Copper may initiate lipid peroxidation through hydroxyl radical formation in a mechanism analogous to the Haber-Weiss cycle (Hanna and Mason, 1992; Stohs and Bagchi, 1995; Britton, 1996).

13.12 Iron (Fe)

The chemical name stems from the Latin word *ferrum*. The use of iron is prehistoric and the element is mentioned in *Genesis:* Tubal-Cain, seven generations from Adam, was 'an instructor of every artefact in brass and iron' (Hammond, 1995). The most common ore is magnetite (Fe_2O_3). Carbon steel is an alloy of iron with carbon, and alloy steels are carbon steels with additives such as nickel, chromium and vanadium. hepatic necrosis and renal failure. Primary haemochromatosis is a disease with autosomal recessive inheritance characterized by an increased absorption of iron in the gastrointestinal tract resulting in chronic iron overload, hepatic portal fibrosis and

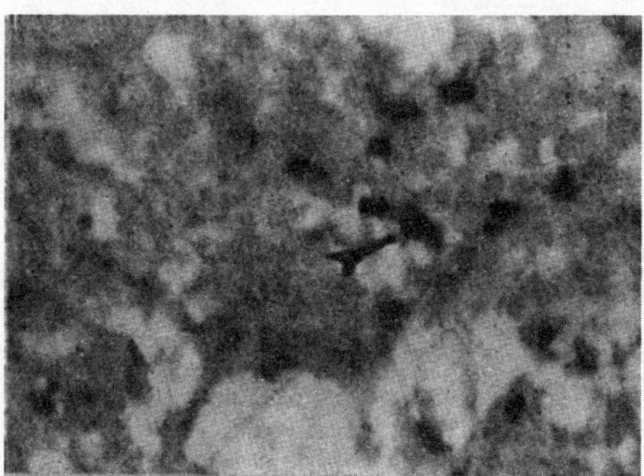

Fig. 13.3 T.S.1 of liver showing hemochromatosis (× 320).

ultimately cirrhosis. The defect has been localized to chromosome 6 closely liked to the HLA locus (Lombard et al. 1990). Redox cycling of iron may lead to hydroxyl radical formation and lipid peroxidation, which have been implicated as crucial events in iron toxicity.

13.13 Selenium (Se)

The name is derived from the word *selene,* the Greek name for moon. Selenium was discovered by Berzelius in 1817 (Hammond, 1995). The element was associated with tellurium, named after *Tellus*, the Earth. Selenium is obtained from byproducts during refining of copper and in sulphuric acid production. The content of selenium in crops, the main human source of exposure, largely depends on the selenium content of the soil, which varies substantially around the world (Jackson, 1988). An average daily requirement of selenium of about 0.4 ìg (kg body weight)$^{-1}$ for adults has been calculated by the WHO (1996).

13.13.1 Toxicity

A number of toxic manifestations due to acute or chronic exposure to selenium have been demonstrated in animals and man. Negative health signs (brittle hair and nails, prolonged plasma prothrombin time, skin lesions and changes in perpheral nerves) have been observed at daily intake exceeding 750 - 850 ìg Se (Yang et at. 1989a,b). The mechanism(s) remain unclear, although several hypotheses including disturbances in sulphur metabolism, glutathione depletion and redox cycling of selenium metabolites have been presented. Based primarily on the studies performed in a seleniferous region in China (Yang and Zhou, 1994; Yang et al. 1989a, b), a maximum safe dietary selenium intake of 400 ìg day^{-1} for adults has been suggested by WHO (1996).

SUGGESTIVE FURTHER READINGS

Underwood E.J. 1977. Trace elements in human and animal nutrition. Academic Press, New York & London.

Ballntyne B. 1999. General and applied toxicology. Grove's Dictionaries Inc. New York.

STUDY QUESTIONS

1. Discuss in details the toxicity of arsenic and cadmium.
2. Write briefly the toxicity of
   a. Lead
   b. Chromium
   c. Beryllium
   d. Bismith
3. Discuss the following
   a. Plumbism
   b. Minamata disease
   c. Itai Itai
   d. Mad as a hatter

Chapter 14

TOXICITY OF PESTICIDES

The word pesticide literally means an agent used to kill an undesirable organism. In the amended US. Federal Insecticide, Fungicide and Rodenticide Act, the definition of an "economic poison" or pesticide was expanded to include.

1. Any substance or mixture of substances intended for preventing, destroying, repelling, or mitigating any pest [insect, rodent, nematode, fungus, weed, other forms of terrestrial or aquatic plant or animal life or viruses, bacteria, or other microorganisms, except viruses, bacteria, or other microorganisms on or in living man or other animals, which the Administrator declares to be a pest and

2. Any substance or mixture of substances intended for use a plant regulator, defoliant or desiccant.

The major classes of pesticides in use today include herbicides, fungicides, rodenticides, insecticides, nematocides, and molluscides. Prior to about 1940, pesticides were primarily inorganic chemicals (e.g., arsenic) and a few natural agents from plant origin (e.g., nicotine and pyrethrum). With the discovery of the insecticidal activity of DDT, however, a burgeoning increase in the development and utilization of synthetic organic chemicals occurred. From about 1940 to 1980, an exponential -increase in the production and use of these synthetic pesticides was evident worldwide. The major chemical classes of pesticides that are used today include inorganic and organic metals, chlorinated hydrocarbons, organophosphorus compounds, carbamates, pyrethroids, substituted phenols, substituted ureas, coumarins, organic acids, organic amides, and triazines. Currently, over 1 billion pounds of about 600 different pesticides (active ingredients) are produced each year in the United States alone, with total worldwide production estimated at around 5 billion pounds.

Insects were the first major focus of pest control, whether to prevent the destruction of food or fiber crops or to limit the spread of insect vectors of disease. There is little doubt the use of insecticides had a profound effect on the further development of civilization. The control of anopheline mosquitoes and malarial infection, as well as vectors for typhus, plague, and yellow fever, by DDT undoubtedly saved millions of lives. Over the past several decades, however, the use of herbicides has dramatically increased and such efforts have markedly altered the methods of modern agriculture. As a result, herbicides now represent the most extensively used class of pesticides in the United States. Some food and fiber crops reportedly increased yields by 300 - 600% after the introduction and widespread use of synthetic herbicides.

While the public health and economic benefits of synthetic pesticide use over the past 50 years are indisputable, these benefits have not been without costs. Widespread environmental contamination by DDT and other organochlorine pesticides reaching global proportions, with concomitant deleterious effects on some members of the food web heralded the end of an era for their extensive use. DDT was banned from use in the United States in 1972 and most other organochlorines were subsequently banned, being replaced by the less environmentally persistent organophosphates and carbamates. While these agents had considerably lower abilities to accumulate in environmental and biological media, they tended to be much more acutely toxic and thus more hazardous to utilize. The pyrethroids are generally regarded as safer than the anticholinesterase organophosphates and carbamates, but still constitute a smaller proportion of total insecticidal use. In general herbicides exhibit markedly lower acute mammalian toxicity than other classes of pesticides. The relative toxicities of these agents are generally scaled, however, on the basis of acute reactions. Knowledge of the long term health consequences of prolonged, low-level exposure to various pesticide classes is still limited. A major challenge for toxicologists in the future is the continued acquisition of data pertaining to the long-term effects of low level pesticide exposures.

14.1 Insecticides

Many insecticides affect the nervous system of insects and since many have some activity against the mammalian nervous system, in man the neurotoxic effects of insecticides are often prominent. The main groups of insecticides are listed in Table 14.1.

There are a number of possible ways in which humans can be exposed to pesticides (Fig 14.1). The effects of pesticide residues in food and water probably cause

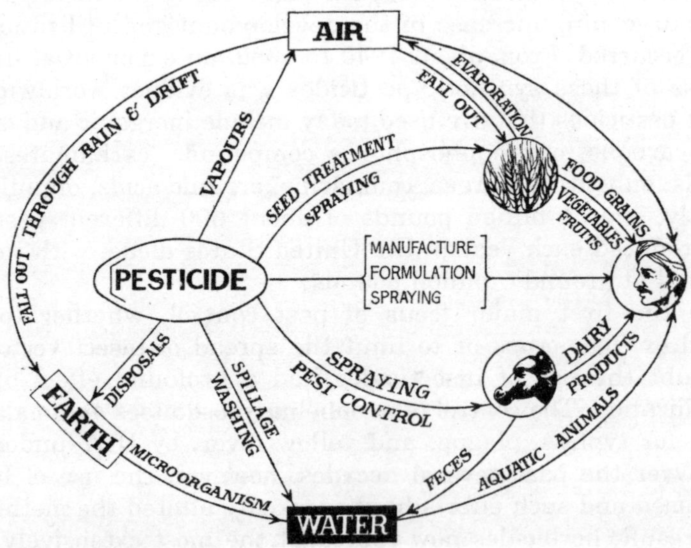

Fig. 14.1: Schematic diagram depicting the pesticide cycle in the environment.

the greatest public concern. However, reports of clinical poisoning by residues seem to be extremely rare. Food-borne pesticide intoxication, especially where clinical signs are non-specific or trivial, would probably pass unnoticed or may not be attributed to pesticides. The problem of consumers who consume high levels of some foodstuffs is taken into account when residue levels are pronounced to be toxicologically acceptable by comparison with the acceptable daily intake (ADI) or the more recently developed concept of the acute reference dose (ARID). On the other hand, simultaneous exposure to more than one pesticide of the same type might occur. Pesticide poisoning might occur from:

a. spillage of pesticides on to food during storage or transport;
b. eating gram or seed potatoes treated with pesticides, where the food article was not intended for human consumption;
c. improper application of pesticides or failure to observe harvest intervals.

Table 14.1: The main groups of pesticides.

Group	Subgroups	Examples
Organochlorines (OCs)	DDT	Endrin
		Aldrin
		Dieldrin
		Endosulfan
		ã-Hexachloro-Cyclohexane (lindane)
Anticholinesterases	Organophosphatase (OPs)	Malathion
		Fenitrothion
		Dichlorvos
		Diazinon
	Carbamates	Carbaryl
	Aldicarb	
Pyrethrins and synthetic pyrethroids	Pyrethrum	
		Permethrin
		Cypermethrin
		Flumethrin
Natural compounds, other than pyrethrins		Abamactin
		Ivermectin
		Rotenone
		Nicotine
Substances which interfere with systems specific to insects	Juvenile hormone analogues	Cyromazine
	Chitin systhesis inhibitors	Diflubenzuron
	Ecdysone agonists	Tebufenozide
Miscellaneous synthetic insecticides	Formamidine	Amitraz
	$GABA_A$	Fipronil

Weighed against the disadvantages of pesticides that accrue from their toxic effects is the fact that insects and fungi are important sources of agricultural loss and give rise to much damage to wooden buildings. Furthermore, many insects carry diseases such as malaria and sleeping sickness, which in the absence of control measures may render land uninhabitable or agriculturally unfit.

The key to a successful pesticide is selective toxicity. An ideal pesticide will interfere with a biological system in the pest, which has no counterpart in non-target species. This is the advantage of the juvenile hormone analogue insecticides. In the case of agricultural insecticides, the pesticide should be toxic to insects, but less toxic to plants, to humans and to other non-target organisms.

14.2 Classification of Pesticides

14.2.1 Organochlorines

This group includes DDT, hexachlorocyclohexane (HCH), the cyclodienes, dieldrin, endrin and heptachlor, and toxaphene. However, the use of organochlorines (OCs) has, in recent years, been severely restricted in most countries. This has mainly been done due to the persistence of OCs in the environment and wildlife and humans. Despite the fact that OCs are less used than before their toxicology is still important because they continue to be present in the environment and in foodstuffs, specially in developed countries. OCs are excreted in breast milk and their persistence is shown by the continuing presence of OCs in human milk.

The most prominent effects of the OCs are on the nervous system, where OCs have an effect on Na^+ channels. DDT produces tremor and incoordination in lower doses and convulsions in high doses. By contract HCH and the cyclodienes may produce convulsions as the first sign of intoxication, as well as fever, by a central effect possibly linked to disturbances in GABA-mediated inhibitory neurotransmission. OCs have been associated with a chronic toxicity syndrome, which includes apathy, headache, emotional lability, depression, confusion and irritability.

OC poisoning is treated symptomatically and diazepam is usually used to deal with the convulsions.

14.2.2 Lindane (γ-HCH)

Lindane is used as a wood preservative, in arable agriculture and in human and veterinary medicine as an ectoparasiticide. It is the ã-isomer of 1,2,3,4,5,6-hexachlorocyclohexane (HCH), which is sometimes known as benzene hexachloride. HCH has eight stereo isomers, of which the á, â- and ã- isomers are the most important. Of these isomers, only lindane (>99% ã-isomer) is still approved in the West, but in some countries preparations of lesser purity continue to be available and these contain other isomers. The effect of lindane on the CNS is stimulatory, causing convulsions, whereas the a-and b-isomers have a depressant effect (Coper et. al., 1951; van Asperen, 1954).

As with other OCs, lindane produces histopathological changes in the liver of experimental animals. The mutagenicity of lindane *in vitro* and *in vivo* has been reviewed and the preponderance of evidence is that the compound is not mutagenic (FAO/WHO, 1990)

14.2.3 Anticholinesterases

Two groups of anticholinesterases, the organophosphates (OPs) and the carbamates are widely used as agricultural insecticides and veterinary medicines. Some anticholinesterase OPs are used as human drugs, for example malathion in the treatment of head lice infestation and metrifonate (trichlorfon) in schistosomiasis (Aden-Abdi et. al., 1990). Several carbamates are used in human medicine, for example pyridostigmine.in myasthenia gravis. The antichlinesterases are often more acutely toxic than the OCs. A variety of cholinergic symptoms and clincal signs occur at parasympathetic effector sites, including bronchorrhoea, salivation, constriction of the pupil of the eye and abdominal colic. Sympathetic effects can also occur, together with signs of/central nervous system involvement, such as confusion and apprehension. Convulsions will occur in severe poisoning. Actions at the neuromuscular junction result in muscle fasciculation and later paralysis. The terminal event in fatal poisonings seems to be respiratory paralysis, which may be of central or peripheral origin.

It is often stated that the main difference between the OPs and the carbamates is that the former produce irreversible inhibition of acetylcholinesterase, whereas the latter produce reversible inhibition.

The numerous cholinesterases in the body show different intensities of sensitivity to inhibitors. Plasma cholinesterase is usually the cholinesterase most open to inhibition and it reactivates slowly thus it is a useful marker of exposure.

14.2.4 Organophosphate

OP anticholinesterases are esters of phosphoric, phosphonic or phosphorothioic or related acids. Their general formula is

$$\begin{array}{c} R^1 \\ | \\ O \\ | \\ R^2 - O - P = O\ (S) \\ | \\ X \end{array}$$

the R groups in pesticides are generally either both methyl groups or ethyl groups. The X or leaving group can be anyone of large variety of moieties.
The OP pesticides have some similarities with the chemical warfare nerve agents, but these are often phosphonofluoridates. Many pesticidal OPs are phosphorothioates and those containing P = S groups such as malathion tend to be of lower acute mammalian toxicity than their corresponding phosphates and phosphonates. Thus paraxon is much more toxic than parathion as is malaxon compared to malathion. These features are called organophosphate-induced delayed polyneurop.athy (OPIDP) and the so called intermediate syndrome. Since the OP pesticides all have the same qualitative anticholinesterase action, and this property is responsible for their acute lethal toxicity.

Intermediate Syndrome: As the syndrome developed before the late effects of OPs, the authors called this syndrome the intermediate syndrome. The syndrome comprises a proximal limb paralysis starting 1-4 days after poisoning. The progression is not altered by atropine or oximes and as the respiratory muscles are affected, respiratory support is necessary. A myopathy has been described *post mortem* in cases of human poisoning (de Rueck and Willems, 1975) and in experimental animals (Preusser, 1967; Wecker et. al.; 1978) and this may be related to the intermediate syndrome. The myopathy appears to be initiated by calcium accumulation in the region of the motor endplate (Inns et al., 1990).

14.2.4.1 Organophosphate-Induced Delayed Polyneuropathy (OPIDP)

OPIDP is a symmetrical sensory-motor axonopathy, tending to be most severe in the long axons, occurring 7-14 days after exposure. It is a polyneuropathy in that there are central and peripheral components. There is degeneration of axons and Schwann cell proliferation in the peripheral nervous system and also changes in the spinal cord and medulla oblongata. Clinically, the most disabling feature is the paralysis of the legs. Less severe cases exhibit a characteristic high-stepping gait, and some recovery may occur, but there is no specific treatment

14.2.4.2 Other Delayed Effects of OPs on the Nervous System

The behavioural toxicity of anticholinesterases, inter alia OPs, has been reviewed (D'Mello, 1993). Since acute intoxication with OPs can cause major effects such as convulsions, respiratory failure and cardiac arrhythmias, all of which can result in anoxia, it is hardly surprising that major intoxication is sometimes associated with long-term CNS changes. More debatable is whether long-term low-dose exposure produces delayed or chronic effects. Such an outcome is biologically less, plausible than delayed effects of acute exposure, but the two problems have frequently been observed.

14.2.5 Other Effects of OPs

It should never be forgotten that OPs may have properties, some of which may be entirely independent of their anticholinesterase effects, including mutagenicity and carcinogenicity as well as organ-specific toxicity to the heart and kidney and other organs.

14.2.6 Carbamates

Carbamates are used as insecticides, herbicides and fungicides, but only the first groups have marked anticholinesterase activity. Methiocarb (see below), an anticholinesterase, is used as a molluscicide.

The anticholinesterase carbamate pesticides are closely related to human drugs such as physostigmine and pyridostigmine. The general structure of most carbamate insecticides is as follows:

$$R-O-\overset{\overset{\displaystyle O}{\|}}{C}-\overset{\overset{\displaystyle CH_3}{|}}{N}H$$

In most cases, the R group is substituted by a phenol or heterocycle, a notable exception being aldicarb, which is 2-methyl-2(methylthio)-propionaldehyde o-(methylcarbamoyl) oxime. Pirimicarb has two N-methyl groups.

In general the carbamates produce toxicity similar to that of organophosphates, but less severe. A major difference is that carbamate-inhibited cholinesterases reactivate more rapidly than enzymes inhibited by OPs, with the result that the effects do not last as long in carbamate poisoning. However, an insecticidal carbamate, aldicarb, is one of the few pesticides that has given rise to poisoning in consumers of treated food. Residues have occurred in cucumbers, watermelons, squashes and similar products sufficient to cause illness, in some cases severe. Outbreaks have occurred in the USA and Canada (Hall and Rumack, 1992) and in the Irish Republic (Department of Agriculture and Food, 1992). The reason for these problems is that aldicarb, in other respects a satisfactory pesticide, has a high mammalian toxicity, with an acute oral LD_{50} of about 1 mg (kg body weight)$^{-1}$. An-opine is effective in carbamate poisoning but oximes less so and there is some evidence that, with certain carbamates, oximes are harmful.

14.2.7 Pyrethroids (and Pyrethrins)

Pyrethrins are natural insecticides produced from, inter alia, pyrethrum, a plant of the Compositae group, and are esters of pyrethric or chrysanthemic acid. Pyrethrins are broken down and deactivated very readily in the environment, particularly by sunlight. The synthetic pyrethroids are structurally similar compounds rendered photostable by various substituent groups, such as chlorine, bromine or cyanide. Some of the newer ones bear a more distant structural relationship to the pyrethrins. Because of their low mammalian toxicity, high insecticidal potency and lack of persistence in the environment, the synthetic pyrethroids have achieved widespread usage in agriculture, as household insecticides and in wood preservation. Synthetic pyrethroids are also used in mosquito control (White, 1993). They have also been widely used on humans to treat such conditions as scabies. However, they are very toxic to aquatic organisms (Zitco et. al., 1979) and their lack o persistence can be a problem when used as wood preservatives. The advantageous properties of the synthetic pyrethroids result from the fact that they hydrolyse relatively easily, both in the mammalian body and in the environment. Consequently, bioaccumulation does not occur, nor do they persist in soils. Pyrethroids can be separated into two classes on the basis of the central neurotoxic syndrome that they produce by parenteral routes in experimental animals (Vijverberg, 1994). Type I synthetic pyrethroids, which include permethrin and resmethrin and also the components of natural pyrethrum, lack an á-cyano group and give rise to the T--syndrome. The T-syndrome is characterized by fine tremor, hypersensitivity to stimuli and aggressive sparring, progressing to coarse whole body tremor; this syndrome is similar to that produced by some of the DCs. Type II compounds, which include detamethrin, flumethrin, cyfluthrin and á-cypermethrin, have an á-cyano group and give rise to the CS-syndrome. The CS-syndrome consists of initial pawing and burrowing and later marked choreQathetosis, salivation, coarse tremor and convulsions. In addition to these central effects in experimental animals, pyrethroids cause peripheral nerve damage with functional impairment when administered repeatedly. Axonal degeneration has been described, but generally at near lethal doses, and there is no evidence that pyrethroids can produce delayed neuropathy of the OP type (Aldridge, 1990).

Specific treatment of the effects of synthetic pyrethroids is rarely necessary as systemic effects are very unusual in humans.

14.3 Toxicity to Fish

The indiscriminate use of pesticide in agriculture and public health operations has increased the scope of disruption of ecological balance as many of the 'non-target' organisms which include important members of food chain that perish in the process thus adversely affecting the secondary and tertiary productivity of fresh water ecosystems. The tragic incidence of 'Handigod syndrome' in Karnataka State was probably due to consumption of pesticide poisoned crabs and fish by the local population. It has been reported that the plankton samples from Arabian Sea between Bombay and Goa contain high levels of DDT. According to a rough estimate, 25% of pesticides used on land may ultimately find their sink in the sea and their cumulative effect to coastal water can be expected to be considerable.

In vitro studies by O'Brien (1967) have revealed that fish appear to be highly susceptible to organochlorine compounds in general and to endrin in particular. This is in accordance with the indication that fish lack mixed function oxidase enzymes responsible for drug oxidations. Presumably fish dispose of foreign lipid-soluble compounds directly by diffusion through gills or skin into the surrounding water without prior conversion to more polar derivatives. However, Potter and O'Brien (1964) showed that trout liver slices converted parathion into paraoxan and since then several trout microsomal enzymes effecting drug oxidation have been described, all requiring NADPH .and oxygen. The oxidizing system was different in one or more components of the electron transport chain as compared to mammalian liver and possibly the level of aldrin epoxidase was also low. The lethal concentration of fenitrothion had an inhibitory effect on the cholinesterase activity in nervous and other tissues of fish with concomitant increase of acetyl choline content and decreased SDH and increased LDH activities indicated favouring of anaerobic metabolism in fenitrothion treated fish.

Toxicity studies in fish reported from India have revealed greater accumulation of organochlorines in tissues. Among the various chlorinated insecticides, viz., DDT, lindane, dieldrin, aldrin, endrin, endosulfan and chlordane, examined for their toxicity, endrin was extremely toxic to fish and endosulfan the least toxic. Thiodan and chlordane were also highly toxic to fish. In cat fish, endrin was reported to be 70,000 times more toxic than carbaryl. Biochemical studies carried out with endrin in fish by some workers revealed that alpha-amylase activity was inhibited by more than 50% which could be regained in the presence of 1 ppm of sodium ion concentration. However results presented by others suggested that phenobarbital and hexabarbital in fish could develop some capacity to tolerate the toxicant, endrin, on the CNS level. During a study on the uptake and metabolism of DDT in *Gambusia affinis,* two metabolites of DDT, viz., p,p'DDE and p,p'DDT were detected. DDE was the major metabolite.

Unlike the chlorinated hydrocarbon insecticides, little accumulation of organophosphorus insecticides in tissues had been reported. Effects of these insecticides on the metabolic fate and activities of different lysosomal enzymes in various tissues have also been noticed. Several defomities were observed in fertilized eggs of *Cyprinus carpio communis* exposed to different concentrations of diazinon, malathion, fenitrothion and phosphamidon. Parathion was reported to be highly toxic and dimethoate least toxic to fish.

14.4 Toxicity to Wildlife

Poultry birds are often exposed to a variety of pesticides either by ingestion of contaminated feed or through the use of pesticides in pent houses. Pesticides are lethal to poultry if ingested at a sufficiently high dose rate. However, the lethal dose of a pesticide is governed by the species of the bird and the nature of the compound. Contamination of poultry with nonlethal levels may impair one or more of a number of production characteristics such as growth, egg production, egg size, shell thickness, fertility and hatchability, etc. Poultry feed contaminated with lindane, DDT, methoxychlor and chlordane when given *ad libitum* for about 24 weeks caused no death in chicken although other reports suggest that DDT and dieldrin contaminated feed, however, produced death in poultry birds.

The metabolism of 'organophosphorus insecticides' in chicken and ducks was accomplished by hepatic microsomal enzyme system. It has been demonstrated that malathion and parathion were converted to their respective and more potent oxygen analogs, malaoxon and paraoxon.

Toxicological studies carried out on poultry with a few organophosphate and chlorinated pesticides in India have shown interesting findings.

Thiometon when sprayed aerially for the protection of crops did not induce any toxic effect in poultry birds. Residues of rabon, an organophosphate insecticide, were found in the fat of all treated then. In the egg yolk, residues appeared two days after initiation of treatment and disappeared five days after the last treatment. Fenitrothion was found to be more toxic than malathion to chick embryos. The effect of fenitrothion on the growth rate was correlated with the growth promoting endocrine glands. The reduced utilization of several metabolites during embryonic development such as nicotinamide, nicotinic acid, try prophan could also be responsible for the toxic manifestation. Inadvertent use of malathion for the control of ectoparasites in poultry has resulted in their large scale killing. Malathion inhibited the brain cholinesterase activity in house sparrow. A correlation has been found with ageing of house sparrow and inhibition of rat brain esterase by different organophosphates. Telodrin produced varying degrees of degenerative haemorrhagic changes in the liver and kidney of cockerels. Metabolic effects of DDT in earthworms and snails, toxic effects of various insecticides to bees and the effect of malathion on enzyme activities of acetylcholinesterase, asparatate and alanine aminotransferases in snail have also been reported.

14.5 Toxicity to Animals

Ingestion of insecticide - contaminated forage crops by animals has shown residues in the body fat of such animals. The chlorinated hydrocarbons are by far the most common residues found in milk since they are concentrated and stored in the fat in animals and translocated to the milk fat. It was reported as early as 1947 that DDT residues appeared in milk of cows that ingested feeds treated with the insecticide. Since that time DDT, dieldrin, BHC, lindane, chlordane, heptachlor, aldrin, endrin and toxaphene have all been detected in milk when the feed of the cow has been treated with these compounds at levels necessary to control the target insects. In contrast, the organophosphates have not shown such residual effects. The work of Cook indicated that the absence of residues in the milk of cows fed with organophosphate compounds is due to the inactivation of the compound by the rumen fluid of the cow.

Insecticides could be detected in blood after exposure. DDT, dieldrin and lindane were identified in the blood of all animals. It was found that rats did not accumulate lindane and it was nearly completely excreted within 4 days of treatment. The metabolism and persistence of lindane have been explored in detail by various workers.

With regard to effects of chlorinated insecticides and the sequence of events which follow the administration of these insecticides in the animal, the mode of action has been studied by various workers. The stimulatory effect on protein biosynthesis altered the general metabolism and mixed function oxidases.

Various concentrations of dieldrin solution when coated on the skin of monkeys and dogs produced loss in body weight and histopathological changes in the skin. Buffalo calves developed signs of poisoning within 16 hours of dieldrin administration. The blood glucose and blood lactic acid levels were increased by 2.5 and 4 fold respectively. BHC (0.27%) increased the weight of liver and altered the microsomal protein content and the level of cytochrome P_{450}. Acute exposure to chlordane produced a significant increase in the serum alkaline phosphatase activity and chronic exposure produced hyper proteinemia, hyperglycemia and significant increase in serum acid and alkaline phosphatase activities in Indian Desert gerbils. Carcinogenic action of DDT in Swiss mice, effect of dietary DDT, effect of DDT on rat adipose tissue lipolytic activity, metabolism of DDT, and acute and chronic toxicity studies of DDT in albino rats have been reported. Significant increases in levels of liver and kidney enzymes by beta isomer of ERC has been observed while gamma ERC produced significant alterations in hepatic aspartate aminotransferase and the hepatic alkaline and acid phosphatases.

Among the organophophorus insecticides, toxicity of malathion in rat has been studied in detail. It was shown that inhibition of acetylcholineesterse activity due to malathion administration in brain and heart was dose-related, while such a relationship was not observed in other tissues. In another study conducted to investigate the effects of malathion on hepatic drug metabolism and lipid peroxidation in young rats, the protein content was found to increase significantly in males and significantly decreased in females and there was an increase in aminopryrine N-demethylase activity and also acetanilide hydroxylase activity in both male and female rats. Other insecticides studied included metasystox, phorate, pyrethrum, dichlorvos and baygon. Recently, the effects of pesticides on reproduction in mammals have been highlighted.

14.6 Toxicity to Man

While assessing the health hazards of pesticide residues, two types of injury have to be considered. First the possibility of an acute illness resulting from ingesting residues on a single day or in a few days and secondly the long-term effects that may occur after ingesting small quantities of residues daily for many years. Residue poisoning appears to cause, only a mild disease manifestation. Gastrointestinal symptoms usually predominate. Residue poisoning also differs epidemiologically from poisoning by direct exposures. In connection with residue poisoning, it has been observed that majority of individuals frequently become sick. Whereas, in poisoning after direct exposure to formulations it is unusual to have more than one or two cases among any group of workers. Of a total of 252 cases reported in California during 1957 of systemic human poisoning attributed to agricultural chemicals, 189 involved the organophosphorus compounds. The accidental poisoning in the state of Karnataka and the involvement of members of food chain like fishes and crabs in the "Handigod Syndrome" have focused attention together caution and control in the spread of pesticides to villages.

Periodical surveys have indicated that humans in India show significantly higher storage levels than their counterparts in U.S.A. Dieldrin has been reported to be present, in the body fat of the U.S. population as a result of extensive application of this chemical. Acute poisoning has often been reported in people exposed to dieldrin.

Signs of intoxication due to aldrin and dieldrin involve the central nervous system and may include electroencephalographic changes, muscle tremors and. convulsions. Levels of liver enzymes, namely, SGOT, SGPT and LDH are greatly enhanced in the group occupationally exposed to pentachlorophenol. Very high levels of alkaline phosphatase are found on exposure to common organochlorines. Levels of albumin and beta and gamma globulins were found also to vary significantly with different types of occupational exposures. Exposure to the chlorinated hydrocarbon pesticides can be estimated from the storage level of the compounds or of their metabolites in adipose tissues or in certain instances from excretion of biotransformation products in urine.

With regard to organophosphorus pesticides, cases of acute poisoning were reported in individuals exposed to these in their occupations. Over 100 deaths in India during the spring of 1985 resulted from eating food accidentally contaminated with parathion during shipment. Most of the organophosphorus insecticides have relatively high acute toxicities and have caused fatal and non-fatal poisonings in man. Men who were occupationally exposed to parathion frequently showed some depression of blood cholinesterase activity, even though, they remained outwardly well. Generally, exposure to the organophosphorus compounds can be determined by measurement of blood cholinesterase activity. Exposure to parathion and other organophosphorus compounds which on hydrolysis, from paranitrophenol or allied metabolites can be estimated by determination of these phenolic compounds in urine. There are two types of biochemical reactions which might play an important role in determining the toxicity of these organophosphorus insecticides, which become significant inhibitors of cholinesterase only after an oxidative biotransformation. The reactions, both of which are catalyzed by microsomal enzymes are: (i) an oxidative biotransformation which converts the original compound into a more potent anticholinesterase (e.g. conversion of parathion to paraoxon) and (ii) the hydrolysis of the original compound or the oxidation product into non-toxic agents.

A number of workers from India have reported the poisoning of humans due to ingestion or exposure to organophosphatic and chlorinated hydrocarbon insecticides. All the organophosphorus pesticides have been reported to cause depression in the cholinesterase activity of blood. The work done in India with special reference to toxic effects of these in human beings is briefly summarized as follows:

Thirty four cases of diazinon poisoning were admitted to the Ruby Hall Hospital, Pune between January 1967 and May 1969. Diazinon was shown to cause rapid and severe depression of serum cholinesterase activity in 17 patients. Similar studies were taken in 158 males of Delhi and Punjab including 84 occupationally exposed to organophosphorus pesticides. The exposed group of individuals showed a significant depression in cholinesterase levels. A significant relationship of blood cholinesterase activity with the total duration of exposure was observed. Electrocardiographic (ECG) changes were observed in 12% of the 86 cases of diazinon poisoning. Neurological manifestations of organophosphorus insecticide poisoning in 200 consecutive cases of suicidal ingestion have also been described. The authors found impairment of consciousness in 10%, fasciculation in 27%, convulsion in 1%, toxic delirium in 50% and paralysis in 26%. A field study on the toxicological hazards to humans of ULV aerial spray of phosphamidon was presented. Maximum percentage of plasma cholinesterase inhibition, which occurred 1-3 days post spray were 0-25% in 11 subjects, 26-50% in 19 subjects and over 50% in 2 subjects. The distribution of insecticides viz. ethyl parathion, malathion, dimethoate, smnithion and phosphamidon in human body tissues such as stomach, intestine, lung, spleen and heart in five different cases of suspected poisoning have been reported.

With regard to the effects of chlorinated hydrocarbon insecticides in humans, the following information is available in the literature.

Data for DDT residues and its metabolites in blood samples from 182 people in Delhi have been reported. The average total DDT concentration in the whole blood ranged from 0.177 to 0.683 mg/l in males and from 0.166 to 0.32 mg/l in females. The DDT metabolites detected were p,p' DDE, p,p' DDD and p,p' DDT. DDE accounted for most of the total DDT. Eleven members of a small village community and five domestic animals developed neurotoxicity in the form of colonic jerks and major motor seizures on being exposed to chlorinated group (aldrin and gammaxene) of insecticides for a period of 6 to 12 months. A report of eight cases with grand mal epilepsy in a village of Uttar Pradesh, secondary to accidental contamination of food grain by BHC was presented. The distribution of endrin in different autopsy tissue from five different victims was assayed by ILC and the workers observed that the largest amounts of the poison remained in the stomach and intestine. The deaths occurred within 1-2 hr of the ingestion of this insecticide. Small amounts of organophosphorus, organochlorine and carbamate insecticides in human biological material by TLC were detected. Toxicological experiments on occupational exposure to pesticides in pest control operators and warehouse workers have been conducted by National Institute of Occupational Health, Ahmedabad in collaboration with Industrial Toxicology Research Centre, Lucknow. A total of 120 exposed workers, sixty each of warehouse and godown workers and pest control agencies workers were examined with the major objective of determining if measurable or observable effects could be detected which were attributable to this occupational exposure. Workers from warehouses and godowns were mostly found to be exposed to celphos, DDVP and malathion, while the PCA workers were mostly exposed to chlordane, aldrin, malathion and hepatochlor. A significant reduction in plasma and RBC cholinesterase level was found. Liver function tests, viz., SGPT and alkaline phosphatase showed no significant difference between control and both the exposed groups.

Besides the field studies from exposure to pesticides, laboratory experiments have been carried out to test the carcinogenicity of BHC and DDT, DDT and BHC in long and short term studies were found to be weak carcinogens in Swiss mice. BHC was found to produce more number of liver tumors of compared to DDT.

SUGGESTIVE FURTHER READINGS

Ballantyne, B. and Marrs, T. C. (1992). *Clinical and Experimental Toxicology of Organophosphates and Carbomates.* Butterworth-Heinemann, Oxford.

Ecobichon, D. J. and Joy, R.M. (1993). *Pessticides and Neurological Diseases.* CRC Press, Boca Raton, FL.

FAO/WHO (series). *Pesticide residues in Food-Evaluations-Toxicology.* Joint FAO/WHO Meeting on Pesticide Residues. World Health Organization, Geneva.

Hayes, W.J. and Laws, E. R. (1991). *Handbook of Pesticide Toxicology.* Academic Press, San Diego.

WHO (series). *Environmental Health Criteria.* World Health Organization, Geneva.

STUDY QUESTIONS

1. Classify different pesticides. Discuss their uses
2. Discuss the toxicity of pesticides in the following with suitable examples.
    a.   In fish
    b.   In animals
    c.   In man
    d.   Wild life
3. Write briefly on the toxicity of following:
    i.   Organochlorine pesticides
    ii.   Organophosphate pesticides
    iii.   Carbamates
    iv.   Pyrethyroids
4. Write short notes on the following:
    i.   Handigod syndrome
    ii.   Organophosphate induced delayed polyneuropathy

Chapter 15

TOXICITY OF ORGANIC SOLVENTS

Organic solvents are a group of organic compounds with low molecular weight. They share properties such as lipophilicity and volatility, although some of them are also hydrophilic or less volatile. The vapour pressure of organic solvents at room temperature is usually high enough for them to represent an inhalation hazard. Organic solvents have been used in large and increasing quantities during the last 50 years in a variety of industrial settings. A wide variety of organic solvents are also used in consumer products.

15.1 Exposure to Organic Solvents

In general, inhalation and skin absorption represent the most significant routes of exposure to organic solvents. The uptake of vapour by inhalation is a simple physical process; the molecules diffuse from the alveolar space into the blood, where they dissolve. The solvent molecules partition between air and blood during the absorptive phase and between blood and other tissues during the distribution phase. The more soluble a vapour is in the blood, the more of it will be absorbed into the blood during each respiratory cycle. The amount of vapour taken up by various tissues depends on the affinity of the organic solvent for each tissue. Thus, the rate of the absorption of organic solvents in the lungs and their distribution to various tissues are variable and depend on the blood-gas and fat-blood partition coefficients, respectively (Fiserova-Bergerova, 1985; Sato and Nakajima, 1987). Hydrophobic solvents have relatively small blood-gas partition coefficients and their fat-blood partition coefficients are much larger than unity. Hydrophilic solvents have blood-gas partition coefficients larger than 200 and fat-blood partition coefficients much smaller than unity (Fiserova-Bergerova, 1985). Differences in the uptake and distribution of various organic solvents have been shown to be, predominantly, a result of differences in their solubilities.

15.2 Toxic Effects of Organic Solvents

The common effects of most solvents include anaesthetic effects on the CNS and irritation of mucous membranes and tissues. These are the dominant signs of short-term exposure at a high concentration. Distinct from general effects are the specific toxic effects of solvents. They are often related to the metabolism of the solvent. Although normally associated with detoxification, metabolism can sometimes result in the formation of reactive metabolites that may cause cellular damage. The specific toxic effects are often not caused by the parent compound, but rather by the metabolites created by this metabolic activation.

15.3 Aliphatic Hydrocarbons

Aliphatic hydrocarbons are open-chain carbon-containing chemicals derived from crude oil. This group can be divided into three subgroups. Alkanes or paraffins (C_nH_{2n+2}) are saturated hydrocarbons. Alkenes or olefins (C_nH_{2n}) contain one or more double bonds. Alkynes of acetylenes (C_nH_{2n-2}) contain one or more triple bonds.

Aliphatic hydrocarbons containing four or less carbon atoms are gases (e.g. methane, ethane, butane, 1,3-butadiene and acetylene). Compounds with a 16-carbon atom chain or longer are solids at room temperature. Aliphatic hydrocarbons containing 5-16 carbon atoms are liquids. C_5-C_8 compounds are very volatile solvents. Gasoline is a complex mixture of liquid hydrocarbons.

Aliphatic hydrocarbons are found in fuels (natural gas, gasoline [petrol], kerosene [paraffin]), propellants, solvents, dyes, inks, plastics and coatings, dry cleaning agents and chemical intermediates. Aliphatic volatile alkanes (e.g. pentane, hexane, heptane, octane and nonane) are CNS depressants in high concentrations. They may also irritate the mucous membrane and the respiratory tract, in addition to causing drying of the skin that may develop into dermatitis. Alkenes, unsaturated aliphatic solvents, are toxicologically very similar to alkanes.

15.3.1 Carbon Tetrachloride (Tetrachloromethane, Perchloromethane)

Carbon tetrachloride (CCl_4) is a well known hepatotoxic solvent (Kefalas and Stacey, 1991). Its use has been declining, and it has been replaced by less toxic chlorinated solvents such as trichloroethylene, methylchloroform (1,1,1-trichloroethane) and perchloroethylene (tetrachloroethylene). However, it is still used, mainly in the chemical manufacture of fluorocarbon refrigerants, solvents and aerosol propellants.

Carbon tetrachloride is well absorbed by the lungs and gastro-intestinal tract. Percutaneous absorption can also occur. CCl_4 is concentrated in fatty tissues. The main excretion route (50 - 80%) is via the lungs as the unchanged compound. Dechlorination occurs in the liver microsomal cytochrome P_{450} system and the formation of free radicals may cause lipid peroxidation and subsequent hepatocellular damage. Only a small part of the dose (4%) is excreted as carbon dioxide via the lungs or kidney.

15.3.2 Chloroform (Trichloromethane, Methenyltrichloride)

Chloroform ($CHCl_3$) is now restricted to industrial uses as a solvent or a chemical intermediate. In 1847, it was introduced as a general anaesthetic because it was less volatile and not as flammable as diethyl ether. Its use as an anaesthetic was banned early in this century. The industrial use of chloroform is also very restricted, because it causes liver cancer in experimental animals. Chloroform is a potent CNS depressant that produces a variety of symptoms, e.g. nausea, headache and coma, which appear soon after exposure (Schroder, 1965).

Chloroform is rapidly absorbed via the lungs and gastro-intestinal tract, the peak blood concentration developing in 1h. A considerable amount (17 - 67%) is expired by the lungs. In the liver, cytochrome P_{450} enzymes dechlorinate chloroform by oxidation to trichloromethanol, which spontaneously dehydrochlorinates to phosgene. Phosgene can react with water to form carbon dioxide or bind covalently to cellular macromolecules. It produces hepatotoxicity characterized by fatty infiltration and necrosis (Larson et al. 1993).

Chloroform-induced cytotoxic effects (necrosis, inflammation and regenerative cell proliferation) can start the carcinogenic process and enhance endogenous and exogenous mutagenic activity (Butterworth et al. 1998). The possible carcinogenicity of chloroform has been of particular concern, because the process of municipal water purification by chlorination results in the formation of traces amounts of disinfection by-products such as trichloromethanes.

Chloroform may cause renal toxicity 24-48 h after exposure, and it is characterized by proteinuria. Like most chlorinated hydrocarbons, chloroform sensitizes the myocardium to endogenous catecheolamines.

15.3.3 Trichloroethylene

Trichloroethylene (TRI) (CHCl = CCl$_2$) has been widely used in metal degreasing, dry cleaning, chemical synthesis and as a cleaning agent in households. It was also used as a non-flammable, narcotic in early 1900s. TRI occasionally produces euphoria, and consequently it has been abused for the purpose. In cases of solvent abuse (glue sniffing), TRI has caused fatal accidents due to overdose, and perhaps through the induction of ventricular fibrillataion.

15.3.4 Tetrachloroethylene

Tetrachloroethylene or perchloroethylene (CCl$_2$ = = CCl$_2$) is used particularly in dry cleaning and degreasing. The toxicity of tetrachloroethylene resembles that of trichloroethylene. It can cause loss of myelin-enriched lipids, which might indicate a persistent loss of myelin membranes (Kyrklund et al. 1990). According to studies with experimental animals; tetrachloroethylene may cause liver cancer (Kyrklund et al. 1990). A cohort study of dry-cleaning workers exposed mainly to tetrachloroethylene has confirmed an increased risk of oesophageal, intestinal, pancreatic and bladder cancers (Ruder et al. 1994).

15.4 Aromatic Hydrocarbons

15.4.1 Benzene

Benzene has mainly been used in chemical processes as a raw material and as a Solvent in rubbers and glues. It is also used in the synthesis of ethylbenzene, styrene, cumene, phenolic resins, ketones and various dyes. However, in most industrial products, benzene has been replaced by other organic solvents because of its myelotoxicity and since it can cause aplastic anaemia and leukaemia (Vigliani and Forni, 1976; Infante et al. 1977). Benzene is a natural constituent of gasoline (petrol) at concentrations of up to 5% and thus becomes constituent of gasoline fumes and automobile exhaust. Tobacco smoke also contains benzene. Benzene is one of the components of volatile organic compounds (VOCs) in the city air, and its concentration is monitored in many cities. Workplace exposures to benzene are limited but both occupational and environmental exposures to benzene still occur (Hricko, 1994; Tompa et al. 1994).

15.4.2 Toluene

Toluene (methylbenzene) is used as a solvent for paints, lacquers, thinners, coatings, glues and cleaning agents. It is also used in the production of other chemicals and in the pharmaceutical industry. The important exposure route is by inhalation, but it is also slowly absorbed through the skin.

The biological half-life of toluene ranges from several hours for white adipose tissue to a few minutes in highly perfused organs (Pyykko et at. 1977). After inhalation exposure of rats to toluene, the distribution to various tissues is very rapid. Toluene accumulates in white adipose tissue, from which the elimination is slow compared with other tissues (Pyykko et al. 1977). Approximately 18% of the absorbed toluene is expired unchanged via the lungs, and only 0.06% is eliminated unchanged in the urine. Toluene is metabolized via cytochrome P450 to benzyl alcohol, which is further oxidized via alcohol and aldehyde dehydrogenase to benzaldehyde and benzoic acid respectively (Pyykko. 1983). Benzoic acid conjugates with glycine to form hippuric acid, which is excreted in the urine. Some other solvents (e.g. n-hexane, benzene, styrene, xylene, and trichloroethylene) can inhibit toluene biotransformation (Sato and Nakajima. 1979; Tabti 1984; Liira et al. 1988). A competitive metabolic inhibition is the plausible mechanism of this toxicokinetic interaction between the individual solvents. Toluene causes induction of microsomal liver enzymes at relatively low concentrations (Pyykko, 1983). Toluene has neither carcinogenic nor mutagenic potency. Kidney damage has been reported in sniffers of glues that may have contained toluene and other solvents.

15.4.3 Xylene

There are three isomers (*ortho, meta* and *para)* of xylene (dimethylbenzene) in the commercial product, *meta* being the major isomer. Xylene is one of the most common solvents in paints and varnishes and it commonly commonly in glues and printing inks. Furthermore, xylene is used as a solvent in the rubber and leather industries and in histological laboratories (Riihimaki at al. 1979). It is also used in the chemical manufacture of insecticides, synthetic fibres, plastics and enamel. Xylenes are widely used to replace benzene as a solvent.

The acute toxicity of xylene is greater than that of toluene, although the symptoms are similar. The major effect of xylene is depression of the CNS, resulting in lightheadedness, nansea, headache and ataxia at low doses. High exposure to xylene produces confusion, respiratory depression and coma. It causes conjunctivitis, nasal irritation, sore throat and respiratory irritation. Xylenes have been shown to induce microsomal enzymes in the liver, although their hepatotoxicity is considered to be low. Concentrations as high as 10.000 ppm produce a reversible and mild increase in hepatic aminotransferase activity and reversible renal failure. Like other aromatic solvents, xylenes may also cause skin and eye irritation.

15.4.4 Styrene

More than 50% of the world's production of benzene is used in the manufacture of styrene (vinylbenzene). Styrene materials account for 20% of all plastic production, e.g. polystyrene, resins and polyesters.

Styrene is readily absorbed via the lungs and gastrointestinal tract, and also to some extent via the skin. Styrene is metabolized to mandelic acid (70%) and phenylglyoxylic acid (30%), which is excreted in the urine. It is also metabolized to styrene oxide, which is more toxic than the parent compound (Mendrala et al. 1993). Hippuric acid is also excreted in the urine following exposure to styrene, and it has been used for biological monitoring of occupational exposure.

15.5 Alcohols

15.5.1 Methanol

Methanol (CH_3OH) is used as a solvent, an antifreeze agent and as a component of car windscreen washer fluid. It is also used in the production of stains, enamels and photographic films.

Recent plans to use methanol as a fuel, or as a component in gasoline, may increase the probability of exposures to methanol. Inhalation and skin absorption are the most important entry routes of methanol in occupational exposure. Ingested methanol is absorbed from the gastro-intestinal tract almost totally (Liesivuori and Savolainen, 1991). The first step in the metabolic pathway of methanol is oxidation to formaldehyde. In humans, alcohol dehydrogenase (ADH) is the enzyme that catalyzes this reaction. The resultant formaldehyde is spontaneously hydrated to methanediol, which is further metabolized by ADH to formic acid (Liesivuori and Savolainen, 1991). The urinary formic acid can be used in biological monitoring of methanol exposure in the work place (Liesivuori and Savolainen, 1987). The effects caused by methanol exposure are dizziness, nausea and vomiting, various types of visual disturbances, headache and metabolic acidosis.

15.5.2 Ethanol

Ethanol (CH_3CH_2OH) is used in the chemical industry and in the production of cosmetics and drugs. Exposure to ethanol by inhalation is rare in industry. Chronic alcohol consumption causes significant CNS toxicity characterized by cerebellar atrophy mid dilatation of cerebral ventricles. Harper and Kril (1990) have proposed a model for alcohol-induced brain damage. It includes loss of dendritic acbour shrinkage of the neuronal body. These changes are potentially reversible and occur in many brain regions. The second phase is neuronal death, which is irreversible and has been demonstrated, primarily in the frontal cortex.

A serious consequence of ethanol consumption is its multiple effects on the developing embryo and foetus (Pratt, 1982). Foetal alcohol syndrome occurs in the children of approximately 10% of women who drink heavily during pregnancy (Sokol et at. 1986). Recent studies have indicated that a possible, explanation for the toxic effects is the enhanced production of acetaldehyde in placenta and foetus due to induction of

CYP2El (Rasheed et at. 1997). One reason for the interindividual variations is that induction varies among heavy drinkers, and may be genetically controlled (Rasheed et al. 1997).

Epidemiological studies have shown that long-term exposures to specific solvents (benzene, trichloroethylene, tetrachloroethylene and carbon tetrachloride) may increase the cancer risks. Occupational exposure to mixtures of solvents in paints has consistently been associated with a 40% increased risk of lung cancer (Lynge et al. 1997). However, addressing these questions becomes ever more demanding when we consider that most long-term exposures are to mixtures rather than to single chemicals at anyone tune. The effects are difficult to detect and often we do not know even what kind of effects to look for.

SUGGESTIVE FURTHER READINGS

Eger, E. I. (1974). Anesthetic Uptake and Action. Williams and Wilkins, Baltimore, MD.

Papper, E.M. and Kitz, R. J. (Eds) (1963). Uptake and Distribution of Anesthetic Agents. McGraw-Hill, New York.

Tahti, H. (1984). Combined effects of organic solvents. A review. In Manninen, O. (Ed.), Combined Effects of environmental Factors, Proceedings of the First international conference on the combined effects of environmental factors, tampere, Finland. Central Printing House, Tampere, pp. 413-429.

Ballantyne B. (1999). General and applied toxicology. Grove's dictionaries Inc. New York.

STUDY QUESTIONS

1. Define exposure. Discuss the toxicity of chloroform and carbon tetrachloride.
2. Give a detailed account the toxicity of aromatic hydrocarbons.

Chapter 16

DETOXICATION MECHANISMS (PHASE - I AND PHASE II REACTIONS)

The majority of xenobiotics that enter the body tissues are lipophilic, a property that enables them to penetrate lipid membranes and to be transported by lipoproteins in the body fluids. The metabolism of xenobiotics, carried out by a complex of relatively nonspecific enzymes, is usually held to consist of two phases. During phase, I, a polar reactive group is introduced into the molecule; its most important effect is to render the xenobiotic a suitable substrate for phase-II reactions. In phase-II, the altered compounds combine with an endogenous substrate to produce a water-soluble conjugation product that is finally excreted.

Although this sequence of events is generally a detoxication mechanism, in some cases the intermediates or final products are more toxic than the parent compound, and the sequence is termed an intoxication mechanism.

Oxidation, reduction, hydroxylation, alkylation, dealkylation are termed as phase-I reactions whereas phase-II reactions mainly involve conjugations.

16.1 Phase - I Reactions

16.1.1 Microsomal Mixed-Function Oxidations

Microsomal mixed-function oxidation reactions are catalyzed by a nonspecific multi enzyme system located in the endoplasmic reticulum of the cell.

The endoplasmic reticulum is a continuous anastomosing network of lipoprotein membranes extending from plasma membrane to nucleus and mitochondria, while the microsomal fraction derived from it consists of membranous vesicles contaminated with free ribosomes and fragments of mitochondria, nuclei, Golgi apparatus, etc. In the hepatocyte the surface area of the endoplasmic reticulum is 37 times that of the plasma membrane and 8.5 times that of the outer mitochondrial membrane.

Mixed-function oxidations are those oxidations in which one atom of molecular oxygen is reduced to water while the other is incorporated into the substrate. The term "monooxygenases" has also been used for enzymes that catalyze the incorporation of one atom of molecular oxygen into the substrate; although this appears to be a logical name, the term of choice is still mixed-function oxidase.

194

Since the electrons involved in the reduction of cytochrome P-450 are derived from NADPH the overall reaction can be written as follows (where RH is the substrate).

$$\text{RH} + O_2 \xrightarrow{\ \ \text{NADPH} + H^+ \quad\longrightarrow\quad \text{NADP}^+\ \ } \text{ROH} + H_2O$$

or, alternatively,

$$\text{RH} + O_2 \xrightarrow{\ \ \text{Reduced cytochrome P-450} \quad\longrightarrow\quad \text{Oxidized cytochrome P-450}\ \ } \text{ROH} + H_2O$$

16.1.1.1 Cytochrome P-450

The cytochrome P-450s are heme proteins and represent a multigene family of isozymes involved in oxidative biotransformation of lipid-soluble compounds to polar metabolites (Fig. 16.1). The key enzymatic components of this membrane bound enzyme system are the flavo protein NADPH-cytochrome-P450-oxydoreductase and cytochrome P-450. The

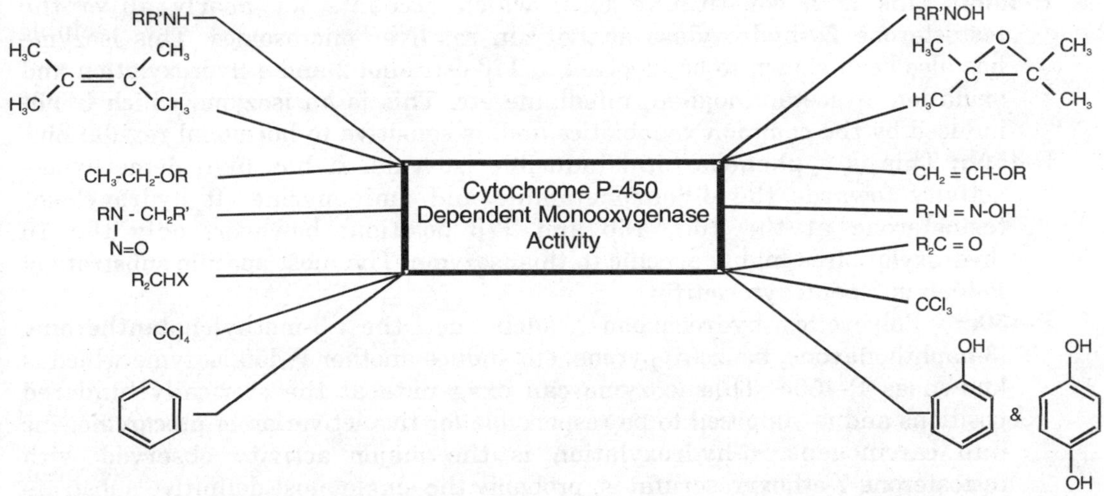

Fig. 16.1 Some important phase I products generated by CYP450-dependent monooxygenase systems.

term cytochrome P-450 has been given to this protein because the visible spectra of reduced protein bound to carbon monoxide shows an absorption maximum at 450 nm as identified by Kilingenberg and Garfinkle in 1958. This pigment was further characterized as a cytochrome P-450 heme protein by Omura and Sato in 1962. Since this protein catalyzes the insertion of one oxygen atom from the oxygen molecule, the enzymes are broadly known as cytochrome P-450 dependant mono oxygenases. They are present in the highest concentration in liver, but in most other tissues, including kidney, lung, gut, skin and nasal epithelium their presence-can also be demonstrated both biochemically and immunohistochemically. Several isozymes are dedicated for the metabolism of specific endogenous molecules such as kidney mitochondrial 1-alpha-hydroxylase is highly substrate specific and appear to metabolize 25-hydroxy cholecalciferol. However, isozymes in the hepatic endoplasmic reticulum show rather broad substrate specificity, catalyzing oxidation of many structurally related chemicals. Furthermore, a single substrate may be metabolizes by many isozymes. Substrate specificities, inducibilities, and amino acid sequenced are highly conserved across species. Oxidative attack occurs at N, S, and C bonds, resulting in the insertion of one atom of molecular oxygen into the substrate, often producing an unstable intermediate that breaks down to yield the final product. For example, the N-alkylated compound N-methylaniline is metabolized to intermediate oxidation product that breaks down to formaldehyde and aniline. Occasionally, the oxidized products are highly toxic unstable electrophiles. A large number of isozymes of cytochrome P-450 have been isolated and many of them have been purified to homogenity. Their encoding genes have been identified and their regulation has been studied. At present, at least 70 different cytochrome P-450s have been purified from mammalian tissues whose protein sequences are known out of which at least 20 distinct P-450s have been identified in liver. The following isozymes in liver are described in a greater detail:

P-450a: This is a constitutive form which accounts for nearly all of the testosterone 2α-hydroxylase ac.tivity in rat liver microsomes. This isozyme has also been shown to be involved in 17β estradiol 2-and 4-hydroxylation and oxidation of acetaminophen, nifedipine etc. This is an isozyme which is not induced by the common xenobiotics and. is sensitive to hormonal regulation.

P-450b: This is a phenobarbital inducible isozyme. It has high demethylase activity towards the d-benzphetamine and aminopyrine. It hydroxylases testosterone at the 16α, 16β and 17β position; however, only the 16-hydroxylation is highly specific to this isozyme. The most specific substrate of P-450b is 7-pentoxyresourfin.

P-450c: Polycyclic hydrocarbons such as the 3-methylcholantherene, {3-naphthoflavone, benzo(a) pyrene, etc. induce another P-450 isozyme which is known as P-450c. This isozyme can oxygenate at the sterically hindered positions and is supposed to be responsible for the activation of procarcinogens into carcinogens. 6-hydroxylation is the major activity observed with testosterone 7-ethoxyresorufin is, probably the single most definitive substrate of this enzyme.

P-450d: The 3-methylcholanthrene and other hydrocarbons besides inducing P-450 also induce this isozyme.

P-450e: Phenobarbital besides inducing P-450b also induces this isozyme.

P-450p: Pregnenolone 16-carbonitrile (PCN) preferentially induces this isozyme. Ethylmorphine N-demethylase activity is preferentially catalyzed by this enzyme.

P-450j: Many solvent like ethanol, acetone, pyridine etc. induce this isoform which is also known to catalyze the conversion of procarcinogens into carcinogenic reactive metabolites.

Besides, a large number of other isozymes are induced in response to both endogenous substrates and exogenous chemicals. Some of the substrates also called as suicidal substrates are the compounds which are metabolized to highly reactive chemical entity that binds irreversibly with the active site. This results in the deactivation of the enzyme protein. This is known as suicidal inactivation. The most important example of suicidal substrates of the cytochrome P-450 are carbon tetrachloride and chloramphenicol.

16.1.1.2 Other Phase I Reactions

1. *Hydrolysis*: The hydrolyzing enzymes are wide spread throughout the body, including the blood plasma, and have been shown to catalyze the hydrolysis of esters and amides. Classification of these enzymes is ill-defined due to the wide substrate distribution and overlapping specificities.

2. *Reduction*: A large number of nitro and azo compounds undergo anaerobic NADPH-dependent reduction reactions that can be inhibited by carbon monoxide, implicating a role of cytochrome P-450 in this process. However, a separate flavin nitroreductase has also been shown to exist. Carbon tetrachloride and halothane are the examples of the chemicals that are metabolized both oxidatively and reductively by cytochrome P-450. Reduction of these compounds results in their bioactivation leading

Aldrin   Dieldrin

to the formation of toxic intermediates, while oxidation reaction is a detoxification step. It has been shown that the oxygen tension in liver determines the route of metabolism. In addition the reduction reactions can be carried out by a number of aldehyde reductases, and alcohol dehydrogenases which are distributed throughout the body. Alcohol dehydrogenase catalyzed conversion of chloral hydrate to trichloroethanol is one such example.

3. *Non-P450 Oxidation*: A number of oxidases other than cytochrome P-450 dependent microsomal enzyme exist in the body, which catalyze various oxidation reactions. Such non-P450-dependant microsomal oxidation system which is present in many tissues including liver but is more versatile for kidney is prostaglandin synthase which is a glycoprotein with a heme containing centre. This enzyme produces prostaglandins from arachidonic acid oxidatively. In this reaction, a peroxide byproduct is also generated. A number of xenobiotics have been shown to be co-oxidized to toxic

metabolites during prostaglandin synthase dependent reactions. The other oxidases present are monoamine oxidases (MAO) and diamine oxidases (DAO). Both of these enzymes are widely distributed. Monoamine oxidases are mitochondrial flavoproteins which are responsible for the catabolism of monoamine neurotransmitters such as dopamine, norepinephrine, epinephrine etc. These enzymes can also oxidize a number of xenobiotic amines. Similarly, diamine oxidases are Vitamin B6-dependent cytosolic enzymes that preferrentially metabolize short chain aliphatic diamines. It may .also convert amines to aldehyde in the presence of oxygen. In addition, alcohol dehydrogenases, another group of cytosolic enzymes can oxidize a variety of alcoholic compounds.

16.1.1.3 Epoxidation and Aromatic Hydroxylation

Epoxidation is an extremely important microsomal reaction since not only can stable epoxides be formed, but arene oxides, the epoxides of aromatic rings, are intermediates in aromatic hydroxylations. In the case of polycyclic hydrocarbons, these reactive intermediates are known to be involved in carcinogenesis.

The expoxidation of aldrin to dieldrin is the best known example of the metabolic formation of a stable epoxide:

The oxidation of naphthalene was one of the earliest understood examples of an epoxide as an intermediate in aromatic hydroxylation.

16.1.2 Aliphatic Hydroxylations

Alkyl side chains of aromatic compounds are readily oxidized, often at more than one position, and provide good examples of this type of oxidation. The n-propyl side chain of n-propyl benzene can be oxidized, in the rabbit, at any of the three carbon atoms to yield 3-phenylpropan-l-ol ($C_6H_5CH_2CH_2CH_2OH$) by ω-oxidation, benzylmethylcarbinol ($C_6H_5CH_2CHOHCH_3$) by α-oxidation, and ethylphenylcarbinol ($C_6H_5CHOHCH_2CH_3$) by á-oxidation. Further oxidation of these alcohols is also possible.

Naphthalene   Naphthalene   1-Naphthol
epoxide

Naphthalene-1,2-dihydrodiol

Alicyclic compounds, such as cyclohexane, are also susceptible to oxidation, in this case first to cyclohexanol and then trans-cyclohexane-1,2-diol.

In compounds with both saturated and aromatic rings, the former appears to be the most readily hydroxylated. For example, the major oxidation products of tetralin (5,6,7,8-tetrahydronaphthalene) in rabbit are 1- and 2-tetralol, whereas only a trace of the phenol, 5,6,7,8-tetrahydro-2-naphthol, is formed:

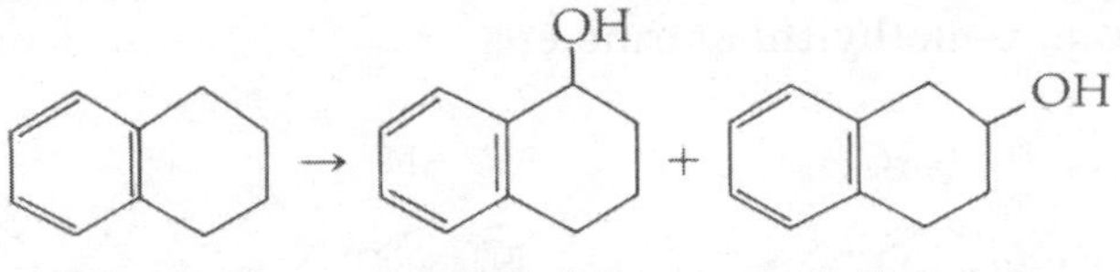

Tetralin   1-Tetralol   2-Tetralol

16.1.3 Dealkylation (O-, N- and S- Dealkylation)

O-Dealkylation Probably the best known example of O-dealkylation is the demethylation of p-nitroanisole. Due to the ease with which the product, p-nitrophenol can be measured, it is a frequently used substrate for the demonstration of mixed-function oxidase activity. The reaction is believed to proceed via an unstable methyl

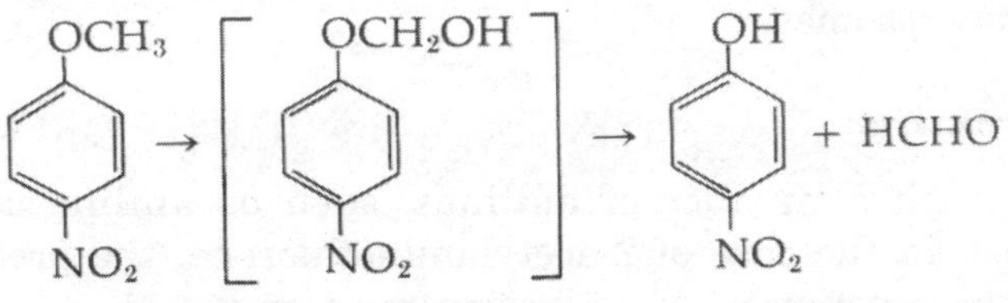

p-nitroanisole   p-nitrophenol

derivative. Other substrates of O-dealkylation include the drugs codeine and phenacetin and the insecticide methoxychlor.

N-Dealkylation: This is a reaction in the metabolism of drugs) insecticides, and other xenobiotics. Both the N- and N,N-dialkyl carbamate insecticides are readily dealkylated and in some cases the methyl intermediates are stable enough to be isolated. N, N-dimethyl-p-nitrophenyl carbamate is a useful model compound for this reaction.

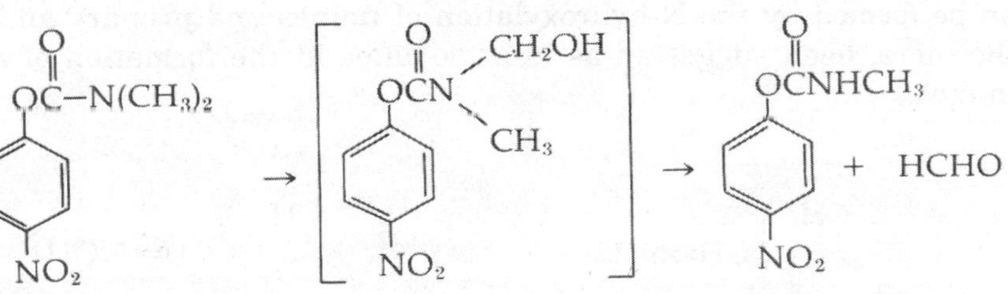

N, N-dialkyl carbamate   Methylol intermediate   N-N-dimethyl-p-nitrophnyl carbamate

Another important example is the insecticide carbaryl, which undergoes several different microsomal oxidation reactions, including an attack on the N-methyl group. In this case the methylol compound is stable enough to be isolated or to be conjugated *in vivo*.

The drug aminopyrene undergoes two N-demethylations to form first monomethyl-4-aminoantipyrene and then antipyrene.

S-Dealkylation: This reaction is known to occur with a number of thioethers such as methylmercaptan, 6-methylthiopurine etc.

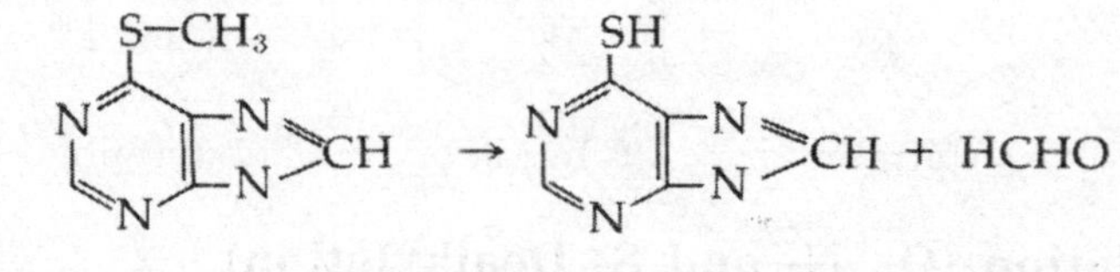

6-Methylthiopurine   6-Mercaptothiopurine

16.1.3.1 N-Oxidation

N-oxidation can occur in a number of ways, including hydroxylamine formation, oxime formation, and N-oxide formation. The latter is primarily dependent on a flavoprotein enzyme found in the microsomes.

Hydroxylamine Formation

This reaction occurs with a number of anilines, such as aniline and many of its substituted derivatives. In the case of 2-acetylaminofluorene, the product is a potent carcinogen and thus the reaction is an activation reaction.

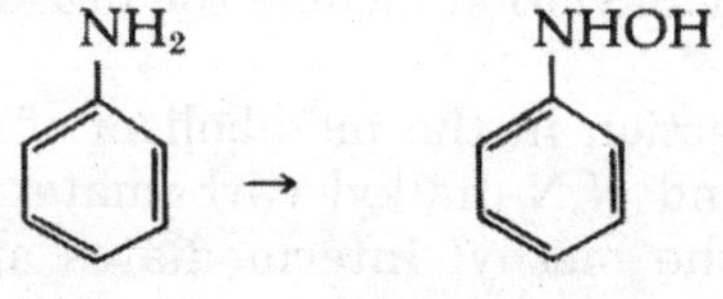

Aniline   Phenylhydroxylamine

Oxime Formation

Oximes can be formed by the N-hydroxylation of imines and primary amines. Imines have, furthermore, been suggested as intermediates in the formation of oximes from primary amines.

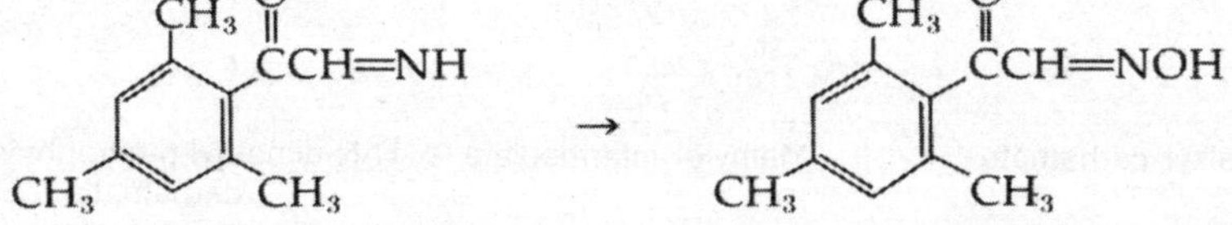

Trimethylacetophenone imine   Trimethylacetophenone oxime

Oxidatire Deamination

Oxidative deamination of amphetamine occurs in the rabbit liver but not to any extent in either the dog or rat, which tend to hydroxylate the aromatic ring.

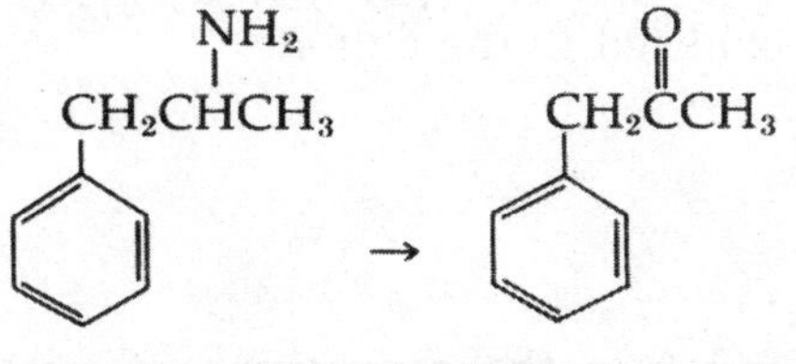

$$\text{Amphetamine} \rightarrow \text{Phenylacetone}$$

Amphetamine   Phenylacetone

A close examination of this reaction indicates that it is probably not an attack on the nitrogen but rather on the adjacent carbon atom, giving rise to a carbinol amine, which produces the ketone by elimination of ammonia.

$$R_2CHNH_2 \xrightarrow{\ O\ } R_2C(OH)NH_2$$

$$R_2C(OH)NH_2 \xrightarrow{\ -NH_2\ } R_2C = O$$

The carbinol, by another reaction sequence, can also give rise to an oxime

$$R_2C(OH)NH_2 \xrightarrow{\ -NH_2\ } R_2C = NH$$

$$R_2C = NH \xrightarrow{\ O\ } R_2CNOH$$

The Oxime can now be hydrolyzed to yield the ketone, which is thus formed by two different routes:

$$R_2CNOH \xrightarrow{\ H_2O\ } R_2C = O$$

S-Oxidation

Thioethers in general are oxidized by microsomal mixed-function oxidase enzymes to sulfoxides, some of which are further metabolized to sulfones. This reaction is very common among insecticides of several different chemical classes, including carbamates,

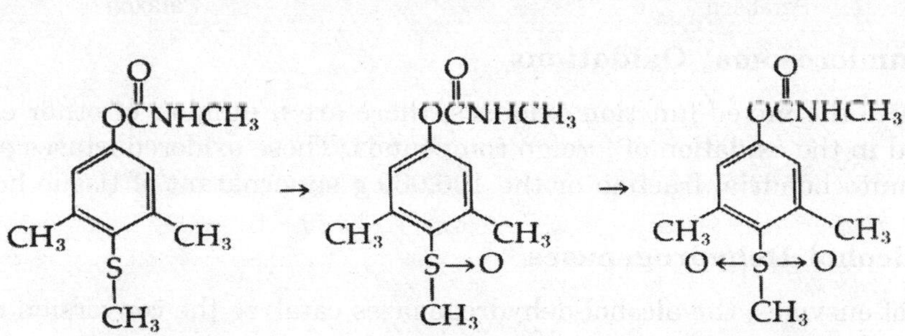

Methiocarb   Methiocarb
sulfoxide   Methiocarb
sulfone

organophosphates, and chlorinated hydrocarbons.

Organophosphates include phorate, demeton, and others, while among the chlorinated hydrocarbons endosulfan is oxidized to endosulfan sulfate and methiochlor to a series of sulfoxides and sulfones to yield eventually the bissulfone.

S-Oxidation is also known among drugs, e.g., chloropromazine, while the solvent dimethylsulfoxide is fhrther oxidized to the sulfone.

$$(CH_3)_2 SO \longrightarrow (CH_3)_2 SO_2$$

P-Oxidation

This little-known reaction involves the conversion of trisubstituted phosphines to phosphine oxides. It is a typical cytochrome P-450-mediated mixed-function oxidation. Known substrates are diphenylmethylphosphine and 3-dimethylaminopropyl-diplumylphosphine.

Diphenylmethylphosphine   Diphenylmethylphosphine oxide

Desulfuration and Ester Cleavage

Phosphothionate and phosphorodithioate insecticides owe their insecticidal activity and their mammalian toxicity to an oxidative reaction in which the P=S group is converted to P=O, thereby converting compounds relatively inactive toward cholinesterases into potent cholinesterase inhibitors. This reaction is known for many organophosphorus compounds but has been studied most intensively in the case of parathion.

Parathion   Paraxon

16.1.4 Nonmicrosomal Oxidations

In addition to the mixed-function oxidases, there are a number of other enzymes that are involved in the oxidation of foreign compounds. These oxidoreductases are located in either the mitochondrial fraction or the 100,000 g supernatant of tissue homogenates.

16.1.4.1 Alcohol Dehydrogenases

As a class of enzymes, the alcohol dehydrogenases catalyze the conversion of alcohols to aldehydes or ketones, e.g.,

$$RCH_2OH + NAD^+ \Leftrightarrow RCHO + NADH + H^+$$

This should not be confused with the mixed-function oxidation of ethanol observed in liver microsomes. This reaction is reversible and carbonyl com-pounds can be reduced to alcohols. Alcohol dehydrogenase is probably the most important dehydrogenase involved in the metabolism of foreign alcohols and carbonyl compounds. The enzyme is found in the soluble fraction of the liver, kidney, and lung and requires NAD or NADP as a coenzyme. The reaction proceeds at a slower rate with NADP. In the intact organism the reaction pro-ceeds to the right, since aldehydes are further oxidized to acids. The oxidation of aldehydes to acids by aldehyde oxidase is a vital detoxication reaction since aldehydes are usually toxic and, because of their lipid solubility, are not readily excreted.

16.1.4.2 Aldehyde Dehydrogenase

The oxidation of aliphatic and aromatic aldehydes to their corresponding acids allows the acids to be excreted or conjugated in phase II reactions:

$$RCHO + NAD^+ \rightarrow RCOOH + NADH + H^+$$

Aldehyde dehydrogenases have been isolated from a variety of sources, but probably the most important source is the liver, the enzyme from which can handle a wide variety of substrates. With a series of linear aliphatic aldehydes, the rate of oxidation increases as the carbonyl carbon becomes more positive.

In addition to liver aldehyde dehydrogenase, a number of other enzymes are present in the soluble fraction of liver homogenates that will oxidize aldehydes and certain N-heterocyclic compounds. Among these are aldehyde oxidase and xanthine oxidase, both flavoprotein enzymes containing molybdenum. These enzymes catalyze the oxidation of aldehydes formed by the deamination of endogenous amines by amine oxidases.

16.1.4.3 Amine Oxidases

The biological function of amine oxidases appears to involve the oxidation of biogenic amines formed during normal biological processes. In mammals, the monoamine oxidases are involved in the control of the serotonin: catecholamine ratios in the brain, which in turn influence sleep and EEG patterns, body tem-perature, and mental depression. Two groups of amine oxidases are involved in the oxidative deamination of naturally occurring amines as well as foreign com-pounds.

16.1.4.3.1 Monoamine Oxidases

Monoamine oxidases (MAD) are flavoprotein enzymes located in the mitochondrial fraction of liver, kidney, and brain. The enzymes have also been found in blood platelets and intestinal mucosa. The MAD exist as a large group of similar enzymes with overlapping substrate specificities and inhibition pat-terns. The number is difficult to assess because of the great variety and variation within each tissue as well as in different animal species.

The general reaction catalyzed is

$$CH_2CH_2NH_2 \xrightarrow[\text{O}]{\text{MAO}} CH_2CHO$$

Phenylethylamine   Phenylacetaldehyde

MAD will deaminate primary, secondary, and tertiary aliphatic amines, the reaction rate being faster with the primary, and slower with the secondary and tertiary amines.

16.1.4.3.2 Diamine Oxidases

Diamine oxidases (DAD) also oxidize diamines to the corresponding aldehydes in the presence of oxygen. The DAD are pyridoxal phosphate proteins contain-ing copper that are found in the soluble fraction of liver, intestine, kidney, and placenta. A typical DAD reaction is the oxidation of putrescine:

$$H_2N(CH_2)_4NH_2 \xrightarrow{\text{DAO}} H_2N(CH_2)_3CHO + NH_3$$
Putrescine

The rate of deamination is determined by the chain length, and with polymethylene diamines, $NH_2 - (CH_2)_n - NH_2$, the maximum rate occurs when $n = 4$ (putrescine) and $n = 5$ (cadaverine) and decreases to zero when $n = 9$ or more. At this stage, MAD become active and can oxidatively deami-nate the diamines. It should be noted that although MAD can deaminate both substituted and primary amines, DAD can only deaminate primary amines.

16.1.5 Reduction Reactions

A number of functional groups, such as nitro, diazo, carbonyls, disulfides, sulfoxides, alkenes, etc., are susceptible to reduction. In many cases it is difficult to determine whether these reactions proceed nonenzymatically, by the action of biological reducing agents such as NADPH, NADH, FAD, etc., or through the mediation of functional enzyme systems.

16.1.5.1 Nitro Reduction

Aromatic amines are susceptible to reduction by both bacterial and mammalian nitroreductase systems:

$$NO_2 \rightarrow NO \rightarrow NHOH \rightarrow NH_2$$

Nitrobenzene   Nitrosobenzene   Phenylhydroxylamine   Aniline

Nitroreductase activity has been demonstrated in liver homogenates as well as in the soluble fraction, while other studies have reported that nitroreductase activity has been found in all liver fractions evaluated. The reductase appears to be distributed in liver, kidney, lung, heart, and brain. The reaction utilizes both NADPH and NADH and requires anaerobic conditions. The reaction can be inhibited by the addition of oxygen. The reaction is stimulated by FMN and FAD, and at high flavin concentrations they can act simply as nonenzymatic electron donors. The reduction process probably proceeds via the nitroso and hydroxylamine intermediates as illustrated above.

16.1.5.2 Azo Reduction

Requirements for azoreductases are quite similar to those for nitroreductases; i.e., they require anaerobic conditions and NADPH and are stimulated by re-duced flavins

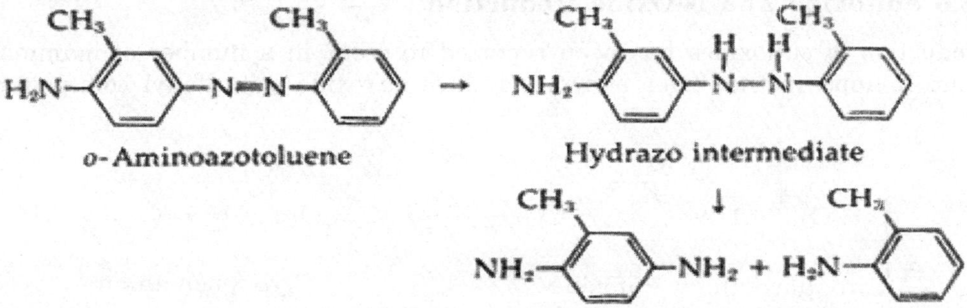

o-Aminoazotoluene

Hydrazo intermediate

The ability of mammalian tissue to reduce azo bonds is rather poor. With *p*-[2,4-(diaminophenyl)azo]benzenesulfonamide (Prontosil), in vivo reduction forms sulfanilamide. However, pretreatment with antibiotics destroys the in-testinal bacteria, which results in a decrease in the formation of the amino compound. Generally, it would appear that both nitro and azo reduction as a function of specific tissues is of minor importance, and intracellular bacteria as well as intestinal bacteria are actually responsible for these reductions.

16.1.5.3 Reduction of Pentavalent Arsenic to Trivalent Arsenic

Pentavalent arsenic compounds such as tryparsamide appear to require reduc-tion to the trivalent state for antiparasitic activity:

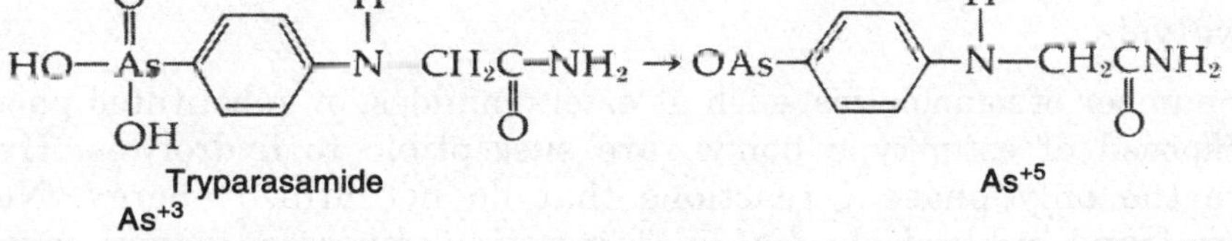

Tryparasamide (As^{+3})   As^{+5}

Various arsenicals containing pentavalent arsenic have only slight in vitro antiprotozoal activity, whereas compounds containing trivalent arsenic are highly active.

16.1.5.4 Reduction of Disulfides

A number of disulfides are reduced in mammals to their sulfhydryl compounds. An example is disulfiram (Antabuse), a drug used for the treatment of alco-holism:

$$(C_2H_5)_2NCSS-SSCN(C_2H_5)_2 \xrightarrow{2H} 2(C_2H_5)_2NCSSH$$

Disulfiram   Diethyldithiocarbamic acid

16.1.5.5 Ketone and Aldehyde Reduction

The reduction of ketones and aldehydes occurs via the reverse reaction of alco-hol dehydrogenases (Section 4.4.1):

$$R_1R_2CO \rightarrow R_1R_2CHOH$$

16.1.5.6 Sulfoxide and N-Oxide Reduction

The reduction of sulfoxides has been reported to occur in a number of mamma-lian systems. It appears that liver enzymes reduce sulYoxide or sulfonyl com-pounds to

Carbophenothion sulfoxide   Carbophenothion

thioether or sulfides under anaerobic conditions

N-Oxides have been reported to be reduced by a bacterial reductase:

$$(R)_3NO \rightarrow (R)_3N$$

16.1.5.7 Reduction of Double Bonds

Certain aromatic compounds have been reported to be reduced by intestinal flora as follows:

$$C_6H_5CH=CHCO_2H \xrightarrow{2H} C_6H_5CH_2CH_2CH_2OH$$

Cinnamic acid

16.1.6 Hydrolysis

A large number of xenobiotics, such as esters, amides, or substituted phos-phates that are composed of ester-type bonds, are susceptible to hydrolysis. Hydrolytic reactions are the only phase I reactions that do not utilize energy. Numerous hydrolases are found in blood plasma, liver, intestinal muscosa, kidney, muscle, and nervous tissue. Hydrolases are present in both soluble and microsomal fractions.

The general reactions are

$$RC\!-\!OR' + H_2O \xrightarrow{\text{esterase}} R\!-\!C\!-\!OH + HOR'$$

$$RCONH_2 + H_2O \xrightarrow{\text{esterase}} RCOOH + NH_3$$

The acid and alcohol or amine formed on hydrolysis can be eliminated directly or conjugated by phase II reactions. In general, amide analogs are hydrolyzed at slower rates than the corresponding esters.

Tissue esterases have been divided into two classes, A-type esterases, which are insensitive, and the B-type esterases, which are sensitive to inhibition by organophosphorus esters. The A esterases include the arylesterases, whereas the B esterases include cholinesterases of plasma, acetylcholinesterases of erythro-cytes and nervous tissue, carboxylesterases, lipases, etc. The nonspecific aryles-terases that hydrolyze short-chain aromatic esters are activated by Ca^{2+} ions and are responsible for the hydrolysis of certain organophosphate triesters, e.g.,

$$(C_2H_5O)_2P\!-\!O\!-\!\!\!\langle\ \rangle\!\!\!-\!NO_2 + H_2O \ \rightarrow\ (C_2H_5O)_2P\!-\!OH + HO\!-\!\!\!\langle\ \rangle\!\!\!-\!NO_2$$

Paraoxon

Certain hydrolases also are present in mammalian plasma and tissue that hydrolyze and subsequently detoxify chemical warfare agents such as the nerve gases tabun, sarin, and DFP:

$$(\text{i-}C_3H_7O)_2PF \ \rightarrow\ (\text{i-}C_3H_7O)_2P\text{-OH} + HF$$
$$\text{DFP}$$

16.1.7 Epoxide Hydration

Epoxide rings of certain alkene and arene compounds are hydrated enzymati-cally by epoxide hydrases to form the corresponding *trans-dihydrodiols:*

Certain labile epoxides that may be responsible for carcinogenesis are deacti-vated by the epoxide hydrases as well as the glutathione epoxide S-transferases (see Chapter 5). The enzyme(s) that mediate the hydration are located in the microsomal fraction of mammalian liver. The mechanism of hydration is opti-mal at alkaline pH and probably involves a nucleophilic attack of the -OH on the 6xirane carbon. With styrene oxide as the substrate, the relative activity in the liver of several species is rhesus monkey> human = guinea pig> rab-bit > rat> mouse. With the guinea pig and rat, low activity was also found in kidney, intestine, and lung, but no detectable activity was found in muscle, spleen, heart, and brain.

There is evidence that more than one epoxide hydrase exists in hepatic micro-somes, based on different purification factors (different relative activities to various epoxide substrates).

16.1.8 DDT-Dehydrochlorinase

In the early 1950s it was demonstrated that DOT-resistant houseflies detoxified DOT mainly to its noninsecticidal metabolite DOE. The rate of dehydrohaloge-nation of DOT to DOE was found to vary between various insect strains as well as between individuals. This enzyme also occurs in mammals but has been studied more intensively in insects.

DDT-dehydrochlorinase, a reduced glutathione (GSH)-dependent enzyme, has been isolated from the high-speed supernatant of resistant houseflies. Al-though the enzyme-mediated reaction requires glutathione, the glutathione levels are not altered at the end of the reaction:

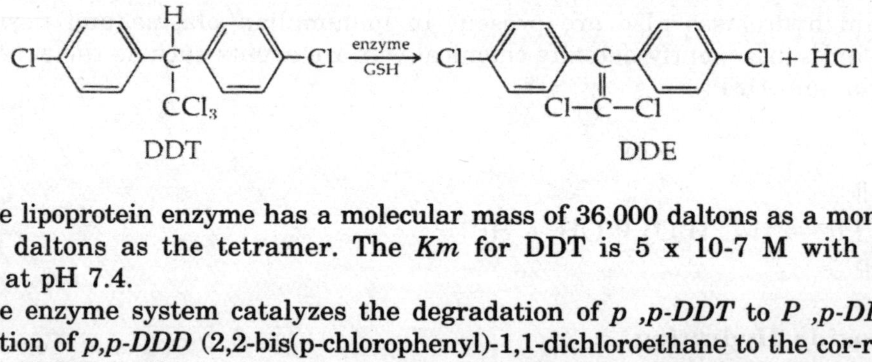

The lipoprotein enzyme has a molecular mass of 36,000 daltons as a monomer and 120,000 daltons as the tetramer. The Km for DDT is 5 x 10-7 M with op-timum activity at pH 7.4.

The enzyme system catalyzes the degradation of *p ,p-DDT* to *P ,p-DDE* or the degradation of *p,p-DDD* (2,2-bis(p-chlorophenyl)-1,1-dichloroethane to the cor-responding ethylene TDEE (2,2-bis(p-chlorophenyl)-1-chloroethylene). *a,p*-DDT is not degraded by DDT-dehydrochlorinase, suggesting a p,p-orientation requirement for dehalogenation. In general, the DDT resistance of house fly strains is correlated with the activity of DDT-dehydrochlorinase, although other resistance mechanisms are known in certain strains.

At one time, it was believed that DDT-dehydrochlorinase, glutathione S-aryltransferase, and enzymes metabolizing the γ-isomer of hexachlorocy-clothexane (γ-BHC) were one enzyme since experimental results suggested the existence of a nonspecific enzyme that catalyzed all these reactions. However, it has since been demonstrated that DDT-dehydrochlorinase is different from glutathione S-aryltransferase based on the following: a difference in elec-trophoretic mobility; a difference in stability and response to inhibitors; results of genetic studies; and purification of a house fly glutathione S-aryltransferase that lacks DDT-dehydrochlorinase activity. The glutathione S-aryltransferase enzyme preparation, however, still has the ability to metabolize γ-BHC

16.2 Phase II Reactions

Foreign compounds containing functional groups such as hydroxyl, amino, carboxyl, epoxide, or halogen, or metabolites of phase-I reactions, as well as many naturally occurring compounds, can undergo biosynthetic reactions referred to as conjugations or Phase II reactions. Biochemical conjugation is the union or coupling of a natural or foreign compound or its phase I metaboliste, with a readily available endogenous conjugating agent derived from a carbohydrate, a protein source, or a sulfur component in general, conjugation products are more polar, less lipid soluble, more readily eliminated from the organism, and less toxic.

Conjugations are generally energy dependent and are linked directly to high-energy intermediates. Williams classified conjugation reactions into two classes based on whether an activated conjugating agent or an activated substrate was utilized in the biosynthesis:

Type I:

Activated conjugating agent + Substrate ⟶ conjugated product
(intermediate)

Type II:

Activated substrate + Amino acid ⟶ Conjugated product

Type I reactions include the formation of glycosides and sulfates as well as methylated and acetylated conjugates, whereas type II reactions result in peptide conjugates.

16.2.1 Glycoside Conjugation

In order for glycoside conjugation to occur, two initial stages are required for both glycoside and glucuronide conjugation. The first is the formation of an activated intermediate, uridine diphosphate glucose (UDPG) and uridine diphosphate glucuronic acid (UDPGA):

$$D-glucose-1-PO_4 + UTP \xrightarrow{\ UDPG\ } UDP-\alpha-D-glu\cos e + P_2O_7^{4-}$$

$$UDP-\alpha-D-glucose-2NAD+H_2O \xrightarrow{\ UDPG\ }$$

$$UDP-\alpha-D-glucuronic\ acid + 2NADH_2$$

Reactions (1) and (2) are associated with the soluble fraction of the liver and, to a lesser extent, the kidney, intestinal tract, and skin.

16.2.2 Glucuronide Conjugation

The three steps of glucuronic acid conjugation involve the reaction of the active intermediate UDPGA with the aglycone:

UDPGA + p-Nitrophenol $\xrightarrow{\text{UDP glucuronyl-transferase}}$ p-Nitrophenyl β-D-glucuronide + UDP

UDPGA is in the β-configuration, and the glucuronide formed due to inversion is in the p-configuration. The enzyme UDP glucuronyl transferase is found in the microsomal fraction of the liver, kidney, intestinal tract, and skin.

A wide variety of reactions are mediated by glucuronyltransferases, depending on the structure of the aglycone. Structurally, O-glucuronides, N-glucuronides, and S-gillcuronides have been isolated and characterized.

16.2.2.1 O-Glucuronides

Ether type. This type of glucuronidation occurs with primary, secondary, and tertiary alcohols and phenols, e.g.

Phenol + UDPGA → Phenyl β-D-glucuronide

Ester type. The conjugates are formed from carboxylic acids: benzoic acid, benzoyl β-D-glucuronide

Benzoic acid + UDPGA → Benzoyl β-D-glucuronide

Hydroxylamino type. When 2-acetylaminofluorene is metabolized by the mixed function oxidase system, the amino group becomes hydroxylated at the nitrogen group and then conjugated with UDPGA:

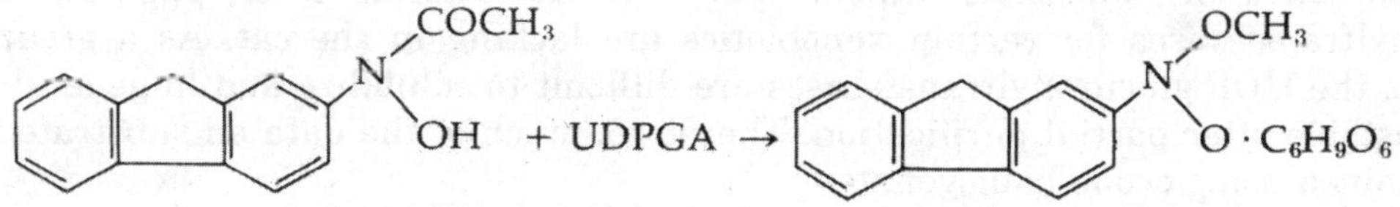

N-Hydroxy-2-acetylaminofluorene + UDPGA → *N*-Hydroxy-2-acetylaminofluorenyl β-D-glucuronide

16.2.2.2 N-Glucuronideation

Several types of N-glucuronides have been identified, in which the glucuronyl moiety is attached to the nitrogen. They include the following functional groups: aromatic amino group, sulfonamide group, carbamyl group and heterocyclic nitrogen group:

Aromatic amino type:

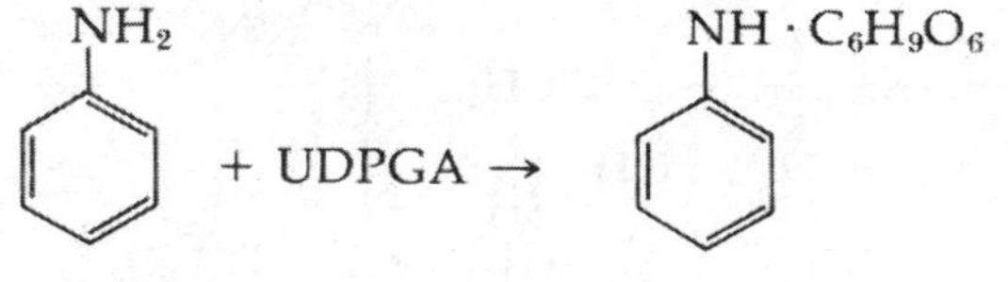

Aniline + UDPGA → Aniline glucuronide

Carbamyl type:

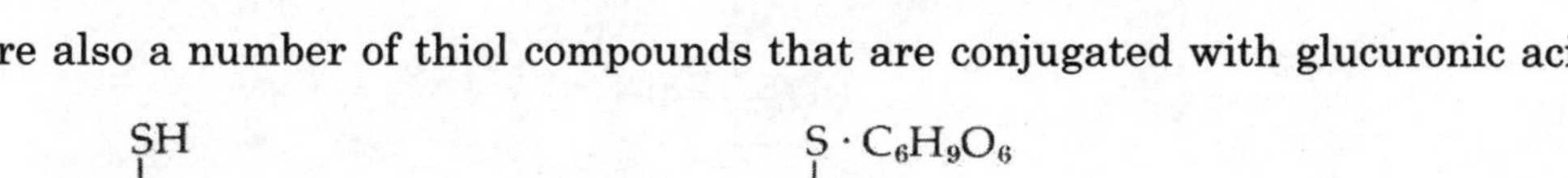

Meprobamate + UDPGA → Meprobamate glucuronide

16.2.2.3 S-Glucuronides

There are also a number of thiol compounds that are conjugated with glucuronic acid:

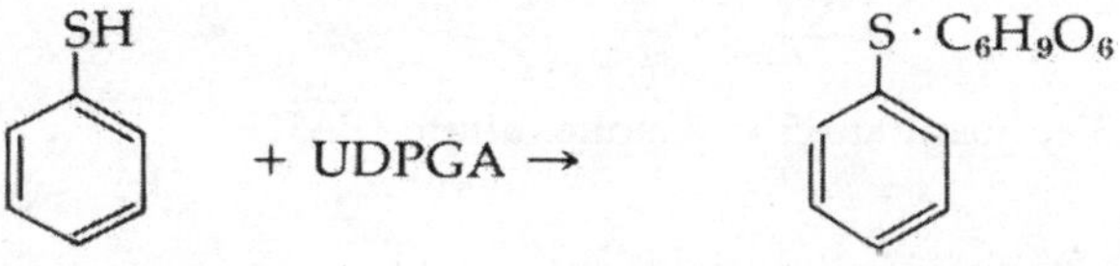

Thiophenol + UDPGA → Thiophenyl β-D-glucuronide

S-glucuronides are not found too often, but are similar to the N-glucuronides in that they are acid labile.

UDP glucuronyl transferases are a group of closely related enzymes with similar properties. The cat is able to form glucuromde conjugated of thyroxine and bilirubin but is unable to form other glucuronides. Cats are able to excrete injected glucuronides and also have a sufficient amount of UDPG. Therefore it appears that glucuronyltransferases for certain xenobiotics are lacking in the cat. As a group of enzymes, the UDP glucuronyltransferases are difficult to solubilize and in general are quite unstable after partial purification. Therefore, much of the data on substrate has been obtained using crude homogenates.

16.2.3 Sulfate Conjugation

Sulfate ester formation is known to occur with primary, secondary, and tertiary alcohols, phenols and arylamines. Sulfate esters, in terms of biological conjugations, are in reality half-esters, i.e., $ROSO_3$, which are completely ionized, highly water soluble, and quickly eliminated from the organism.

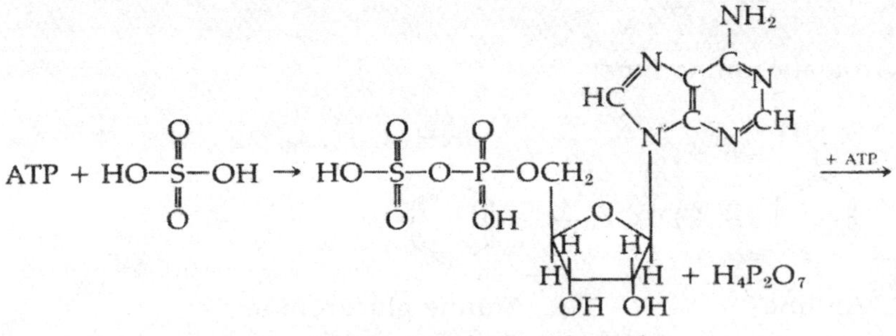

Adenosine 5′-phosphosulfate

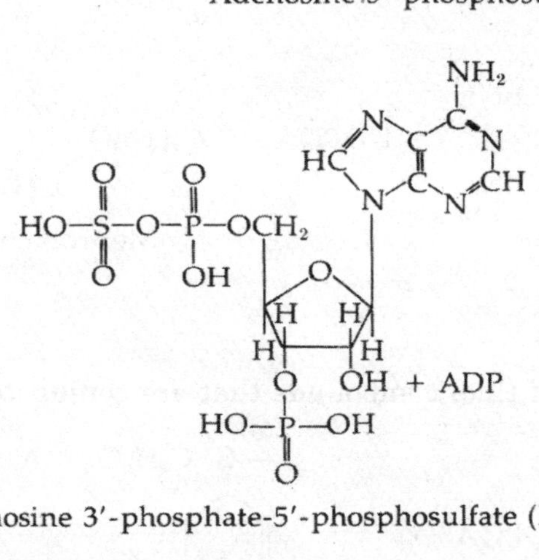

Adenosine 3′-phosphate-5′-phosphosulfate (PAPS)

Conjugation via the sulfate route requires a great deal of energy. The activation of the sulfate ion by ATP kinase has an absolute requirement for Mg^{2+}.

The enzymes responsible for PAPS formation have been found in the soluble fraction of the adrenal, brain, intestinal mucosa, kidney, liver, muscle, ovaries, and testis.

The sulfating agent (PAPS) with a group of sulfotransferases will form sulfate conjugates with phenol or ethanol:

Aryl Sulfates:

$$\text{Phenol} + \text{PAPS} \rightleftharpoons \text{Phenyl sulfate} + \text{ADP} \qquad \text{(an aryl sulfate)}$$

Alkyl sulfates:

$$C_2H_5OH + \text{PAPS} \rightleftharpoons C_2H_5OSO_3H + \text{ADP} \qquad \text{(an alkyl sulfate)}$$

$$\text{Ethanol} \qquad\qquad \text{Ethyl sulfate}$$

It appears that a number of sulfotransferases exist. No sulfotransferase has been obtained in a pure state, and the enzymes have usually been studied as crude tissue preparations. Therefore, it is not certain how many separate enzymes actually exist. With xenobiotics, phenol sulfotransterases, alcohol sulfotransierases, and arylamine sulfotransterases appear to be the most important. Phenol sulfotransferase is found in the soluble fraction of mammalian liver, kidney and intestinal mucosa, whereas some alcohol sulfotranslerases have only been found in mammalian livers. A number of other sulfotransferases are known, such as steroid sulfotransferase, choline sulfotransterase, mucopolysaccharide sulfotransferase and cerebroside sulfotransferase. However, of these probably only the steroid sulfotransferase may play a role in detoxication of foreign compounds.

16.2.4 Methylation

Methylation is a common biochemical reaction involving the transfer of methyl groups from three major coenzymes to amines, phenols, and thiol compounds to form N-, O-, and S-methyl conjugates. The coenzymes that provide methyl groups for transfer and S-adenosylmethionine, N^5-methyltetrahydrofolate derivatives, and vitamin B_{12} (methylcorrinoid) derivatives of these coenzymes, S-adenosylmethionine is probably the most important for xenobiotic methylation. S-Adenosylmethionine is synthesized from L-methionine and ATP. In all cases involving methylation, products are less water soluble than the parent compound, except for the tertiary amines. Even with the decrease in water solubility methylation generally is considered a detoxication reaction. Transmethylation reactions can be classified according to substrates.

16.2.4.1 N-Methylation

There are several enzyme systems that catalyze the N-methylation of natural and foreign amines. A highly specific enzyme, imidazole-N-methyltransferase, methylates histamine

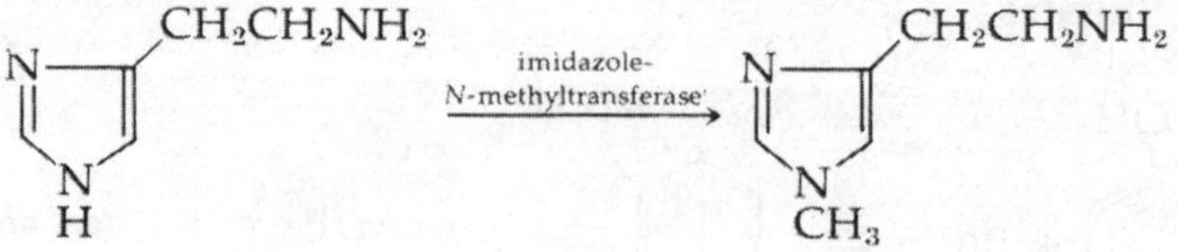

$$CH_2CH_2NH_2 \xrightarrow[\text{N-methyltransferase}]{\text{imidazole-}} CH_2CH_2NH_2$$

Histamine   N-Methylhistamine

Phenylethanolamine-N-methyltransferase, located in the soluble fraction of the adrenal, medulla, heart and brain, catalyzes the methylation of noradrenaline and other phenylethanolamine derivatives but cannot methylate phenyl ethyl amine:

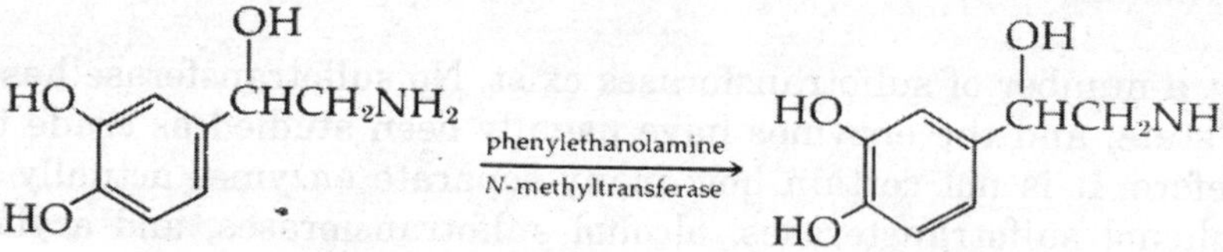

Noradrenaline   Adrenaline

Nonspecific-N-methyltransferases have been isolated in the soluble fraction from rabbit lung. Other tissues that have activity include the adrenals, kidney, and to a lesser degree, the liver and the spleen. The nonspecific N-methyltransferase methylates serotonin, tryptamine, and foreign amines such as normorphine, nornicotine, and norcodeine. Certain heterocyclic tertiary amines such as pyridine are also methylated to form quaternary salts:

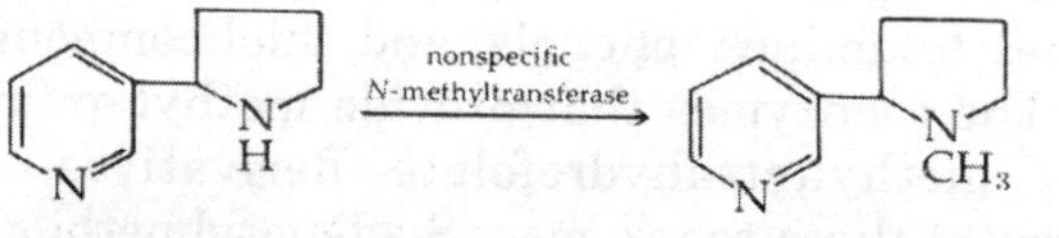

Nornicotine   Nicotine

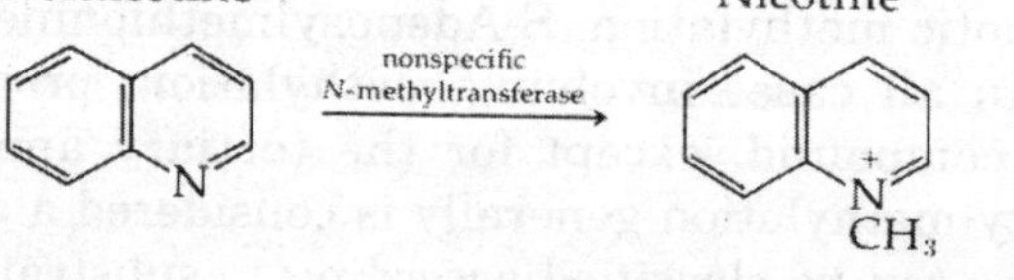

Quinoline   N-Methylquinoline

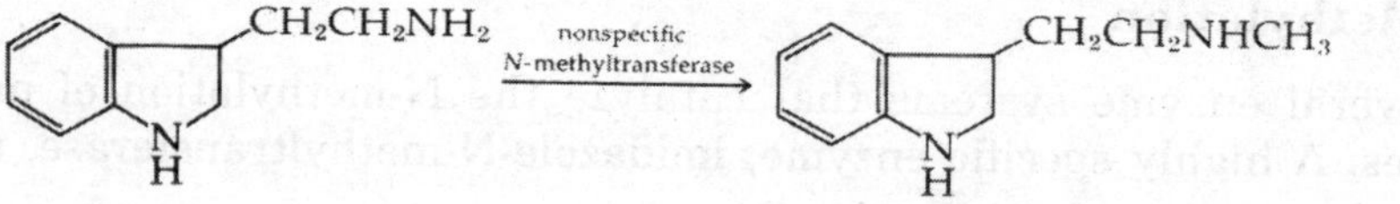

Tryptamine   N-Methyltryptamine

16.2.4.2 O-Methylation

Catechol-O-methyltransferase is localized in the soluble fraction of rat liver, kidney, glandular tissue and nervous tissue. The reaction requires S-adenosylmethionine and Hg^{2+} or other divalent ions:

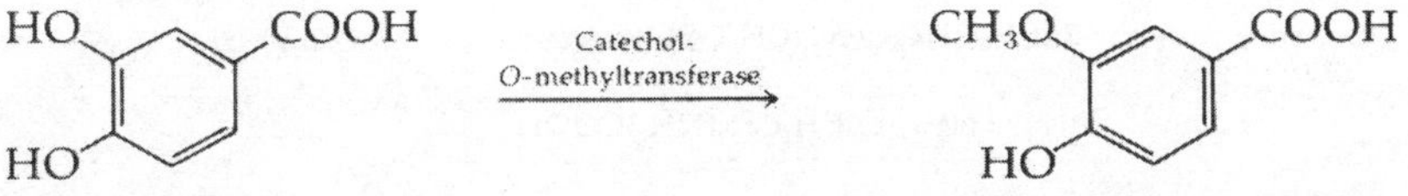

3,4-Dihydroxybenzoic acid   3-Methoxy-4-hydroxybenzoic acid

This enzyme catalyzes the methylation of epinephrine, norepinephrine, and other catechol derivatives. With this enzyme, methylation generally occurs at the *meta* position, but in some cases it will occur at the *para* position.

16.2.4.3 S-Methylation

Methyl groups are also transferred to thiol groups of certain xenobiotics, but the reaction appears to be of minor importance. S-methylation has been demonstrated with methyl and ethyl mercaptan, mercaptoethanol, thiouracil, mercaptoacetic acid and dimercaprol (BAL).

$$HSCH_2CH_2CH_2OH \xrightarrow{\;S-methyltransferase\;} CH_3SCH_2CH_2OH$$
Mercaptoethanol   S-Methylthioethanol

16.2.4.4 Biomethylation of elements

Biomethylation of toxic elements in the environment is an important biotransformation mechanism. A variety of metals i.e., mercury, lead, tin, pallidium, platinum, thallium and gold, as well as the metalloids, i.e., arsenic, selenium, tellurium, and sulfur, are methylated. The coenzymes reported to be the methyl donors for biomethylation are S-adenosylmethione and vitamin B_{12} (methylcorrinoid derivatives).

$$Hg^{2+} \longrightarrow CH_3Hg^+ \longrightarrow (CH_3)_2Hg$$
Inorganic   Monomethyl   Dimethyl
mercury   mercury   mercury

16.2.4.5 Formation of Mercapturic Acid

A large number of foreign compounds are metabolized into mercapturic acids, conjugates of L-acetylcysteine, and eliminated via biliary excretion. Mercapturic acids were first reported in 1875, but it was not until the 1950s that it was demonstrated that the cysteine portion of the conjugate came from reduced glutathione (GSH). The general scheme of mercapturic acid biosynthesis is given in Fig 16.1.

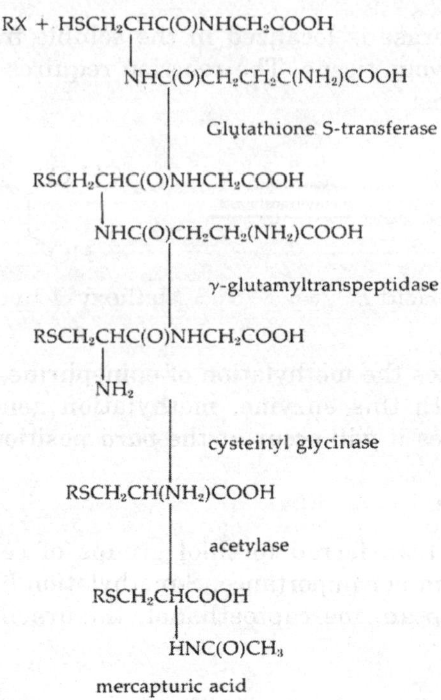

Fig. 16.1: Mercapturic acid biosynthesis.

The overall reaction involves the conjugation of foreign compounds possessing eletrophilic centers with GSH followed by transfer of the glutamate group. Loss of glycine and finally acetylation. As can be seen, a number of enzymes are involved in the mercapturic acid pathway. The GSH conjugation, the initial reaction, does not require the use of a high-energy intermediate. However, the synthesis of various amino acid components and the N-acetylation of the cysteine conjugate does require ATP.

The initial step is mediated by a group of enzymes referred to as glutathione S-transferases, which are involved in the conjugation of GSH -with certain foreign compounds.

The role of the glutathione S-transferases is to conjugate toxic electrophiles with endogenous GSH and therefore protect critical nucleophiles such as proteins and nucleic acids. GSH conjugates are anionic and have the necessary properties for biliary excretion, especially with molecular weights greater than 300.

The glutatlnone S-transferases seem to be distributed quite widely, e.g. in mammals, insects, fish, aquatic invertebrates, plants, and possibly bacteria. In mammals, the highest concentration of the enzymes is found in the soluble fraction of the liver. No sulfur-containing compounds other than GSH, will mediate this reaction.

A sufficient amount of GSH appears to be present in various mammalian tissues, plants and insects to act as a component of functional glutathione S-transferase system. However, depletion of hepatic GSH, which results when compounds that are reactive with GSH are administered, can have an important toxicological effect, such as a dramatic increase in toxicity when the organism is exposed to additional foreign compounds.

Glutathione-S-transferases were initially classified according to their substrate specificity and pH optima. Some substrates used for the classification were as follows:

Glutathione *S*-alkyltransferase:

$$CH_3I \quad + GSH \rightarrow CH_3\text{-}SG$$

Methyl iodide

Glutathione *S*-aryltransferase:

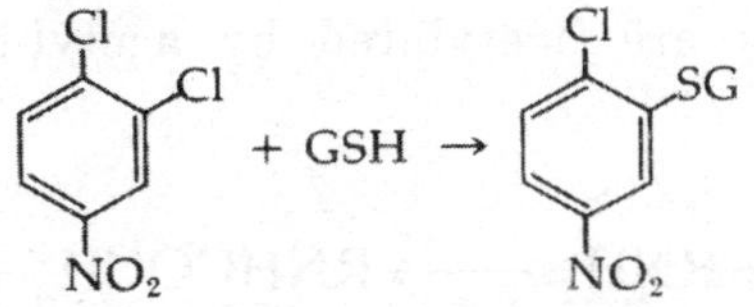

3,4-Dichloronitrobenzene (DCNB)

Glutathione *S*-aralkyltransferase:

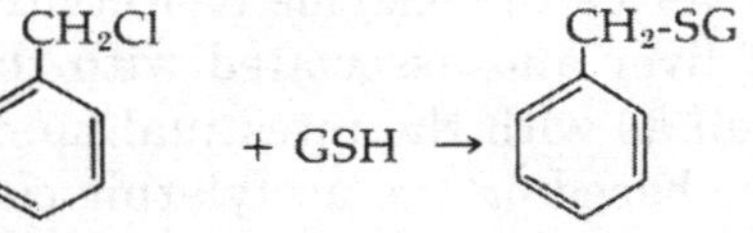

Benzyl chloride

Recently, the presence of multiple forms of glutathione S-transferases have been reported from livers of the rat, mouse, and man and in insects, Five forms of glutathione S-transferases have been purified to apparent homogeneity from the soluble

Glutathione *S*-alkenetransferase:

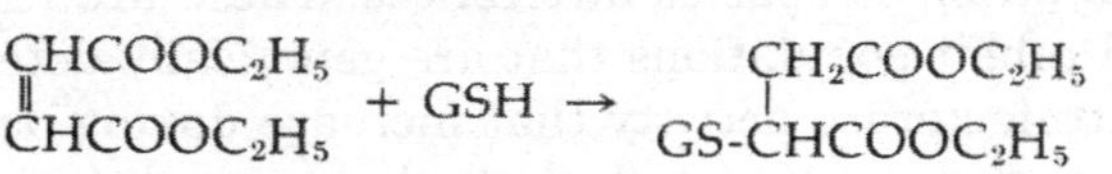

Diethyl maleate

Glutathione *S*-epoxidetransferase:

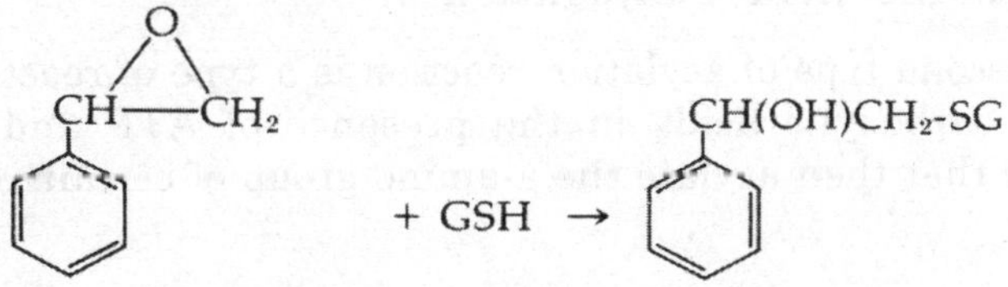

1,2-Epoxyethylbenzene

fraction of both the rat liver and human liver. The molecular masses of the transferases were 45,000-50,000 daltons and they were composed of two subunits with a molecular mass of 23,000-25,000 daltons.

16.2.4.6 Acylation

Foreign carboxylic acids and amides undergo biological acylation in mammalian species to form amide conjugates before excretion. This reaction renders the compound less water soluble than the parent compound but generally decreases toxicity.

Acylation reactions are of two general types. The first involves an activated conjugating intermediate (type I reaction) and the foreign compound.

16.2.4.7 Acetylation

Foreign amines are acetylated by acetyl-coenzyme A (CoA) to form acetylated derivatives.

$$CH_3COSCoA + RNH_2 \longrightarrow RNHCOCH_3 + CoASH$$

Acetyl-CoA   Amine   Conjugated - CoA
                                   amine

The acetylation involves the enzyme N-acetyltransferase, which has been found in the soluble fraction of liver and associated with the reticuloendothelial cells of liver, spleen, and lung as well as with the intestinal mucosa. Evidence exists for more than one N-acetyltransferase based on the acetylating capacity of mammalian tissue.

Acetylation of amino, hydroxy, and sulfur compounds occurs quite readily *in vivo*, Acetylation of -OH groups has been demonstrated with choline, while the acetylation of -SH groups is known to occur in the formation of acetyl-CoA. At the present time, the acetylation of hydroxy or thiol group present in foreign compounds is not known. Aromatic amines sulfonamides and certain amino acids are N-acetylated while aliphatic and phenyl-substituted aliphatic amines are not acetylated.

The ability to acetylate foreign compounds is influenced by developmental and genetic factors in addition to species differences. There are rapid and slow acetylators in both human and rabbit populations that are genetically determined. New borns have low levels of acetyltrafesterase activity that increase during development.

It has also been demonstrated that slow acetylators are likely to be more susceptible than fast acetylators to adverse effects from xenobiotics that are inactivated by acetylation.

16.2.4.8 Amino Acid Conjugation

The second type of acylation reaction is a type of reaction involving the activation of foreign carboxylic acids in the presence of ATP and CoA to form acyl S-CoA derivatives that then acylate the a-amino group of certain amino acids to form peptide conjugates:

$$CoASH + ROOH \xrightarrow{ATP} RCOSCoA$$

$$RCOSCoA + H_2NCH_2COOH \longrightarrow RCONHCH_2COOH$$

Acyl-CoA   Glycine   Glycine conjugate

Peptide conjugation has been reported to occur with glycine and glutamine in mammals and certain primates. In other organisms, different amino acids are utilized in peptide conjugation. Ornithine is utilized by reptiles and some birds, while ticks utilize arginine and glutamine, insects use alanine, glycine, serine, and glutamine, and fish use taurine.

The activating enzyme system is present in human and rat liver mitochondria and is presumably identical to the fatty acid-activating enzyme system. The acyltransferase has been found in both mitochondria and the soluble fraction. An acyltransferase has been purified from liver mitochondria that is able to mediate the transfer of various aliphatic (C_2-C_{10}) and aromatic acyl-CoA derivatives. The molecular mass of the 53--fold purified enzyme was estimated to be 190,000 daltons. Tests for various acyl acceptors showed that it was specific for glycine. Another purified rat liver glycine acyltransferase preparation had a different pH optimum than the beef liver enzyme.

Evidence would indicate that more than one acyltransferase may be present in various tissues.

16.2.4.9 Deacetylation

Enzymes that deacetylate N-acetylated xenobiotics have been found in the liver and kidney of many species. The aromatic de acetylase is known as arylacylamidase, which cleaves the conjugate and forms the starting compound:

$$NHCOOCH_3 \longrightarrow NH_2$$

Acetanilide   Aniline

Deacetylation occurs in various species, and there is a large difference in the amount of deacetylase activity among species, within different strains of the same species, and among individuals of the sample population.

Acetylation and deacetylation reactions are independent of one another and act in opposite directions. Therefore, it is difficult to evaluate the importance of acetylation *in vivo* in certain animal species because of the two competing systems.

16.2.4.10 Phosphate Conjugation

The biosynthesis of phosphate esters is a general phenomenon found in all intermediary metabolisms. However, phosphate conjugates of xenobiotics are rarely encountered. At only major group of animals that utilize this present it appears that insects are the mechanism as a phase II reaction.

An active phosphotransferase preparation has been obtained from gut tissue of the Madagascar cockroach. The enzyme requires ATP and Mg^{2+} for phosphorylation of 1-naphthol and *p*-nitrophenol.

16.3 Phase III Reactions

Drug transport has been designated as phase-III reactions. Drug transport influences not only the therapeutic efficacy but also the absorption, distribution and elimination of a drug (Zhang et al. 2003). The drug transporters are located in epithelial and endothelial cells of the liver, gastrointestinal tract, kidney, blood-brain barrier and other organs. They are responsible for the transport of most of the drugs across cellular barriers and thus concentrate on the target or biotransformation site. Multidrug resistance proteins (MRP, *p*-glycorprotein and others) have been shown to be important in explaining the pharmacokinetics of a drug in man. Thus, the elucidation of the influence of drug transporters on the adsorption, disposition, metabolism and elimination of a drug is essential in the early developmental and elimination of a drug.

<h2 style="text-align:center">SUGGESTIVE FURTHER READINGS</h2>

Arias, I. M., Jakoby, W.B. (Eds.). Glutathione, Metabolism and Function. New York: Raven Press, 1976.

Jakoby, W.B., Habig, W.H., Keen, J.H., Ketley, J.N., Pabst, M.J. Glutathione-S-transferases: Catalytic aspects. p. 189.

Brodie, B.B., Gillette, J.R., Ackerman, H.S. (Eds.). Handbook of experimental pharmacology, vol. 28 part 2, Concepts in Biological pharmacology. Berlin: Spinger, 1971.

<h2 style="text-align:center">STUDY QUESTIONS</h2>

1. Discuss in details the Phase I reactions.
2. Write briefly on the role of CYP450 in the metabolism of a xenobiotic.
3. Describe in details the Phase II reactions as detoxication mechanisms.
4. Discuss in the significance of the following enzymes
   a. CYP450
   b. Glutathione transferases
   c. Glutathione peroxidase
   d. Epoxy-hydrolase

Chapter 17

ENVIRONMENTAL CARCINOGENESIS

It has been established now that a high proportion of human cancers are caused by environmental agents - mainly environmental chemicals, but also viral and physical agents. It is unclear to what extent genetic factors may contribute to individual susceptibility to chemically induced carcinogenesis. However, perhaps three-fourths of human cancers are due to environmental chemicals. In a series of reviews of the carcinogenicity of approximately 300 substances by International Agency for Research on Cancer, 21 substances are listed as carcinogenic in humans and an additional 150 as carcinogenic in experimental animals. Others have claimed that 30 identified compounds are definite human carcinogens. Much of the current research is directed toward the identification of carcinogenic agents among environmental substances, industrial chemicals, and drugs and the development of methodology to predict the carcinogenicity of a tested substance.

The distribution of potential carcinogens in the environment is ubiquitous. Water may contain carbon tetrachloride and other chlorinated compounds or metallic salts that may be potentially carcinogenic. Laboratory and industrial solvents such as benzene and carbon tetrachloride may also be carcinogens. Alternatively, during the cooking of food, the nitrites can react with the amines to yield carcinogenic nitrosamines. Approximately three-fourths of the total 100-120 nitroso compounds tested are carcinogenic to experimental animals. Certain foods may also be contaminated with the aflatoxins, potentially carcinogenic compounds produced by some *Aspergillus* strains. Epidemiological investigations in man have suggested an association of increased incidence of hepatic tumors with increased dietary contamination by aflatoxins. Various polynuclear aromatic hydrocarbons, such as the carcinogen benzopyrene formed in the combustion of organic substances are potential carcinogens.

17.1 Historical Background

The environmental basis of carcinogenesis was first put forth by an English Surgeon, Percivall Pott (1775) when he observed high incidence of scrotal cancer amongst chimney sweeps exposed to soot. It is, however, now known that soot contains a group of a potentially hazardous carcinogenic compounds termed as polyaromatic

221

hydrocarbons. During the 16th century it was also realized that miners were frequent victims of what is recognized as lung cancer, which was rare in the rest of the population. In the latter half of 19th century 'coal tar' derivatives were found to induce carcinoma of skin and vulva. Later on, Japanese scientists Yamagiwa and Ichikawa successfully induced papilomas by painting rabbit ear with 'tar'. Kennaway (1930) induced tumors in mouse skin with a polyaromatic hydrocarbon benzo(a)pyrene.

17.2 Environmental Pollution and Carcinogens

In an urge for providing sufficient food and better life to its citizens, every country undertook industrial expansion alongwith large scale use of pesticides and chemical fertilizers which resulted in environmental pollution. Moreover, introduction of heavy metals, polyaromatic hydrocarbons, noxious gases, synthetic chemicals, dyes, food additives and a host of other pollutants also caused pollution in the wake of heavy industrialization, modernization and changed life style. As a result of all these activities our environment was polluted by a wide variety of chemicals (Ray, 1986a). There are approximately more than 700,000 chemicals in the environment and to these 1000 to 2000 new chemicals are being added every year (Ray, 1985a).

Cancer has become one of the leading causes of death in many parts of the world. The exact cause of human cancer is not well understood, however, some of the aetiological agents of environmental origin (environmental carcinogens) include polyaromatic hydrocarbons, tobacco, some metals, mycotoxins, certain pesticides and drugs etc. It is now well documented that 90% of human cancers have environmental origin (Higginson, 1978).

Epidemiological surveys have revealed that many cancers originate due to occupational exposure of workers to carcinogenic pollutants. Examples include high incidence of lung cancer among workers exposed to asbestos dust and coke oven emission, liver angiosarcoma among workers exposed to vinyl chloride, bladder cancer in aniline dye workers and many others (Lassiter, 1980).

Various epidemiological surveys and experimental studies have shown that most of the environmental carcinogens can cause immunotoxicity resulting in altered immune function of the exposed individual developing thereby a state of either partial or complete immunosuppression. The prolonged immunosuppression makes an individual extremely susceptible to various infections and even may predispose the host to develop cancer (Ray and Prasad 1987; Ray 1986a).

17.3 Types of Carcinogens

17.3.1 Direct Acting Carcinogens

These are in general, synthetic chemicals developed primarily for laboratory research or those used in industry. Added to these are the anticancer drugs. These carcinogens do not require the host metabolic activation by mixed function oxidases (MFOs) or other enzymatic activation in generating key reactive metabolites. To name a few of them are: p-propionlactone, 1,2,3,4-diepoxybutane, ethyleneimine, dimethyl sulphate, bis (2-chloroethyl) sulphide (mustard gas), nitrogen mustard, melphalan (sacrolysin),

2-naphthyl amine mustard (chlomaphazine), bis (chloromethyl) ether, benzyl chloride and dimethyl carbamyl chloride (Weisburger, 1975).

17.3.2 Procarcinogens

The parent compounds of this class of carcinogens are mostly inert in the microsomal fraction of liver. This system of enzymes contains CYP450(s). They require metabolic activation by mixed function oxidases (MFO). It is now known that there are at least six forms of cytochrome P-450 with different steroids and biphenyl substrates. The various forms of cytochrome P-450 show preferential hydroxylation at specific positions.

17.3.3 Co-carcinogenes

These are agents which increase the effect of directly acting carcinogens and/or pro carcinogens e.g. dilute solution of croton oil greatly enhances the effect of PAH (3-methyl cholanthrene) in causing skin tumors in mice. Tobacco tar and tobacco smoke are the co-carcinogenes that contain strong co-carcinogenic agents (Weisburger, 1975).

17.4 Mechanism of Chemical Carcinogenesis

There is little doubt nowadays that carcinogenesis in general and chemical carcinogenesis specifically is usually a multistage process, comprising different stages from the conversion of a normal somatic to a tumor cell (transformation) and ultimately, after a long latency period, to a clinically manifest malignant tumor.

While many of the later stages of carcinogenesis (promotion and prog-ression) are little understood at present and are fields of intensive research, especially from a biochemical and molecular-biological point of view, the stage of initiation (i.e the transformation of a normal to a poten-tial tumor cell) has been extensively studied in the past two decades and has finally resulted in a theory of chemical carcinogenesis that has been and is widely accepted. Some theories of initiation, especially the one by Miller and Miller is based on the concept of somatic mutation as the first step in carcinogenesis (Miller, J.A. et al. 1966, 1970, 1977).

Any theory of carcinogenesis has to incorporate the fact that the initiation process leads to irreversible changes at the cellular level; thus, mechanistic theories must explain that the induced effect becomes permanent. The only plausible explanation at present is that the initiation step is a genetic change, probably a somatic mutation. Therefore, the direct interaction of the carcinogen or its activated *metabolite/s* with DNA as a promutational step has received much attention.

However, it should not be forgotten that reasonable models do exist, which discuss macromolecular cellular targets other than DNA (Miller, J.A. et al. 1977). Thus, chemical carcinogenesis might also result from a primary alteration of a cellular RNA, since Temin (Temin, H.M. et al. 1970) has shown that RNA's can be transcribed intracellularly as DNA and the resulting DNA can be integrated into the host genome. Dickson and Robertson (Kickson, E. and Robertson, H.D. 1976) discuss a mechanism of the potential regulatory role of specific RNA's in cellular development.

Alterations of specific proteins could also have a potential for permitting the development of cellular clones with altered genomes (Miller, J.A., 1977). A possible mechanism is the greatly increased error rates of certain DNA polymerases (Loeb, L.A. et al. 1974; Springgate, G.F. et al. 1973). Another possible molecular mechanism of carcinogenesis, not involving direct genomic change in the cells, is based on the repressor--derepressor systems that control the expression of genomes of viruses or bacteria. Monod and Jacob (1962) and Pitot and Heidelberger (1963) both have argued that loss or modification of protein repressors or parts of the genome that are expressed, could result in near stable states of cellular differentiation.

This historical "review" however should by no means distract from the fact that direct interaction of chemical carcinogens with DNA and the ensuing mutation is no doubt the most probable as well as the most plausible mechanism of initiation. The groups Lawley (1957) and Brookes (1960) first showed the reaction of direct alkylating carcinogens with nucleic acids and proteins, with covalent binding to these substrates. They also showed that DNA interaction correlated best with carcinogenicity data.

The indirectly acting carcinogen 4-dimethyl-aminobenzene (butter yellow) was first shown by Miller and Miller (1947) to bind to proteins in 1947 and, later, by the same group, also to form adducts with nucleic acids (Lin, J.K., et al. 1975),

Identification of formed adducts invariably showed that cellular nucleophiles were the sites to which binding occurred, thus indicating that reactive intermediates of carcinogens were electrophilic chemical species.

17.4.1 The Miller and Miller Theory

These and similar results thus formed the basis of a more generally accepted theory of chemical carcinogenesis by Miller and Miller (1966, 1970; 1977). It postulates that chemical carcinogens are either electrophiles (directly acting agents) or form electrophilic reactive metabolites in vivo from per se chemically nonreactive carcinogens (so-called procarcinogens). The reactive electrophilic species then interacts with cellular nucleophiles, especially in DNA. DNA adduct formation is then considered as a promutagenic step which after cell replication, may lead to a mutation and therefore to a "fixation" of the change. DNA interacting carcinogens are therefore also called geno-toxic carcinogens, they are usually mutagenic in appropriate test systems. To summarize: genotoxic carcinogens bind covalently to DNA and tumor initiation is therefore the consequence of mutation(s) resulting from such interactions. This theory, based on new scientific results of carcinogenesis research, also lead to a revival of the somatic mutation hypothesis, first suggested by Boveri (1919) in 1914 and later emphasized again by K.H. Bauer (1949).

In a critical discussion of the theory, of course, it must be strongly emphasized that it explains plausibly many of the known facts in carcino-genesis, but that it is an idealized scheme that does not explain all facts. It effectively explains the initiation mechanism of genotoxic carcinogens, but it certainly cannot easily explain carcinogenesis with proven human car-cinogens, such as inorganic carcinogens (arsenic, nickel, chromium) and fibrous inorganics such as asbestos. Hormonal agents such as 17-â-estradiol or diethylstilbestrol are nonmutagenic, but carcinogenic in animal bio-assays. Similarly many other chemical compounds are carcinogenic in certain animal bioassay systems, but are nonmutagenic and probably do not interact

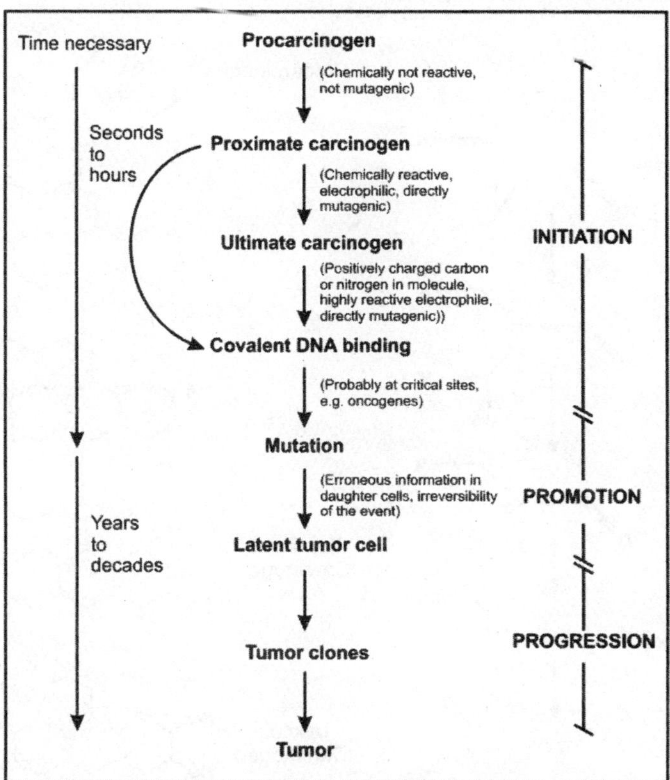

Fig. 17.1 Scheme of mechanism of carcinogenesis according to Miller and Miller (Miller, J.A., 1966; 1970; 1977).

with DNA directly. It is now clearly evident that the title "Carcinogens are Mutagens" in the paper by B. Ames (1973) is wishful thinking and in this gen-eralized form wrong. However, some carcinogens formally do not fit into Miller's scheme.

17.4.2 Inorganic Carcinogens

Arsenicals, some nickel and chromium compounds are recognized human carcinogens and have also induced cancer in animal bioasssays. The usual short-term tests for assaying mutagenicity usually give negative results with inorganic compounds; however mammalian cell systems are better suited to test for DNA and chromosomal damages: Carcinogenic metal compounds sometimes but not always induce chromosomal aberrations in different test systems such as human lymphocytes or in Chinese hamster embryo cells where they also induce transformations.

Asbestos and similar fibers, also negative in simple mutation test systems, induce chromosomal damage (Brown, R.C. et al. 1980) and also transform Syrian hamster embryo cells (DiPaolo, J.A., 1983; Barrett, J.C. et al. 1985).

Thus, for inorganic carcinogens there is some, but often not sufficient evidence for genotoxicity in a broad sense but they probably do not form DNA- adducts; their DNA interaction, mechanistically not understood, is of a more complex nature.

17.5 Molecular Mechanism of Carcinogenesis

Recent researches oriented to investigate the mechamism of carcinogenesis reveal that alterations in genes of normal cells are responsible for initiation of carcinogenesis and cancer is genetic in origin (Weinberg, 1988; Paul, 1984). It is now known that 'tumor producing genes' or 'oncogenes' are not only present on genetic locus of reteroviruses but also in normal cells-cellular oncogenes (c-oncogenes) (Barbacid, 1986).

Various environmental carcinogens including radiation can activate or cause enhanced expression of c-oncogenes, resulting in the formation of 'oncogene encoded proteins' in large amounts. Several carcinogens lead to initiation of carcinogenesis by activating c-oncogenes through deletion and/or substitution mutational mechanism (Tabin et. al. 1982). A good amount of evidence is now available, which indicates that carcinogens can cause substitution at specific focus of c-ras oncogene leading to initiation of carcinogenesis. Weinstein (1988) has documented the role of 'Suppressor genes' (S-genes) in the initiation of carcinogenesis. In a normal cell S-genes control and regulate the cellular growth. The activation of c-oncogenes by carcinogens or radiation decreases the activity of S-genes. The elimination or lesser expression of S-genes results in the removal of normal constraint on cell growth, and such genetically depleted cells, grow uncontrolled and become wild or malignant. The involvement of S-genes in carcinogenesis is evident from the experimental observations of Weinstein (1988) who noticed that the deletion of S-gene-Rb gene (located on 13th human chromosome) results in the conversion of normal retinoblastomas (Weinstein, 1988, Weinberg, 1988).

Carcinogens/radiation may cause activation of oncogenes which can disturb signal transduction responsible for normal growth and differentiation involving protein kinase C (PKC) (Hunter, 1984). PKC is a membrane bound enzyme often associated with regulatory functions and signal transduction for cellular growth and differentiation. The activation of c-oncogenes alongwith other events leading to carcinogenesis and tumor development are shown in Fig 17.2.

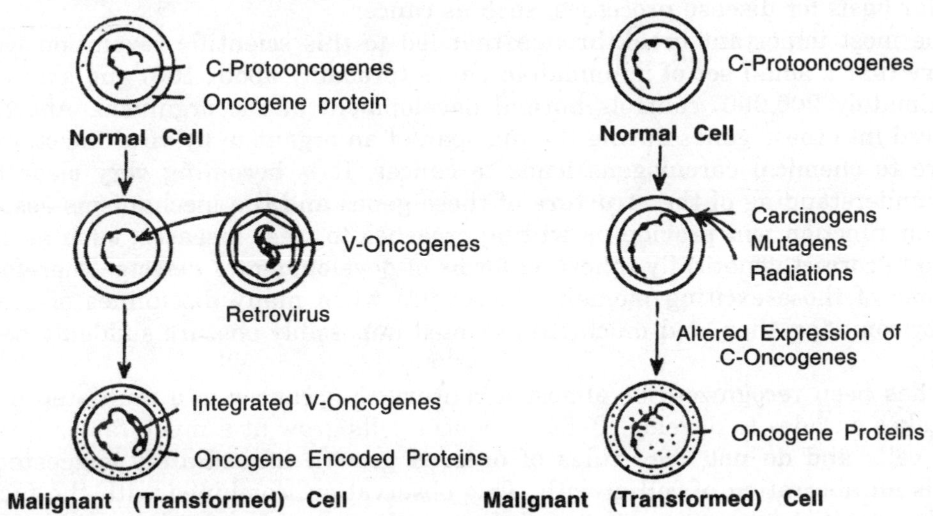

Fig. 17.2: Oncogenes and tumor development by tumor viruses, carcinogens and radiations.

Oncogenes encode several proteins, many of which are now characterized. 'Src', oncogene encodes a protein kinase having mol. wt. 60,000 daltons. Several oncogene encoded proteins are produced following activation of *'fms'*, *'raf'*, *'ras'* and *'mos'* oncogenes and have unique property of phosphorylating cell membrane proteins and assist in bringing about carcinogenic transformation (Hunter, 1984). Some of the 'oncogene encoded' proteins are shown in the Table 17.1.

Table 17.1 Some of the oncogene coded proteins

Oncogene	Retrovirus	Tumor	Property of organic proteins
sar	Chicken sarcoma	-	Tyrosine specific kinases bound to cellmembrane
abl	Mouse leukemia	Human leukemia	Plasma membrane bound protein
Ha-ras	Rat sarcoma	Human lymphoma	Class 4 nuclear
myb	Chicken leukemia	Human lymphoma	Nucleoprotein
Ki-ras	Rat sarcoma leukemia and carcinoma	Human sarcoma,	GPT binding class 3cytoplasmic protein

Sar, Part of Rous Sarcoma virus genome; abl, Abelson murine leukemia oncogene; Ha-ras. Haryey murine Sarcoma virus oncogene; myc, Avian klian MC 29 myelocytomatosis virus oncogene; myb, Avian myeloblastosis vims oncogene.

In the past 25 years a major revolution has occurred in understanding the molecular basis of cancer. The long drive to understand how normal cells grow and contribute to the ordered development of an organism and the nature of genetic events that disrupt this ordered process and result in cancer has become a major focus of biological research. The explosions of knowledge from basic research on cancer has unified several disciplines of science and led to a clear understanding of the mechanisms associated with normal cell growth, genetic events that produce defects in regulatory mechanisms that lead to aberrant growth and differentiation, and finally the molecular basis for disease processes, such as cancer.

The most important breakthrough that led to this scientific revolution was the discovery that a small set of mammalian genes (probably about 100) and avian genes (approximately 200,000) controls normal development of the organism. Aberrations introduced into these genes during the life span of an organism by viral infection or by exposure to chemical carcinogens leads to cancer. It is becoming very clear that a precise understanding of the structure of these genes and the mechanisms associated with their function will provide us with approaches to treat diseases, such as cancer, and also to correct genetically inherited forms of developmental defects. Therefore, we are at one of those exciting moments in science when many disciplines of scientific endeavor come together, and much that seemed impossibly obscure suddenly becomes clear.

It has been recognized for almost a century that cancer is a multistep process, which takes decades to develop. Because cancer cells grow at a much faster rate than normal cells and do not obey rules of ordered growth immediately suggested that cancer is an aberration of cell growth. This observation, combined with the fact that cancer is a multistep process, suggested that involvement of multiple genetic events of the growth deregulation seen in cancer cells. Research during the past decade has shown that the genetic events include activating a group of genes, termed "oncogenes", and inactivating another group of genes named "growth suppressor genes". By definition, oncogenes are genes that promote cell growth, and growth suppressor genes are those that block cell growth. A fine balance between the activities of these two groups of gene products dictates normal cell growth, and disrupting this balance provides a cell with a growth advantage that ultimately results in a neoplastic state.

17.6 Discovery of Viral and Cellular Oncogenes

Oncogenes were initially discovered by studying transforming retroviruses that produce tumors in animals. Several decades ago, it was observed that chickens die predominantly of cancer, suggesting that these animals have a genetic predisposition to this disease. This observation led to examining the tumors derived from these animals for a disease transmitting agent, which led to the identification of retroviruses. In 1991 Peyton Rous at the Rockefeller Institute in New York first isolated a retrovirus from a spontaneous chicken sarcoma. He established the viral etiology of the tumor by demonstrating that extracts derived from the tumor could be filtered through membranes that retain bacteria and nevertheless induce cancer in animals injected with the filtered extracts. The virus isolated by Rous has been named the Rous sarcoma virus in honor of its discoverer. Most interestingly, it was found that these viruses contain RNA as their genetic material and hence was called "RNA tumor viruses" or "retroviruses". Later studies revealed that extracts from animal tumors often contain two types of RNA tumor viruses, which were termed acute transforming viruses and leukemia viruses. These viruses differ from each other in a number of properties, which are listed in Table 17.2. The most important biological difference between these two classes of viruses is their ability to induce tumors *in vivo* as a function of time. Although acute transforming viruses induce tumors in susceptible, hosts in a very short time (1 to 2 weeks), the leukemia viruses require several months to years. A second important difference between the two classes of viruses is their ability to transform cells in tissue culture. The acute transforming viruses readily transform cells in tissue culture, whereas the leukemia viruses fail to do so. Finally, several of the acute transforming viruses (with the exception of Rous sarcoma virus) are replication-incompetent, whereas leukemia viruses replicate readily *in vitro* and *in vivo*.

The advent of recombinant DNA techniques in late 1970s allowed the molecular cloning and sequence analysis of the two types of viruses, which provided a molecular explanation of their biological differences. These studies showed that the leukemia viruses encode three polypeptides, named **gag, pol** and **env.** Of these, the *gag* gene encoded a polypeptide chain that is cleaved proteolytically into smaller polypeptides that from the core proteins of the virion (Fig. 17.3). The *pol* gene encodes a polypeptide

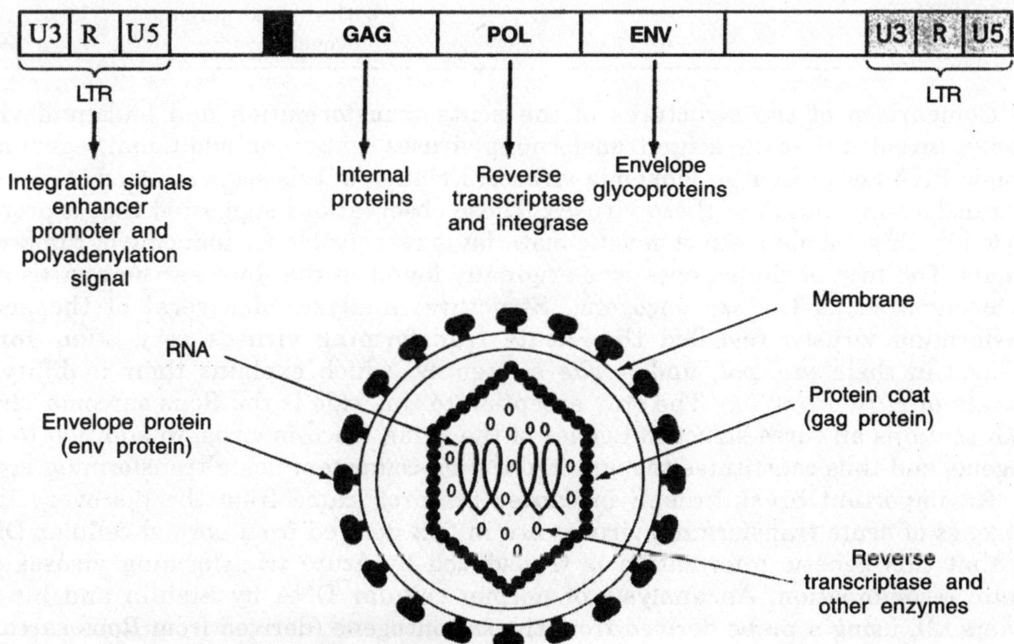

Fig. 17.3 Genetic elements of retroviruses and their gene products. A typical retroviral provirus encodes three genetic elements, gag, pol, and env, that are translated into pol gene codes for reverse transcriptase (an enzyme that coverts RNA into DNA) and the viral integrase (an enzyme that facilitates the integration of viral sequences into the host genome), and the env gene product forms the viral envelope protein. The LTR sequences direct high levels of viral gene expression within the host cell. Ψ denotes the packaging signal that enables the viral proteins to recognize the viral RNA that is to be incorporated into the mature virion.

chain that is cleaved into two polypeptides, one of which is the reverse transcriptase that enables the virus to convert its RNA into DNA. The second polypeptide fractions as an integrase that allows the proviral DNA to integrate into the host viral genome. The *env* gene encodes the viral envelope protein that plays an important role in virus-host cell membrane interactions. In addition to these three genes, the retroviral genome contains a stretch of sequences that are duplicated at the 5' and 3' ends of the provirus, termed Long Terminal Repeats (LTRs). The LTRs contain sequences that constitute some of the most potent eukaryotic promoter/enhancer elements in mediating high-level transcription of viral RNA.

Table 17.2. Retroviruses or Type C RNA Tumor Viruses

Properties	Chronic Leukemia Viruses	Acute Transforming Viruses
Induce	Lymphomas	Sarcomas, carcinomas, and hematopoietic tumors
Tumor latent period	Months to years	Weeks
Transformation of cellsin culture	No	Fibroblasts and/or hematopoietic cells
Replication-competent	Yes	Generally not
Mechanism of transformation	Integration next to oncogene	Discrete cell-derived gene within viral genome (oncogene)

Comparison of the structures of the acute transformation and leukemia viral genomes revealed that the acute transforming viruses contain an additional segment of genomic RNA not present in leukemia viruses. Deletion of this sequence leads to loss of the transforming ability of these viruses. These observations suggested that a protein encoded by this unique piece of genetic material is responsible for inducing of tumors in animals. The first of these genes was originally found in the Rous sarcoma virus and was designated as the *src* oncogene. Structural analysis of several of the acute transforming viruses revealed that acute transforming viruses very often suffer deletions in their *env, pol,* and/or *gag* sequences, which explains their inability to replicate *in vitro* or *in vivo*. The only exception to this rule is the Rous sarcoma virus, which contains all three structural genes of the avian leucosis virus, in addition to the *Ere* gene, and thus constitutes the only replication-competent acute transforming virus.

An important breakthrough in cancer research came from the discovery that oncogenes of acute transforming viruses are in fact derived from normal cellular DNA and that this genetic information is transduced by acute transforming viruses via genetic recombination. An analysis of normal cellular DNA by Stehlin and his co-workers (2), using a probe derived from the *src* oncogene (derived from Rous sarcoma virus), revealed the presence of endogenous oncogene-related sequences in normal chicken DNA Following the lead provided by the Rous sarcoma virus, retrovirologists have isolated more than 200 different tumor-producing, acute transforming viruses from animal tumors of different species. Table lists some of the acute transforming viruses and the oncogenes transduced by them. An examination of this table shows that several of the acute transforming viruses isolated from different animal species contain the same oncogene, suggesting that there are only a finite number of transforming genes in the avian and mammalian genomes and that these often undergo recombination with replicating retroviruses, leading to the formation of acute transforming viruses. Since the first identification of the *v-src* oncogne, a number of different approaches have been used to identify genes responsible for the altered growth properties of tumor cells. Now, the term "oncogene" is used more broadly to include any gene whose expression is associated with enhanced growth of tumor cells.

17.7 Carcinogenic Risk Assessment

Assessment of carcinogenic risk involves a number of steps: (1) hazard identification: (2) mechanism elucidation: and (3) quantification of risk based upon the mechanism.

It as has been suggested, all substances are hazardous, then the methods used for what we refer to as 'hazard identification' are, in fact, telling us something about the potency of the chemical. With all its difficulties, the 'hazard identification' process remains the easiest part of the process leading to quantitative risk assessment. Mechanism of action is the next, extremely important undertaking and must involve a broad range of disciplines, including pathologists, toxicologists, metabolic chemists, pharmacokinetics specialists and molecular biologists. Mechanistic studies have always been interesting, but they have now achieved a very high level of importance in the evaluation of the toxicology of a compound. Today it would be difficult to overestimate their importance. There are few general principles which can be applied to understanding mechanisms of action, each carcinogenic entity being studied on a case-by-case basis, although genetic toxicity has been a high-profile consideration in most mechanistic studies. Nevertheless, each chemical structure is unique and it is a serious error to attempt to group chemicals without experimental evidence – on the basis of either their structural similarities or similarities in their effects. Chemicals sharing particular active groups can have very different effects and particular effects may be achieved by different mechanisms.

Quantification of risk is the ultimate step, although investigators in this field do not pretend that a solution to the problem is at hand. The best that they can offer at the moment is mathematical models which fit the observed data and may have a foundation in biology.

Active efforts need to be constantly directed towards the reduction of environmental pollution by carcinogens and other immunosuppressive toxicants. This will reduce the chances of human cancer by environmental carcinogens. Judicious adoption and control of changed life style (smoking, drinking, using sophisticated foodstuff with preservatives etc.) is also needed to prevent environmental pollution. Environmental pollution is caused by human beings which is preventable. Therefore, environmental linked cancer too (90% of human cancer) can be kept well under control.

The use of anticarcinogens and host immunostimulation are the promising areas and should be clinically tried to prevent carcinogenesis.

SUGGESTIVE FURTHER READINGS

E.P. Reddy, A.M. Skalka, and T. Curran (1988). *The Oncogene Handbook,* Elsevier, New York.

G.M. Cooper (1995). *Oncogenes,* Jones and Bartlett, Boston.

B. Vogelstein and K.W. Kinzler (1995). *The Genetic Basis of Human Cancer,* McGraw-Hill, New York.

STUDY QUESTIONS

1. What do you understand by environmental carcinogenesis. Discuss the problem of environmental pollution and carcinogenesis.
2. Classify environmental carcinogens.
3. Discuss the mechanism of environmental carcinogenesis.
4. Write short notes on the assessment :
   i.   Carcinogenic risk assessment
   ii. Viral & cellular oncogenes

Chapter 18

ENVIRONMENTAL HEALTH RISK ASSESSMENT

With few exceptions such as veterinary and agricultural products, toxicological studies are not conducted solely to assess the toxic effects of chemicals in animals but to identify the effects which might occur in humans. *Risk assessment is the process of evaluating the toxic properties: of chemicals and the conditions of human exposure to ascertain the likelihood that humans: will be adversely affected and to characterize the nature of the effects which may be experienced.*

The risk assessment process, such as performed by the U.S. Environmental Protection Agency, can be divided into four steps; ***hazard identification, dose-response assessment, exposure assessment, and risk characterization*** (fig. 18.1). In hazard identification, a determination is made of whether a substance of concern, be it a pharmaceutical, industrial chemical, environmental pollutant etc., can be linked to an adverse effect. Dose-response assessment establishes relationships between the magnitude of exposure and the occurrence of the adverse effect. The major activities of toxicologists are concentrated in these two steps. In exposure assessment, human exposure to the substance of concern is identified through characterization of the exposed population, routes of exposure, and magnitude of the exposure under various conditions. All the information derived in these three steps of the risk assessment process is used in the risk characterization. In this fourth and final stage of the risk assessment process, a determination is made of the likelihood that humans may experience the identified adverse effect under actual or plausible hypothetical conditions of exposure.

Based on the risk characterization, the need for and the degree of risk management will be determined. A number of options are available to the risk manager, including education and communication of risk, exposure monitoring and controls, limitation in the use of the substance of concern or a total ban of the chemical. Risk management decisions are influenced by economic, political, and social concerns. Risk management is considered separately from the risk assessment process.

An overview of some of the major concepts that form the foundation of the risk assessment process, as well as assumptions and factors considered and/or used in risk assessment is presented in the following table and figures. A review of this information will provide a basic assessment of risk assessment.

233

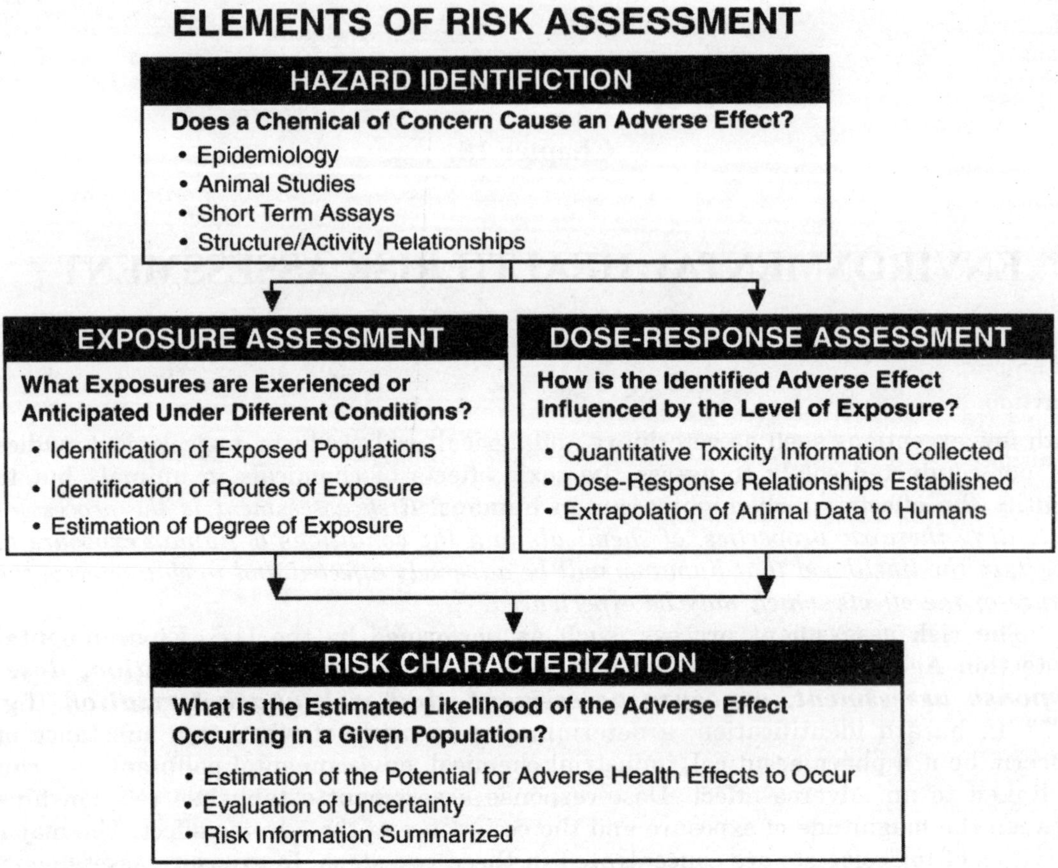

ELEMENTS OF RISK ASSESSMENT

HAZARD IDENTIFICTION

Does a Chemical of Concern Cause an Adverse Effect?

- Epidemiology
- Animal Studies
- Short Term Assays
- Structure/Activity Relationships

EXPOSURE ASSESSMENT

What Exposures are Exerienced or Anticipated Under Different Conditions?

- Identification of Exposed Populations
- Identification of Routes of Exposure
- Estimation of Degree of Exposure

DOSE-RESPONSE ASSESSMENT

How is the Identified Adverse Effect Influenced by the Level of Exposure?

- Quantitative Toxicity Information Collected
- Dose-Response Relationships Established
- Extrapolation of Animal Data to Humans

RISK CHARACTERIZATION

What is the Estimated Likelihood of the Adverse Effect Occurring in a Given Population?

- Estimation of the Potential for Adverse Health Effects to Occur
- Evaluation of Uncertainty
- Risk Information Summarized

Table 18.1 Hierarchy data selection for risk assessment

- Human data *lifetime exposure* via the more appropriate route (inhalation or oral)
- Human data *less than lifetime exposure* with lifetime observation (exposure via the more appropriate route)
- Human data *less than life time exposure* with less-than-lifetime observation (exposure via the more appropriate route)
- Human data *lifetime exposure* via the less appropriate route if reasonable toxicologically
- Human data *less than lifetime exposure* with lifetime observation (exposure via the less appropriate route)
- Human data *less than lifetime exposure* with less-than-lifetime observation exposure via the less appropriate route)
- Animal data the same sequence as for human data. Animal studies of less than 90 days of exposure and/or less than 18 months of observation from first exposure should not be used.

Table 18.2 Major factors that influence a Risk assessment

Factor	Effect
Low dose extrapolation	Can involve as many as 50 or more assumptions each of which introduce uncertainty. Often considered the greatest weakness in risk assessment.
Population variation	The use of standard exposure factors can understimate actual risk to hypersensitive individuals. Addressing the risk assessment to the most sensitive individuals can overestimate risk to the population as whole
Exposure variation	The use of modeling and measurement techniques can provide exposure estimates that diverge widely from reality.
Environmental variation	Can affect actual exposures to a greater or lesser degree than assumed to exist.
Multiple exposures	Risk assessment generally deal with one contaminant for which additive, synergistic, and antagonistic effects are unaccounted. Can result in underestimate or overestimate of risk.
Species differences	It is generally assumed that humans are equivalent to the most sensitive species. Can overestimate or underestimate risk.
Dose based on body weight	Toxicity generally does not vary linearly with body weight but exponentially with body surface area.
Choice of dose levels	Use of unrealistically high dose levels can result in toxicity unlikely to occur at actual exposure levels. The number of animals being studied may be insufficient to detect toxicity at lower doses.
Uncertainty factors	The use of uncertainty factors in attempting to counter the potential uncertainty of a risk assessment can overestimate risk by several orders of magnitude.
Confidence intervals	The upper confidence interval does not represent the true likelihood of an invent can overestimate risk by an order of magnitude or more.
Statistics	Experimental data may be inadequate for statistical analysis. Statistical significance may not indicate biological significance, and a biologically significant effect may not be statistically significant. Statistical significance does not prove causality. Conversely, lack of statistical significance does not prove safety.

Table 18.3 Criteria Defining "High Exposure" Chemicals

- Production greater than 100,000 kg
- More than 1000 workers exposed
- More than 100 workers exposed by inhalation to greater than 10 mg/kg/day
- More than 100 workers exposed by inhalation to 1 - 10 mg/day for more than 100 days/year

(Contd.........)

Table 18.3 Contd.........

- More than 250 workers exposed by routine dermal contact for more than 100 day/yea r
- Presence of the chemical in any consumer product in which the physical state of the chemical in the product and the manner of use would make exposure likely
- More than 70 mg/year of exposure via surface water
- More than 70 mg/year of exposure via air
- More than 70 mg/year of exposure via ground\1\later
- More than 10,000 kg/year release to environmental media
- More than 1000 kg/year total release to surface water after calculated estimates of treatment

Source: U.S. Environmental Protection Agency, 1988

Table 18.4 Factors that influence risk management decisions

Decreases degree of risk management	Increases degree of risk management
Risk assumed voluntarily	Risk borne involuntarily
No alternatives available	Safer alternatives available
Exposure is essential	Exposure considered a luxury
Exposure primarily occupational	Exposure non-occupational
Exposure primarily to average individuals	Exposure involves hypersensitive individuals
Intended use can be reasonably guaranteed	Potential for misuse high
Toxic effects reversible	Toxic effects permanent
Toxic effects not inheritable	Toxic effects inheritable
Risk perceived acceptable	Risk perceived unacceptable

Adapted from Lowrance, Of Acceptable Risk: Science and the Determination of Safety, 1976

18.1 Dose-response Relationships

The dose-response relationship forms the basis of the most fundamental concept of toxicology, that the toxicity of any material is defined by its dose-response curve. According to Paracelsus, "the dose makes the poison". Dose-response assessment is one of the four steps in the risk assessment process.

The dose-response curve generally takes two forms: the first displays the distribution of an effect within a population as a function of changing exposure, and the second indicates the degree of change of an effect in an exposed individual of a population as a function of changing exposure. The demonstration of dose-response suggests causality.

The typical dose-response curve is sigmoidal, but can also be linear, concave, convex, or bimodal. The shape of the curve can offer clues to the mechanism of action of the toxin, indicate multiple toxic effects, and identify the existence and extent of sensitive subpopulations. Analysis of the dose-response curve can demonstrate average response, the degree of susceptibility within a population, and the range of exposure affecting hyperreactive individuals. The slope of the dose-response curve categorizes the potency of the toxin and indicates the magnitude of effect associated with incremental increases in exposure. The curve can display the degree of confidence (and conversely uncertainty) associated with the data.

The purpose of this section is to provide examples of some of the ways dose-response relationships are used in risk assessment and the type of information that can be obtained from the dose-response curve. Risk assessment takes advantage of the ability of the dose- response curve to quantitate the susceptibility of individuals in an exposed population to a substance of concern. Dose-response assessment determines the relationship between the extent (magnitude) of exposure and the probability of occurrence of health effects. For pharmaceuticals, safety ratios such as the therapeutic index and margin of safety are derived. "Safe levels" of exposure such as reference dose (RID), acceptable daily intake (AD1), and permissible exposure limits (PEL) for pollutants, food additives, and industrial chemicals, respectively, are estimated from dose-response.

The above values are determined for substances demonstrated to have a threshold below which an effect of concern is not expected to occur. Where thresholds are assumed not to exist, such as for carcinogens, virtually safe doses (VSD) or exposures are determined by extrapolation of the dose-response curve to levels of risk deemed acceptable by society. The reliability of this approach is compromised by the uncertainty associated with modeling the low-dose region of the dose-response curve.

The quality of a dose-response assessment is only as good as the data with which the dose-response curve was generated. Dose-response assessment, coupled with exposure assessment, allows the risk of exposure to a chemical to be characterized, which ultimately is the purpose of the risk assessment process. The more accurate the dose- response and exposure assessment, the more realistic the risk characterization will be.

18.2 Epidemiology

Epidemiology, the study of the distribution and determinants of disease in human populations, is one of the tools used in the hazard identification step of the risk assessment process. Despite the many problems inherent to epidemiological studies including various biases, confounding factors, and inadequate quantization of exposures, these studies offer a major advantage over those conducted with animals: the direct observation of effects in humans.

Much of the uncertainty associated with risk assessment results from the extrapolation of animal data to humans. Quality epidemiology studies can significantly reduce or eliminate such uncertainty. For this reason, human data are preferred over animal data for risk assessment. Usually however, the availability of quality epidemiological studies is limited, and human, animal and *in vitro* data are used together in a "weight-of-evidence" approach to the risk assessment process.

Most toxicologists will encounter epidemiological data at some point in their careers. Epidemiology has its own unique methodology, measurement indices, and terminology. Much of the information from these studies is reported as relative risks, rates, ratios, and proportions. Knowledge of such information is useful for evaluating the design and interpreting the results of human studies.

SUGGESTIVE FURTHER READINGS

U.S. Environmental Protection Agency, *General Quantitative Risk Assessment Guidance for Non-cancer Health Effects,* ECAP-CIN-538M, cited in Hooper et al.

National Research Council, *Risk Assessment in the Federal Government,* National Academy Press, Washington D.C., 1983.

Van Ryzin, J., Quantitative risk assessment, *J. Occup. Med.,* 22, 321, 1980.

Weisburger, E.K., Industrial and environmental cancer risks, in *Dangerous Properties of Industrial Materials,* 6th ed., Sex, N.I. Ed., Van Nostrand Reinhold, New York, 1984, section 3.

STUDY QUESTIONS

1. Describe different elements of risk assessment.
2. Write briefly on the following
   i. Dose response relationship
   ii. Epidemiology

Chapter 19

HUMAN TOXICOLOGY

Nearly two million human poisonings are reported to poison information centers each year however, there are an estimated 2 to 3 million additional unreported exposures. The purpose of this chapter is to review the epidemiological characteristics of human poisonings, clinical toxicology research designs and general management techniques of the actually poisoned patient. In addition the role of the poison control centers in managing these patients and several contemporary issues in clinical toxicology are discussed.

Although several excellent texts contain valuable information on the toxicity and treatment of poisoned patients, the most up-to-date Information on both human and animal poisonings is provided by Poisindex® (MICROMEDEX, Medical Information Systems, Denver). It contains data on the chemical composition, toxicity, and the current medical management of more than 750,000 drugs, household chemicals, industrial and environmental toxins, and biologicals (including plant and animal toxins). Poisindex® also facilitates the identification of manufactured drugs by providing a description of the tablet/capsule shape, color, and the symbols imprinted on them. It also provides slang terminology, color, and shape for street drugs. Poisindex is edited and updated very 3 months.

Another valuable source of information is a regional poison control center. Currently there are more than 100 regional poison control centers located throughout the United States; 38 have been certified by the American Association of Poison Control Centers (AAPCC).

19.1 Designing of Clinical Research

Predicting the effects of toxic agents in animals are critical and mandatory facts of human risk anagement. The three main tasks of experimental animal studies in toxicology are listed in Table 19.1. Basic to understanding human toxicities is the assumption that information gained from animal models can be extrapolated to analogous human situations. This places great importance on the validations of extrapolations from animal data. However, extrapolation from an animal requires specification of the effects that will serve to test the validity of the model. A hierarchy of five criteria, shown in Table 19.2, can be used to determine the validity of animal models. Other relevant information, from epidemiology studies and clinical research, is then integrated with the animal data to make regulatory decisions regarding human safety.

The principle of research design in clinical toxicology involves assessing causation in disease- exposure associations, evaluating the appropriateness of the research design, and evaluating the validity of a particular research study.

The various research designs available to study clinical problems include the randomized clinical trial (RCT), cohort studies, case-control studies, cross-sectional studies, case series, and case reports.

Table 19.1 The Three Main Tasks of Experimental Toxicology

1. Spectrum of toxicity
   Detection of adverse effects of chemicals in selected laboratory animal species and description of the dose-effect relationship over a broad range of doses.
2. Extrapolation
   Prediction of adverse effects in other species, particularly in man.
3. Safety
   Prediction of safe levels of exposure in other species, particularly in man.

From Zbinden (1991) Reprinted with permission.

Table 19.2 Five Criteria that Test the Validity of Animal Models

1. Face validity
   A model is superficially similar to the human condition.
2. Content validity
   Examining the characteristics shared in common by an animal model and the human condition it seeks to simulate, to determine whether the model represents the specific content which a study is designed to measure
3. Concurrent validity
   Multiple measures of toxic reactions within subjects can provide profiles distinctive to particular toxins or classes of toxins.
4. Construct validity
   Utilizing a theoretical model of the nature of living organisms and how they interact with their biosphere
5. Predictive validity
   Demonstration that extrapolation from animal models compares or can be predictive of human toxicity

From Russell (1991) Reprinted with permission.

Table 19.3 The three main tasks of experimental toxicology

1. **Spectrum of toxicity**
   Detection of adverse effects of chemicals in selected laboratory animal species and description of the dose-effect relationship over a broad range of uses.

(Contd..........)

Table 19.3 Contd..........

2. **Extrapolation**
   Prediction of adverse effects on these species, particularly in man.
3. **Safety**
   Prediction of safe levels of exposure in other species, particularly in man.

19.2 Epidemiology of Human Poisonings

19.2.1 General Characteristics

The AAPCC estimates that 4.4 million poisonings occurred nationwide in 1991. Of the 1.8 million exposures reported, 764 resulted in death.

Although nonpharmaceuticals were involved in more than 50% of all poisonings (Table 19.4), pharmaceuticals were most frequently involved in fatalities.

Table 19.4 Substances most frequently involved in human exposure

Substance	No.	%[a]
Cleaning substances	191,830	10.4
Analgesics	183,013	10.0
Cosmetics	153,424	8.3
Plants	112,564	6.1
Cleaning Cough and cold preparations	105,185	5.7
Bites/envenomations	76,941	4.2
Pesticides (includes rodenticides)	70,523	3.8
Topicals	69,096	3.8
Antimicrobials	64,805	3.5
Foreign bodies	64472	3.5
Hydrocarbons	63,536	3.5
Sedatives/hypnotics/antipsychotics	58,450	3.2
Chemicals	53.666	2.9
Alcohols	50,296	2.7
Food poisoning	46,482	2.5
Vitamins	40,883	2.2

Note: Despite a high frequency of involvement, these substances are not necessarily the most toxic, but rather often represent only ready availability.
[a] Percentages are based on the total number of human exposures rather than the total number of substances.

Ingestion was the most common route of exposure. Accidental exposures accounted for more than 87% of all poisonings; they were most common in children younger than 6 years of age and in the elderly. Intentional exposures were most common in adolescents (>14 years of age) and the most frequent cause of death in the adult population. Most accidental poisonings involved only one substance; however, 50 to 60% of intentional poisonings in adults were poly-drug exposures.

19.2.2. Pediatric Poisonings

According to the 1991 statistics, approximately 60% of exposures involved children younger than 6 years of age. The number of pediatric deaths due to poisonings increased from 25 deaths in 1990 to 44 deaths in 1991.

A 5-year (1985-1989) retrospective analysis of pediatric deaths was conducted to aid poison prevention and educational efforts, guide new product formulation and aversive agent use, reassess over-the-counter status for selected pharmaceuticals, and identify areas of research for the treatment of pediatric poisonings. A hazard factor was devised to assess the risk of each agent to produce a major (residual disability) or life-threatening outcome when a child is involved in an overdose This factor indicates a substance's relative pediatric hazard by evaluating its packaging, accessibility (as a reflection of common storage practices in the home») availability (as a reflection of marketing), formulation, and closure types. The hazard score ranking allows for a comparison among categories.

Of the 3.8 million pediatric exposures reported between 1985 and 1989, 2117 patients experienced a major outcome and its fatalities occurred. Table 19.5 shows the substance categories implicated in pediatric exposure calls, the total number of pediatric exposures that occurred in each category, the number of major effects, the number of fatal exposures, and the hazard factor.

Table 19.5 Reported Poison Exposures in Children Younger than 6 Years *of* Age, 1985 through 1989

Categories and Subcategories of Poisonous Substances	Total Pediatric Exposures (1985-1989)	No. of Major Effects	No. of Deaths	Hazard Factor
Non pharmaceutical exposures				
Adhesives/glues	37,986	15	0	0.7
Alcohols	80,443	46	5	1.0
Ethanol (beverage)	2622	11	2	8.0[c]
Arts/crafts/office supplies	80,294	3	0	0.1
Auto/aircraft/boat products	12,019	11	4	2.0
Ethylene glycol	2321	4	1	3.5[c]
Methanol	1883	3	3	5.1[c]
Batteries	12,753	7	0	0.9
Bites/envenomations	48,821	100	0	3.3[c]
Copper head	44	4	0	146.4[c]
Rattle snake	125	19	0	244.8[c]
Unknown snake	651	10	0	24.7[c]
Other unknown reptile	682	3	0	7.1[c]
Scorpion	1585	43	0	43.7[c]
Black widow spider	1240	4	0	5.2[c]
Brown recluse spider	455	2	0	7.1[c]
Building products	13,721	2	0	0.2
Chemicals	87,463	72	3	1.4[c]
Acid: hydrochloric	784	3	0	6.2[c]
Alkali	10,267	24	0	3.8[c]
Dioxin	11	1	0	146.4[c]

Table 19.5 Contd..........

Categories and Subcategories of Poisonous Substances	Total Pediatric Exposures (1985-1989)	No. of Major Effects	No. of Deaths	Hazard Factor
Ethylene glycol	1002	6	0	9.6 [c]
Strychnine	32	1	0	50.3 [c]
Cleaning substances	386,052	205	4	0.9
Acid: drain cleaner	208	6	0	46.5 [c]
Acid: industrial cleaner	438	3	0	11.0 [c]
Aklali: drain cleaner	1474	19	I	21.9 [c]
Alkali: industrial cleaner	938	17	0	29.2 [c]
Alkali: oven cleaner	4619	10	0	3.5 [c]
Oven cleaner: other unknown	224	4	0	28.8 [c]
Cosmetics/personal care	395,985	57	3	0.2
Deodorizers (nonpersonal)	39,408	1	0	0.0
Dyes	10,369	0	0	0.0
Essential oils	6557	5	0	1.2
Fertilizers	23,581	1	0	0.1
Fire extinguishers	860	0	0	0.0
Food products/poisoning	49,500	7	0	0.2
Foreign bodies/toys	163,722	21	0	0.2
Fumes/gases/vapors	8436	55	21	14.5 [c]
Carbon monoxide	3103	42	18	31.1 [c]
Chlorine gas	2208	6	0	4.4 [c]
Hydrogen sulfide	228	2	0	14.1 [c]
Methane and natural gas	700	3	1	9.2 [c]
Gas:other	1138	1	2	4.2 [c]
Fungicides (nonmedicinal)	2714	1	0	0.6
Heavy metals	11,926	14	0	1.9 [c]
Other	1094	4	0	5.9 [c]
Unknown	39	1	0	41.3 [c]
Herbicides	6488	3	0	0.7
Paraquat	66	1	0	24.4 [c]
Hydrocarbons	*129,024*	168	5	2.2 [c]
Kerosene	10,751	51	2	7.9 [c]
Lighter fluid/naphtha	8865	24	0	4.4 [c]
Mineral seal oil	6564	13	0	3.2 [c]
Insecticides/pesticides	100,105	122	6	2.1 [c]
Chlorinated hydrocarbon alone	9694	29	0	4.8 [c]
Organophosphate alone	16,560	56	2	5.6 [c]
Organophosphate with other pesticide	1806	4	0	3.6 [c]
Rotenone	284	2	0	11.3 [c]
Lacrimators	4779	1	0	0.3
Matches/fireworks/explosive	11,655	1	0	0.1
Moth repellants	19,548	6	0	0.5

(Contd..........)

Table 19.5 Contd..........

Categories and Subcategories of Poisonous Substances	Total Pediatric Exposures (1985-1989)	No. of Major Effects	No. of Deaths	Hazard Factor
Mushrooms	32,724	9	0	0.4
Paints/stripping agents	47,114	10	0	0.3
Photographic products	2688	1	0	0.6
Plants	375,649	33	1	0.1
Polishes/waxes	12,249	4	0	0.5
Rodenticides	41,261	2	1	0.1
Sporting equipment	2134	0	4	3.0 [c]
Gun bluing compound	100	0	3	48.3 [c]
Swimming pool/aquarium	8067	1	0	0.2
Tobacco products	36,742	14	0	0.6
Unknown nondrug substance	20,318	14	0	1.1
Pharmaceutical exposures				
Analgesics	325,539	119	8	0.6
Acetaminophen with propoxyphene	2171	6	0	4.5 [c]
Aspirin: unknown formulation	10,002	18	1	3.1 [c]
Methadone	127	4	2	76.1 [c]
Morphine	164	3	0	29.5 [c]
Propoxyphene	514	1	1	6.3 [c]
Ofher unknown narcotic	732	4	0	8.8 [c]
Anesthetics	14,025	17	4	2.4
Anticholinergic	6516	13	0	3.2 [c]
Anticoagulants	1860	I	0	0.9
Anticonvulsants	9198	106	4	19.3 [c]
Carbamazepine	4113	81	2	32.5 [c]
Phenytoin	3619	19	2	9.3 [c]
Valproic acid	1197	5	0	6.7 [c]
Other anticonvulsant	54	1	0	29.8 [c]
Antidepressants	12,003	125	7	17.7 [c]
Amitriptyline	2897	39	2	22.8 [c]
Amoxapine	200	4	0	32.2 [c]
Desipramme	935	13	3	27.6 [c]
Doxepin	887	8	0	14.5 [c]
Imipramine	2503	29	2	20.0 [c]
Maprotiline	209	3	0	23.1 [c]
Nortriptyline	347	2	0	9.3 [c]
Other cyclic antidepressant	179	2	0	18.0 [c]
Unknown cyclic antidepressant	142	5	0	56.7 [c]
Cyclic antidepressant with benzodiazepine	288	2	0	11.2 [c]
Cyclic antidepressant with phenothiazine	832	10	0	19.4 [c]
Lithium	1054	7	0	10.7 [c]

(Contd..........)

Table 19.5 Contd..........

Categories and Subcategories of Poisonous Substances	Total Pediatric Exposures (1985-1989)	No. of Major Effects	No. of Deaths	Hazard Factor
Antihistamines	38,390	20	4	1.0
Antimicrobials	122,686	28	3	0.4
Antimalarials	204	2	2	31.6 [c]
Isoniazid	219	3	0	22.1 [c]
Rifampin	84	2	0	38.4 [c]
Antineoplastics	570	0	1	2.8
Asthma therapies	20,502	43	3	3.6 [c]
Aminophylline	8622	38	3	7.7 [c]
Cardiovascular drugs	37,385	182	7	8.1 [c]
Antiarrhythmics	1203	3	0	4.0 [c]
Antihypertensives	8099	139	0	27.6 [c]
Cardiac glycosides	3846	24	2	10.9 [c]
Nitroprusside	47	I	1	68.5 [c]
Cough/cold preparations	249,038	72	4	0.5
Diagnostic agents	697	1	0	2.3
Diuretics	11,175	8	0	1.2
Electrolytes/minerals	50,751	57	8	2.1 [c]
N Iron	11 ,234	52	7	8.5 [c]
Eye/ear/nose/throat preparations	32,805	17	0	0.8
Glaucoma medications	74	1	0	21.8 [c]
Gastrointestinal preparations	99,636	48	3	0.8
Antidiarrheals: diphenoxylate/ atropine	2500	18	1	12.2 [c]
Hormones and antagonists	6357	20	1	0.5 [c]
Insulin	217	2	1	22.3 [c]
Oral hypoglycemics	2609	11	0	6.8 [c]
Miscellaneous drugs	17,650	10	2	1.1
Neuromuscular blocking agents	8	2	0	402.7 [c]
Muscle relaxants	3165	7	0	3.6 [c]
Methocarbamol	341	2	0	9.4 [c]
Other	1012	3	0	4.8 [c]
Sedative/hypnotics/antipsychotic	33,048	153	2	7.6 [c]
Barbiturates: long acting	4475	37	0	13.3 [c]
Barbituarates: short acting	1051	4	0	6.1 [c]
Chloral hydrate	443	18	0	65.5 [c]
Ethchlorvynol	81	1	0	19.9 [c]
Glutethimide	30	2	0	107.4 [c]
Methaqualone	66	1	A	24.4 [c]
Phenothiazines	7451	57	2	12.8 [c]
Other	330	2	0	9.8 [c]
Serums, toxoids, vaccines	448	0	0	0.0
Stimulants/street drugs	21,260	84	1	6.4 [c]

(Contd...........)

Table 19.5 Contd..........

Categories and Subcategories of Poisonous Substances	Total Pediatric Exposures (1985-1989)	No. of Major Effects	No. of Deaths	Hazard Factor
Amphetamines	6409	18	0	4.5 [c]
Cocaine	546	20	0	59.0 [c]
Lysergic acid diethylamide	117	2	0	27.5 [c]
Marijuana	694	3	0	7.0 [c]
Mescaline/peyote	325	2	0	9.9 [c]
Phencyclidine	177	30	0	273.0 [c]
Topicals	175,378	55	2	0.5
Silver nitrate	44	1	0	36.6 [c]
Miscellaneous veterinary	4630	0	0	0.0
Vitamins	145,872	33	1	0.4
Unknown drugs	21,020	38	0	2.9 [c]
Total	**3,852,618**	**2,270**	**122**	**1.0**

[c] Subcategories with hazard factors 23 and statistical significance are listed under each substance category; [b] See text for explanation; [c] $p < .05$, Fisher's Exact Test comparing each individual category (or subcategory) with all other cases.

The ingestion hazards parallel the substances required to have child-resistant closures by the Poison Prevention Act. This demonstrates, however, that the requirement of child-resistant closures does not render a product "child-proof". In 1989, King and Palmisana made an epidemiological study to identify the risk factors responsible for the ineffectiveness of child-resistant closures. Although the Poison Prevention Packaging Act of 1970 has resulted in a 65% decline in the ingestion of products packaged in child-resistant containers, ingestion of prescription drugs by children has declined by only 36%. Reasons for these data include.

1. Availability of non-child-resistant packaging upon consumer request (i.e, consumer noncompliance)
2. Misuse of child-esistant closures by the consumer in the home
3. Transferring medicines :from child-resistant packages to unsafe containers, or using no container at all
4. Unsafe storage practices, such as leaving containers within easy reach; and
5. Violations of the ACT by the dispensing pharmacist, physician, or Health Care Facility. (King and Palmisana, 1989).

Pharmacies and health care facilities have also been shown to be noncompliant with the packaging act of 1970. Reports indicate that between 14% and 44% of pharmacies surveyed were in violation of federal packaging standards.

Table 19.6 demonstrates that 2 year olds accounted for the majority (53 7%) of ingestions of prescription drugs. Table demonstrates that in more than 75% of cases surveyed, non-child-resistant packages or no containers were involved in the ingestion. Table indicates the leading solid prescription drugs ingested that resulted in hospitalization. Table demonstrates that the owner of each prescription was able to be identified in 80% of the cases surveyed. Although parents prescriptions accounted for 53.6% of the ingestions, nearly 30% involved grandparents medications. Noncompliance with child-resistant packaging was a major reason for the exposure as indicated in Table 19.7.

Table 19.6 Ingestions of Solid Prescription Drugs, by Age of Victim

Age Categories (months)	No.	(%)
Infant (< 12 mo)	1	(0.1)
(12-23)	205	(24.2)
(24—35)	456	(53.7)
(36-47)	124	(14.6)
(48-59)	37	(4.4)
(60-71)	25	(2.9)
Unknown	1	(0.1)
Total	849	(100.0)

Table 19.7 Container Type and Ownership of Medication [a]

	Father	Mother	Grand father	Grand mother	Totals
Child-resistant	13	82	7	27	129
No container	23	75	13	40	151
Non-child-resistant	25	79	25	56	185
Totals (%)	61 (13.1%)	236 (50.8%)	45 (9.7%)	123 (26.4%)	465
Total noncompliance (%)	48 (78.6%)	154 (65.3%)	38 (84.4%)	96 (78.0%)	
Parent's noncompliance	202 (68.0%)				
Grandparent's noncompliance	134 (79.8%)				

[a] Excludes other owners of medication (i.e., neighbor, aunt, sibling, other).

Other barriers to pediatric poisonings include the use of warning stickers designed to deter children from getting into containers. However, studies have failed to demonstrate any benefit from their use and, in fact, the warning stickers may attract children who otherwise would hay ignored the product. Woolf and Lovejoy discuss the epidemiology of drug overdose in children and the determinants that result in a high risk of drug poisoning in this group.

19.2.3 Adolescent Toxic Exposures

Pool Bon (1988) indicated that intentional poisoning in adolescents is one of the 10 leading causes of death and potentially productive years of life lost in the U.S. Alcohol use and abuse plays a large role in fatal injuries in this age group.

Approximately 8% of all intentional exposures resulting in death were in children aged 13 to 17 years as illustrated in table 19.9. Drug-related fatalities by drug class adolescents 11 to 17 years old reported to the AAPCC from 1989 to 1991 are listed in table. Of 764 fatalities reported in 1991, 52 (6.8%) were in the adolescent age group; 92% were 13 to 17 years old.

Table 19.8 Drug-Related fatalities by drug class among adolescents 11 to 17 years old reported to the American Association Poison Control Centers in 1989, 1990 and 1991.

Drug class	1989	1990	1991	Total
Cyclic antidepressants	12	7	11	31
Amitriptyline	1	3	1	
Desipramine	4	2	7	
Doxepine	1	0	2	
Imipramine	5	1	0	
Maprotaline	1	0	0	
Nortriptyline	1	1	1	
Calcium channel blockers	2	2	3	
Nifedipine	0	0	1	
Verapamil	2	2	2	
Salicylates	1	0	5	6
Theophylline	3	1	1	5
Propranolol	3	2	1	5
Methamphetamine	1	0	0	1
Cocaine	0	0	1	1
Street drug (unknown type)	1	0	0	1
Paracetamol (acetaminophen)	1	0	2	3
Carbamazepine	1	0	0	1
Haloperidol	0	1	0	1
Glipizide + cyclobenzaprine	0	1	0	1
Isoniazid	0	0	1	1
Lidocaine (lignocaine)	0	1	0	1
Methocarbamol	0	0	1	1
Chlorpromazine	0	0	1	1
Temazepam	0	0	1	1
Colchicine/allopunnol/ibuprofen	0	0	1	1
Amfebutamone (bupropion) + lithium	0	1	0	1
Phenylpropanolamine/chlorphenamine	0	1	0	1
Paracetamol + doxylamine + detromethorphan + pseudoephedrine	0	1	0	1
Totals	26	18	29	72

19.2.4 Poisoning in the Elderly

Poisoning in the elderly is a continuing public health problem. Accidental poisoning, due to dementia and confusion, improper use or storage of a product, and therapeutic errors account for the most exposures of patients >64 years of age. However, approximately 11% were intentional with suicidal intent. The mortality rate from poisoning is much higher in the elderly than in other age groups. Of the 764 fatalities reported in 1991, 18% were 264 years old.

Woolf et al. analyzed poisoning-related hospitalization and mortality rates among older adults in Massachusetts from 1983 to 1985.

19.3 Management Trends

Table outlines the frequency of initial decontamination procedures, measures used to enhance toxin elimination, and antidotes administered to patients reported to the 1991 AAPCC database. The use of Syrup of Ipecac as an initial method of gastrointestinal decontamination has declined in the emergency department setting. It is most often administered to the pediatric patient in a non-health care facility environment. This trend was reversed in adult patients. Table demonstrates a continued decline from 1983 to 1991 in the use of emesis induced by Syrup of Ipecac and an increase in the sole use of activated charcoal administration in the emergency department.

Table 19.9 Therapy Provided in Human Exposure Cases

Therapy	No.
Initial decontamination	
Dilution	653,687
Irrigation/washing	366,557
Activated charcoal	129,203
Cathartic	107,556
Ipepac syrup	95,124
Gastric lavage	58,677
Other emetic	4239
Measures to enhance elimination	
Alkalinization (with or without diuresis)	7092
Hemodialysis	692
Forced diuresis	500
Hemoperfusion (charcoal)	124
Exchange transfusion	106
Acidification (with or without diuresis)	99
Hemoperfusion (resin)	38
Peritoneal dialysis	11
Specific antidote administration	
Naloxone	7136
N-Acetylcysteine (oral)	7075
Atropine	923
Deferoxamine	901
Antivenim	579
Ethanol	540
Hydroxocobalamin	384
N-Acctylcysteine (IV)	250
Pralidoxime (2-PAM)	248
Fab fragments	241
Pyridoxine	227
Physostigmine	226
Dimercaprol (BAL)	146
Methylene blue	117
Cyanide antidote kit	99
EDTA	94
Penicillamine	86

SUGGESTIVE FURTHER READINGS

Goldfrank, L. et al., Goldfran's Toxicologic Emergencies, 5th ed., Appleton & Lange, Norwalk, CT, 1994

Zbinden G., Predictive value of animal studies in toxicology, Regul. Toxicol. Pharmacol., 14, 167, 1991.

Fletcher, R., Clinical Epidemiology: The Essentials, 2nd ed., Williams & Wilkins, Baltimore, MD, 1988.

Klein-Schwartz, W. and Oderda, G., Poisoning in the elderly: Epidemiological, clinical and management considerations, Drugs, and Aging, 1, 67, 1991.

STUDY QUESTIONS

1. Discuss briefly the scope & history of human toxicology.
2. Write in details on the epidemiology of poisoning in children
3. Write briefly on –
   a. Adolescent toxic exposures
   b. Poisoning in the elderly

Chapter 20

NEW TRENDS IN TOXICOLOGY

Science of Toxicology, deals with the study of interactions between chemicals and biological systems in order to quantitatively determine the adverse effects in living organisms and investigate the nature, incidence, mechanism of production, factors influencing their development and reversibility of such adverse effects (Ballantyne, 1989). In formal terms, it appears to be a young science, however, it was conceptualized by a physician known as Paracelsus. Borzellaca (2000) honoured Paracelsus as herald of modern toxicology.

Paracelsus, Philippus Theophrastus Aureolus Bombastus von Hohenheim, the "father of Chemistry and the reformer of "materia medica", the "Luther of Medicine", the "godfather of modern chemotherapy", the founder of medicinal chemistry, the founder of modern toxicology, a contemporary of Leonardo da Vinci, Martin Luther, and Nicholas Copernicus, was born near or in the village of Einsiedeln near Zurich, Switzerland on 10 or 14 November, 1493. Paracelsus studied at a number of Universities in Europe & conferred doctorate in 1516 from the University of Ferrara. It was at this time that he assumed the name Paracelsus (Para: beside, beyond and celsus: a famous Roman physician).

Paracelsus discounted the humoral theory of Galen who postulated that balance amongst four humors in the body (blood, phlegm, yellow and black bile) is essential for health. Paracelsus believed in three humors: salt (representing stability), sulfur (representing combustibility) and mercury (representing liquidity). He defined disease as a separation of one humor from the other two. He propounded the principle of simplitude meaning that "a poison in the body will be cured by a similar poison". He introduced chemistry into medicine. He extended his interest in chemistry and biology to what we now consider toxicology. Basic tenets of Paracelsianism were summed up by Temkin and Coworkers (1996).

Since then, toxicological developments have witnessed new heights. War and prospects of war played a great role in the development of toxicology. In "world war I", a variety of chemicals were used in the battlefields of France. Occupational toxicology originated in 19th century as a product of industrial revolution. Development of chemical and pharmaceutical industries in 19th and 20th century gave birth to regulatory toxicology. Increasing concerns for consumer and environmental health during last three decades brought toxicology to the age of Science. World war II offered stimulus to the evolution of Environmental Toxicology.

20.1 Earlier Developments in Toxicology

Soon the nature and magnitude of toxic effects were studied. Factors viz: physicochemical properties of the substance, its bioconversion, the conditions of exposure, and the presence of bioprotective mechanisms subsequently dominated the scene. Morphological and biochemical injury produced by a toxin was classified as inflammation, necrosis, enzyme inhibition, biochemical uncoupling, lethal synthesis, lipid peroxidation, covalent binding, receptive interaction, immune mediated hypersensitive reactions, genotoxicity, developmental and reproductive toxicity and pharmacological effects.

In 1848, Blake in the United States published his opinion that the biological activity of a salt was due to its basic or its acidic component and not to be whole salt; as with lead nitrate it was the lead moiety and not the acetate or nitrate part. This was, for Ist time, a daring thought, because it was not until 1884 Arrhenius introduced his theory of electrolytic dissociation. The Scottish authors CrumBrown and Fraser (1869) made a major discovery. They wrote, *"there can be no reasonable doubt that a relation exists between the physiological action of a substance and its chemical composition and constitution"*, understanding by the latter term, the mutual relations of the atoms in the substance. This discovery was the first to show structure-action relationship at the turn of present century. Ernest Overton and Hans Meyer independently put forward a, "Lipoid theory of cellular Depression". This stated that chemically inert substances exert depressant properties on cells (particularly those of central nervous system) that are rich in lipids and that higher the partition coefficients, the greater the depressent action. The idea that drugs act upon receptors began with John Langley in (1878) in Cambridge. Later, Langley coined the term 'receptive substances'. Paul Ehrlich was already using the term receptor in Germany. In his Noble prize address Ehrlich outlined the receptor as a small chemically defined area, which was normally occupied with the cell's nutrition and metabolism but which could take up specific antigens or drugs instead (Ehrlich, 1908). First the idea of receptor was received with skepticism because of repeated failure to isolate any such substance. However, the idea of receptors became more firmly established by the work of Alfred Clark who showed that combination of drug with a receptor quantitatively followed the law of mass action. He summed up his work in a monograph (Clark, 1937), a few years before his death in 1941.

The period of second worldwar (1939-1944) was a turning point in the study of structure action relationship.

There was a period when dose response relationships were highly predominant. Development of physiology and biochemistry also influenced the growth of toxicology. Metabolism of substances was conveniently classified as phase – I and phase – II reactions.

20.1.1 Concept of QSAR

The concept of QSAR (quantitative structure activity relationships) was applied to study the toxicity of inorganic cations. While molecules of organic compounds reflect their properties as a whole, the inorganic compounds dissociate in various degrees and properties have thus to be attributed to anions, cations or undissociated molecules (Free and Wilson, 1964). Inorganic cations can form complexes with inorganic or organic ligands contributing new properties to the complex. Components of this system (cations, anions, undissociated molecules) could mutually influence each other depending upon the ratio amongst components. Quantitative relationships between a chemical structure of the complex and the biological activity formed a new line of action in toxicology (Tichy & Krivucova, 1976).

QSAR studies generated the concept of molecular connectivity in order to characterize the organic biologically effective substances. The index of connectivity is deduced from the numeric evaluation of the extent of branching of chemical bonds in the section of the molecule. There exists a correlation between the connectivity index and toxicity of cations.

20.2 Modern Toxicology

In recent years, toxicology has developed from an activity relying principally on the tools of classical pathology to observe and classify harmful effects so as to become a discipline of increasing ability to explain the effects of toxic compounds in molecular and mechanistic terms.

Over the last several years, it has become apparent that many environmental toxicants exert their effects by the action or disruptron of specific signaling pathways, ultimately resulting in alterations in gene expression. With the completion of the human genome project and the advent of many powerful new technologies, there has been a revolution in our understanding of these mechanisms at molecular level. Toxicant-induced alterations in gene expression depend on receptors. Four receptors namely the Ah receptor (AhR, the constitutive Androstone Receptor (CAR), the Pregnane X Receptor (PXR), and the peroxisome Proliferator Activated Receptor (PPAR) mediate the toxicity of four broad classes of chemicals. In contrast to these specific receptor mechanisms, metals exert their toxicity through both stress response pathways and specific metal responsive transcription factors. Role of tissue selective transcription factors on the expression of xenobiotic metabolizing enzymes is now being investigated in several laboratories.

Recent developments show that toxicology is not merely a study of the effects of a variety of poisons in animals, plants and man but a multidisciplinary science embracing pathology, pharmacology, cell biology, biochemistry and public health. While dwelling with the subject for about three decades, I could witness the sustained progress made by science of toxicology. Earlier developments got smoothly integrated into modern concepts of toxicology. This communication is an attempt to review the present status of toxicology.

20.2.1 Toxicogenomics

It is broadly defined as gene and protein expression technology that addresses pertinent issues of toxicology. The term genome has been traditionally used to define the haploid set of chromosomes in the nuclei of multicellular organisms. The study of genome is referred as genomics. The patterns by which genes and their protein products act in concert to affect function is known as functional genomics. Certain environmental stimuli will perturb the normal cellular functions of proteins and cause changes in gene expression. These kinds of environmental factors can also lead to the pathology of disease. Often the development of disease will be the result of complex mix of factors including inherent genetic susceptibility and a series of environmental changes or challenges.

The term proteome was coined in 1994 by Mark Wilkins (Garrels., 2001). It refers to total protein repertoire able to be expressed from a given genome. A proteome of a cell, tissue, or organ is not only different, it can express differently under particular set of conditions. Thus toxicogenomics appear challenging.

20.2.2 Metabonomic Technology

Toxicants by definition, disrupt the normal composition and flux of endogenous biochemicals in, or through, key intermediate cellular metabolic pathways. These disruptions, either directly or indirectly, alter the blood that percolates through the target tissues. The diagnostic utility of any one trace biomolecule is limited by the number of variables affecting its concentration *in situ* and by the common biochemical processes disrupted by toxicants. However, if a significant member of trace molecules are monitored, the overall pattern or "figerprint" produced may be more consistent and protective than any other marker. This comprehensive information can be obtained from high field nuclear magnetic resonance (NMR) spectroscopy coupled with pattern recognition technology. Magic angle spinning NMR technology enables similar information to be garnered from tissues as well. Temporal evaluation of metabolic consequences of toxicity, coupled with genomic and proteomic technologies and metabonomics permit complete assessment of toxicity from genotype through phenotype.

20.2.3 Pharmacogenetics

Pharmacogenetics, a term originally coined in the 1950s may now be viewed as the study of correlations between an individual's genotype and the same individuals ability to metabolize an administered drug or compound. Genotypic variations, often in the form of single nucleotide polymorphism (SNPs), exists for many of the enzymes that metabolize drugs/chemicals. Extensive metabolism of a drug is a general characteristic of the normal population. Poor metabolism which typically is associated with excessive accumulation of specific drugs or active metabolic products is a recessive trait requiring a functional change, such as frameshift or splicing defect in both copies of the relevant gene. *Ultraextensive metabolism* which may have the effect of diminishing a drug, apparent efficacy in an individual, is generally an autosomal dominant state derived from a gene duplication or amplification. Some representative example of genetic polymorphism that affect drug metabolism are shown in table 20.1. For example in cancer chemotherapy, several common drugs show wide polymorphism related metabolic variations with 30 fold or greater interindividual variability.

Table 20.1 Model organisms, gene names, and human disease

Human disease	Mice	Humans
Congenital heart defects Tinman	Csx/NKx2.5	Csx/NKx2.5
Huntington's polyglutamine repeats cause neural degeneration	Polyglutamine repeats	Polyglutamine repeats
Alzheimer's APPL	APP	APP
Nmyotrophic lateral sclerosis SOD	SOD1	SOD1
Cancer Cdx(2) caudal	cdx2	cdx2
Diabetes insulin receptor	Insulin receptor (knockout mice)	Insulin receptor
Retinitis pigmentosa rhodopsin	Rhodopsin	Rhodopsin
NF MERLIN	NF2	NF2
Ataxia telangiectasia mei-41	ATM	ATM

APP, amyloid precursor protein: SOD, superoxide dismutase, NF, Neurofibromatosis.

20.2.4 Molecular Toxicology

Apoptosis is a natural consequence *in vivo* and there is now substantial evidence that apoptosis plays an important role in the toxic effects of a number of drugs and chemicals. Numerous coherent pathways regulate cell suicidal process. Target organ toxicities, target organ apoptogenic drugs and chemicals, regulation of apoptosis at organs, cellular and subcellular and molecular levels emerges a new discipline. Since oxidative stress, caspases, caspase activated DNAse, reactive oxygen species, mitochondrial and cell cycle related events are known to modulate this process, their respective roles are under investigation in several laboratories.

In-excitable and non-excitable tissues are the direct and indirect targets of many xenobiotics that produce apoptotic and necrotic cell death. Determination of the temporal and sequential relationships between the opening of the mitochondrial permeability transition (PT) pore, mitochondrial depolarization and swelling, cytochrome C release and caspase activation during cell death are critically important. In toxicology, understanding the role and molecular mechanisms of PT pore opening will allow the development of pharmacological and genetic strategies to prevent inappropriate apoptosis as well as to initiate and control the apoptotic process for therapeutic purposes. The current evidence suggests that the PT pore is a complex of the voltage dependent anion channel (VDAC), adenine nucleotide translocase (ANT) and cyclophilin-D (CyP-D), formed at connect sites between the inner and outer mitochondrial membranes.

20.2.5 Concept of Biomarkers

The emergence of specific biomarkers offer the promise of being able to measure signals and/or events that reflected more accurately the biology associated with exposure, effects, and susceptibility. The 1983 NRC publication formalized human health risk assessment into a four component process namely exposure, assessment, hazard identification, dose response assessment and risk characterization.

20.2.6 Chronotoxicology

Biological rhythms are toxicologically important because they have a positive or negative effect on all measures of normal physiological functions, and health of the individual. Circadian patterns affect the absorption of drug/toxin. Once the drug/toxin has been absorbed, it is transported to its tissue/target sites and its elimination sites. Large circadian variations have been shown in humans and rats in plasma proteins binding of a variety of drugs. In addition, drug/toxin transport can occur by their binding to red blood cells. In general, lipophilic materials pass into blood cells more rapidly than hydrophilic materials. Circadian variation with respect to drug permeability of the blood-brain barrier is of interest. Circadian rhythms in heavy metal toxicology have been described for mercury and cadmium (Camber et al. 1992a). The level of liver microsomal benzene hydroxylase activity is highest at a particular time of the day. Carcinogenicity and teratogenicity of chemicals have also been found to be affected by circadian rhythms (Senerbier, 1992). Circadian susceptibility in mammals to organophosphate pesticides have been well described (Mayersbach, 1974).

Circadian time structure is not routinely considered in toxicity testing in human or preclinical (rodent) models. Actually, they are not the only rhythms which modulate the outcome following drug or chemical exposure. Other cycles, such as fertility cycle and seasonal cycles markedly and reproducibly alter toxicity profiles.

In summary, considering toxicology in the absence of these three factors within the biological time structure of living animals seems to be uneconomical, misleading and unwise.

20.2.7 Superinteractions

It is now well recognized that human environmental exposures are not to single chemical. Rather humans are exposed concurrently or sequentially to multiple chemicals, by various routes of exposure and from a variety of sources. The process of carcinogenesis can be modified significantly by other chemicals. The term cocarcinogenesis was initially defined as the enhancement of neoplasm induction brought about by new carcinogenic factors, which act in conjunction with an initiating carcinogen (Sivak, 1979). Whereas the additive or synergistic effects of two or more carcinogens in neoplasm production has been defined as syncarcinogenesis (Nakahara, 1970). When the toxic responses grossly exceed the expected response after an ideal substantial, the process is called as superinteraction. An important example of an environmental 'superinteraction' is that of chlordecone (CD) and CCl_4. In this case a prior 15 days exposure to CD enhanced the acute toxicity of CCl_4 in male rats by 67-fold (Mehendale, 1991). This physiological/biochemical framework within which extremely potent interactions could occur is important in planning screening programmes or to predict superinteractions in toxicology/pharmacology.

A brief review of studies made by the science of *Toxicology* from the times of its founder *Paracelsus* to modern times as presented in this communication might attract young workers to this wonderful discipline of science.

SUGGESTIVE FURTHER READINGS

Ballantyne, B. (1989). *Toxicology*. In "Encyclopedia of Polymer Science and Engineering Vol. 16. pp. 879-930. Wiley, New York.

Temkin, C.L., Rosen, G., Zilboorg, G., and Sigerist, H.W. (1996). Four treatises of Theophrastrus von Hohenheim called Paracelsus (H.E. Sigerist eds.) The John Hopkins University Press, Baltimore Manyland.

Adrien Albert (1848). Selective toxicity. Chapman and Hall, London & New York 1985.

STUDY QUESTIONS

1. Define toxicology and discuss earlier developments in toxicology.
2. Write briefly on the following –
   a. Toxic genomics
   b. Pharmacogenetics
   c. Molecular toxicology
   d. Chronotoxicology

Chapter 21

CHEMICAL WARFARE AGENTS

United Nations (1969) report defines chemical warfare (CW) agents as "....chemical substances, whether gaseous, liquid or solid, which might be employed because of their direct toxic effects on man, animals and plants ...".

The Chemical Weapons Convention defines chemical weapons - including not only toxic chemicals but also ammunition and equipment for their dispersal. Toxic chemicals are stated to be "... any chemical which, through its chemical effect on living processes, may cause death, temporary loss of performance, or permanent injury to people and animals".

Toxins produced by living organisms and their synthetic equivalents, may be called as chemical warfare agents if they are used for military purposes. However, they have a special position since they are covered by the Biological and Toxin Weapons Convention, 1972. This convention bans the development, production and stockpiling of such substances not required for peaceful purposes.

Today, thousands of poisonous substances are known but only a few are considered suitable for chemical warfare. About 70 different chemicals have been used or stockpiled as CW agents during the 20th century. Today, only a few of these are considered of interest owing to a number of demands that must be placed on a substance if it is to be used as a CW agent. These are:

- A presumptive agent must not only be highly toxic but also "suitably highly toxic" so that it is not too difficult to handle.
- The substance must be capable of being stored for long periods in containers without degradation and without corroding the packaging material.
- It must be relatively resistant to atmospheric water and oxygen so that it does not lose effect when dispersed.
- It must also withstand the heat developed when dispersed.

21.1 War Gases

CW agents are frequently called war gases and a war where CW agents are used is usually called a gas war. These incorrect terms are a result of history. During the First World War chlorine and phosgene which are gases at room temperature and normal atmospheric pressure were used. The CW agents used today are only exceptionally gases. Normally they are liquids or solids. However, a certain amount of

258

the substance is always in volatile form (the amount depending on how rapidly the substance evaporates) and the gas concentration may become poisonous. Both solid substances and liquids can also be dispersed in the air in atomized form, so-called aerosols. An aerosol can penetrate the body through the respiratory organs.

Some CW agents can also penetrate the skin. Solid substances penetrate the skin slowly unless they happen to be mixed with a suitable solvent.

21.1.1 Classification of War Gases

CW agents can be classified in many different ways. There are, for example, volatile substances, which mainly contaminate the air, or persistent substances, which are involatile and therefore mainly cover surfaces.

CW agents mainly used against people may also be divided into lethal and incapacitating cathegories. A substance is classified as incapacitating if less than 1/100 of the lethal dose causes incapacitation, e.g., through nausea or visual problems. The limit between lethal and incapacitating substances is not absolute but refers to a statistical average. In comparison, it may be mentioned that the ratio for the nerve agents between the incapacitating and lethal dose is approximately 1/10. Chemical warfare agents can also be classified according to their effect on the organism.

In order to achieve good ground coverage when dispersed from a high altitude with persistent CW agents the dispersed droplets must be sufficiently large to ensure that they fall within the target area and do not get transported elsewhere by the wind. This can be achieved by dissolving polymers (e.g. polystyrene or rubber products) in the CW agent to make the product highly-viscous or thickened.

Although it may appear that a CW agent can be "custom-made" for a certain purpose, this is not the case. Instead, there is always some uncertainty about the persistence time, the dispersal and the effect.

Toxicities of mustared gas, nerve agents, lewisite, phosgene and hydrogen cyanide are discussed below:

21.2 Mustard Gas

Mustard gas is a chemical warfare agent belonging to the blister agent/vesicant class. It is a cytotoxic alkylating compound similar to the other type of vesicants or blister agents such as nitrogen mustard, lewisite, and phosgene oxide. Mustard causes blistering of the skin and mucous membranes.

Actually, mustard gas is not a gas, but a dense, yellow to brown oily liquid with a low vapor pressure and relatively high melting point. Mustard can also be released in the air as a vapor and thus exposure of skin, eyes, and respiratory tract can be to vapor or liquid. If sulfur mustard is released into water supplies, exposure can occur from drinking contaminated water or getting it on the skin. Although it currently has no medical use, it was available for use as a treatment for psoriasis. Mustard may smell like garlic, onions, mustard, or have no odor. The chemical properties are summarized in the following box:

> · CHEMICAL NAME: bis (2-Chloroethyl) sulfide
> · CHEMICAL ABSTRACTS SERVICE REGISTRY NUMBER: C A S
>   505-60-2
> · SYNONYMS: Sulfur mustard; 1,1-Thiobis(2-chloro-ethane); 1-Chloro-
>   2-({3-chloroethylthio) ethane; 2,2' -Dichlorodiethyl sulfide; Distilled
>   mustard; S mustard; S-lost; Schwefel-Iost; Yellow cross liquid;
>   Yperite; Kampstoff lost; HD; HT; H
> · CHEMICAL/PHARMACEUTICAL/OTHER CLASS: Thio-ether
> · CHEMICAL FORMULA: C_4H_gChS
> · CHEMICAL STRUCTURE:
> CL-CH_2-CH_2-S-CH_2-CH_2Cl

21.2.1 Toxicokinetics

Mustard gas is lipophilic and will accumulate in brain and fatty tissue. Mustard gas is soluble in water to less than 0.1 % and hydrolyzes in water with a half-life of 3-5 min. It is freely soluble in organic solvents. It has been detected in the blood after dermal or inhalation exposure. Although mustard dis-solves slowly in aqueous solution, it must first dissolve in sweat or extracellular fluid to be absorbed. Following dissolution, mustard molecules rapidly rearrange to form extremely reactive cyclic ethylene sulfonium ions that immediately bind to intracellular and extracellular enzymes, proteins, and other cellular components. Mustard binds irrever-sibly to tissues within several minutes after contact. If decontamination is not done immediately after ex-posure, injury cannot be prevented. However, later decontamination might prevent a more severe lesion.

The biological half-life of the chemical has not been published, but products are excreted in the urine for several days after acute exposure. Traces of the agent are exhaled and excreted in the feces. Several hours can pass before symptoms become manifest but this is attributed to the mechanism of action and not to direct effects of residual levels of the agent. Mustard gas is a greater threat in hot and humid climates.

21.2.2 Mechanism of Toxicity

Although the exact mechanism by which mustard pro-duces tissue injury is not known, the mechanism of action has been suggested to be its ability to directly alkylate DNA. This DNA alkylation and cross-linking in rapidly dividing cells such as basal keratinocytes, mucosal epithelium, and bone marrow precursor cells leads to cellular death and inflammatory reactions. Systemic effects with extensive exposures include bone marrow inhibition, with a drop in the white blood cell count, and gastrointestinal tract damage.

21.2.3 Human Toxicity

Mustard gas is a powerful irritant and vesicant. Dermal effects range from itching to erythema, blistering, corrosion, and necrosis. Dermal blistering is delayed in onset and slow to heal. Ocular irritation, conjunctivitis, and blindness can occur. Respiratory symptoms include a harsh and painful cough, bronchitis, sneezing, rhinorrhea, and sore throat. Mustard gas is not acutely lethal; 2-3% of soldiers exposed to mustard gas during World War I died of its direct effects. Death is generally due to respiratory collapse, shock, and secondary infections. Survivors may be susceptible to bronchitis and pneumonia, and may be at an increased risk to develop tumors of the respiratory tract. Mustard gas produces tumors in animal studies.

The International Agency for Research on Can-cer has classified sulfur mustard as carcinogenic to humans (Group I). Epidemiological evidence indicates that repeated exposures to sulfur mustard may lead to cancers of the upper airways. There is limited evidence that repeated exposures to sulfur mustards may cause defective spermatogenesis years after exposure. Sulfur mustard has been implicated as a potential developmental toxicant because of its similarity to nitrogen mustard; however, data are inconclusive.

21.2.4 Chronic Toxicity

Production workers exposed to mustard gas had increased rates of bronchitis and pneumonia. Increased incidence of tumors of the respiratory tract, lung cancer, bladder cancer, and leukemia were also found. Mustard gas is classified as a human carcmogen.

21.2.5 Clinical Management

It is important that the agent be washed from the skin with soap and water as soon after exposure as possible. The eyes should be thoroughly flushed. The onset of symptoms typically is delayed and an ab-sence of immediate effect does not rule out toxicity. Although ingestion is unlikely, due to the sources of mustard gas, an emetic should not be administered because of the extreme caustic nature of the chemical. If the patient is not comatose, dilution of stom-ach contents with milk or water, prior to gastric lavage, may be attempted. Application of a solution of sodium thiosulfate to the skin and inhalation of a nebulizing mist of sodium thiosulfate may speed inactivation of mustard gas. Animal studies have shown that administration of corticosteroids (e.g., dexamethasone) and antihistamines (e.g., prom-ethazine) may prove beneficial.

21.2.6 Animal Toxicity

The irritant, vesicant, and respiratory effects of mustard gas are the same in animals and humans, although the hair coat and lack of extensive sweat glands may somewhat protect the animal from the dermal effects of the agent.

21.3 Nerve Agents

Casualties are caused primarily by inhalation; however, they can occur following percutaneous and ocular exposure, as well as by ingestion and injection.

- PREFERRED NAME: Nerve agents. Also known as nerve gases
- SYNONYMS: Tabun (GA; CAS 77-81-6); Sarin (GB; CAS 107-44-8); Soman (GD; CAS 96-64-0); Cyclosarin (GF; CAS 329-99-7); G agents; Organophos-phates (OP): VX (CAS 20820-80-8); V agents; Chemical warfare agents; Irreversible cholinesterase inhibitors; Anticholinesterase compounds
- DESCRIPTION: Nerve gases are clear liquids; therefore, the term 'gas' is a misnomer. The preferred term is 'nerve agents'. Because of the chiral phosphorus in their structure, the nerve agents contain various steroisomers. So man, sarin, tabun, and VX contain equal amounts of (+) and (-) enantiomers. The G agents are volatile and thus present both vapor and liquid hazard. In decreasing order of volatility are sarin, soman, tabun, and VX. VX presents a negligible vapor hazard, but its volatility increases with increasing temperature. At temperatures above 4^0 C it also presents a vapor hazard
- CHEMICAL STRUCTURES:

$$R_1\diagdown\quad \diagup\!\!\diagup O$$
$$P$$
$$R_2O\diagup\quad \diagdown X$$

21.3.1 Toxicokinetics

Nerve agents are absorbed both through the skin and via respiration. Because VX is an oily, nonvolatile liquid it is well absorbed through the skin (persistent nerve agent), although it can also be absorbed by inhalation. Thus, VX is more of a percutaneous threat than by inhalation, whereas the G agents (nonpersistent), which are also liquids, pose more of an inhalation hazard because of their vapor pressure. Sarin (GB) is the most volatile, but evaporates less readily than water, while cyclosarin (GF) is the least volatile of the G agents.

Nerve agents are hydrolyzed by the enzyme organophosphate (OP) hydrolase. The hydrolysis of GB, soman (GD), tabun (GA), and diisopropyl fluorophosphate occurs at approximately the same rate. The isomers of the asymmetric OPs may differ in overall toxicity, rate of aging, rate of cholinesterase inhibition, and rate of detoxification. The rates of detoxification differ for different animal species and routes of administration. The onset of effects from nerve agents depends on the route, duration, and amount of exposure. The effects can occur within seconds to several minutes after exposure. There is no latent period following inhalation exposure of high concentrations where loss of consciousness seizures has occurred within 1 min. At low concentrations; however, miosis, rhinorrhea, and other effects may not begin for several minutes. Maximal effects usually occur within minutes after contamination ceases.

21.3.2 Mechanism of Toxicity

The nerve agents inhibit the enzymes butrylcholi-nesterase in the plasma, acetylcholinesterase on the red blood cell, and acetylcholinesterase at cholinergic receptor sites in tissues. These three enzymes are not identical. Even the two acetylcholinesterases have slightly different properties, although they have a high affinity for acetylcholine. The blood enzymes reflect tissue enzyme activity. Following acute nerve agent exposure, the red blood cell enzyme activity most closely reflects tissue enzyme activity. However, during recovery, the plasma enzyme activity more closely parallels tissue enzyme activity.

Following nerve agent exposure, inhibition of the tissue enzyme blocks its ability to hydrolyze the neurotransmitter acetylcholine at the cholinergic receptor sites. Thus, acetylcholine accumulates and continues to stimulate the affected organ. The clin-ical effects of nerve agent exposure are caused by excess acetylcholine.

The binding of nerve agent to the enzyme is con-sidered irreversible unless removed by therapy. The accumulation of acetylcholine in the peripheral and central nervous systems (eNS) leads to depression of the respiratory center in the brain, followed by pe-ripheral neuromuscular blockade causing respiratory depression and death.

The pharmacologic and toxicologic effects of the nerve agents are dependent on their stability, rates of absorption by the various routes of exposure, distri-bution, ability to cross the blood-brain barrier, rate of reaction and selectivity with the enzyme at specific foci, and behavior at the active site on the enzyme.

Red blood cell enzyme activity returns at the rate of red blood cell turnover, which is ~ 1 % per day. Tissue and plasma activities return with synthesis of new enzymes. The rates of return of these enzymes are not identical. However, the nerve agent can be removed from the enzymes. This removal is called reactivation, which can be accomplished therapeutically by the use of oximes prior to aging. Aging is the biochemical process by which the agent-enzyme complex becomes refractory to oxime reactivation. The toxicity of nerve agents may include direct ac-tion on nicotine acetylcholine receptors (skeletal muscle and ganglia) as well as on muscarinic acetyl-choline receptors and the central nervous system.

These include the effects of nerve agents on y-amino-butyric acid neurons and cyclic nucleotides. In addition, changes in brain neurotransmitters, such as dopamine, serotonin, noradrenaline, as well as acetyl-choline, following inhibition of brain cholinesterase activity, have been reported. These changes may be due in part to a compensatory mechanism in response to overstimulation of the cholinergic system or could result from direct action of nerve agent on the enzymes responsible for noncholinergic neurotransmission.

21.3.3 Human Toxicity

Rhinorrhea may precede miosis as the first indication of exposure to even small amounts of nerve agent vapor. After exposure to high concentrations/ doses by any route, rhinorrhea occurs as part of the generalized increase in secretions. Direct ocular contact to nerve agents may cause miosis, conjunctival injection, pain in or around the eyes, and dim or blurred vision.

Acute exposure of 3 mg min m^{-3} of GB vapor will produce miosis in most of the exposed population. Other routes of exposure may not cause any eye effects or cause a delayed onset of them, but will cause vomiting, sweating, and weakness.

The onset of miosis is within seconds to minutes following aerosol or vapor exposure but may not be maximal for up to 1 h, especially at low concentrations. The duration of miosis varies and is dependent on the extent of exposure. The ability of the pupil to dilate maximally in darkness may not return for up to 6 weeks. There is no correeiation between miosis and blood cholinesterase levels.

Respiratory distress also occurs within seconds to minutes following vapor exposure. The symptoms include tightness of the chest, shortness of breath, and gasping and irregular breathing leading to apnea. Bronchoconstriction and bronchial secretions contribute to this. With larger concentrations, cyanosis and audible pulmonary changes occur, which can only be relieved by therapeutic intervent-ion. Death due to nerve agent intoxication is attributable to respiratory failure resulting from bronchoconstriction, bronchosecretion, paralysis of skeletal muscles, including those responsible for respiration, and failure of the central drive for respiration. Nerve agent intoxication causes skeletal muscles to fasciculate, twitch, and fatigue prior to paralysis.

The cardiovascular effects of nerve agent exposure are variable. Bradycardia may occur via vagal stimulation, but other factors such as fright, hypoxia, and adrenergic stimulation, secondary to ganglionic stimulation may produce tachycardia or hyperten-sion. Following inhalation exposure to large amounts of nerve agent, the CNS effects will cause loss of consciousness, seizure activity, and apnea within 1 min.

Following skin contact with large amounts of liq-uid, the dermal effects may be delayed up to 30 min. Long-term exposure to an OP, diisopropyl phosphorofluoridate, used in the treatment of myasthenia gravis, caused side effects including nightmares, con-fusion, and hallucinations.

21.3.4 Animal Toxicity

Small doses of nerve agents can produce tolerance. The cause of death is attributed to anoxia resulting from a combination of central respiratory paralysis, severe bronchoconstriction, and weakness or paral-ysis of the accessory muscles for respiration.

Signs of nerve agent toxicity vary in rapidity of onset and severity. These are dependent on the specific agent, route of exposure, and dose or concen-tration. At the higher doses or concentrations, convulsions, apnea, and neuropathies are indications of CNS toxicity. Following nerve agent exposure, animals exhibit hypothermia resulting from the cho-linergic activation of the hypothalamic thermoregulatory center. In addition, plasma levels of pituitary, gonadal, thyroid, and adrenal hormones are increased during OP intoxication. The nerve agents are anticholinesterases and as such inhibit the cholinesterase enzymes in the tissues resulting in the accumulation of acetylcholine at its various sites of action in both the autonomic nervous system and the CNS. These include the endings of the parasympathetic nerves to the smooth muscles of the iris, ciliary body, bronchial tree, gastrointestinal tract, bladder, blood vessels, the secretory glands of the respiratory tract, the cardiac muscles, and the endings of sympathetic nerves to the sweat glands. Accumulation of acetylcholine at these sites results in characteristic muscarinic signs and symptoms, while the accumu-lation at the endings of the motor nerves to voluntary muscles and in the autonomic ganglia results in nicotinic signs and symptoms. The accumulation of acetylcholine in the brain and spinal cord results in the characteristic CNS signs and symptoms.

Nerve agents inhibit the activity of acetylcholine-sterase by attaching to its active sites so that it cannot hydrolyze the neurotransmitter acetylcholine into choline, acetic acid, and regenerated enzyme. Thus, acetylcholine cannot attach to the enzyme, is not hydrolyzed, and continues to produce action poten-tials until the mechanism is fatigued. The biological effects of the nerve agents result from the excess of acetylcholine.

Although there is a lack of information on the general toxicological effects of low-level, and sublethal repeated exposures, there are studies on the behavioral effects of such exposures to nerve agents in animals free of observed signs of intoxication. These were conducted in an effort to determine whether behavioral studies can provide markers of early neurotoxicity that are more sensitive than ne-urochemical and neuropathological changes.

21.3.5 Clinical Management

Following exposure the victim should be removed from the area to avoid further contamination and decontaminated (water/hypochlorite) by adequately protected (protective clothing and gas mask) and trained attendants. Contaminated clothing should be removed carefully so as to avoid further contamina-tion. Respiration should be maintained and drug and supportive therapy instituted. If exposure is anticipated, pretreatment with carbamates (pyridostigmine bromide) may protect the cholinesterase enzymes before GD and possibly GA exposures, but not for GB and VX exposures. The three types of therapeutic drugs to be administered following nerve agent ex-posure are (1) a cholinergic blocker, anticholinergic or cholinolytic drug such as atropine; (2) a re-activator drug to reactivate the inhibited enzyme, such as the oxime pralidoxime chloride; and (3) an anticonvulsant drug such as diazepam or benzodiaze-pine. Oxygen may be indicated in respiratory failure.

21.4 Lewisite

Lewisite was synthesized in 1918 by Dr. Wilford Lee Lewis as a vesicant for chemical warfare. Its production was too late to use in World War 1. It can be used with mustard to lower the freezing point of the mixture for ground dispersal and spraying.

Other organic arsenical chemical warfare agents are methyldichlorarsine (MD), phenyldichloroarsine (PD), and ethyldichloroarsine (ED). These plus lewi-site (L), mustard agents and phosgene oxime make up the vesicant chemical warfare agents.

> · CHEMICAL ABSTRACTS SERVICE REGISTRY NUMBERS:
> CAS 541-25-3 (Lewisite 1: 2-chlorovinyldichlo-roarsine); CAS 40334-69-8 (Lewisite 2: (2-chlorovinyl)chloroarsine); CAS 40334-70-1 (Lewisite 3: Tris(2-chlorovinyl)arsine)
> · SYNONYMS: Arsine; Arsonous; Dichloride; Arsine, L
> · CHEMICAL/PHARMACEUTICAL/OTHER CLASS: Blister agent
> Nesicant class of chemical warfare agents
> · CHEMICAL FORMULA: $C_2H_2AsCl_3$
> · CHE IICAL STRUCTURE: CICH = HC-AsCb

21.4.1 Exposure Routes and Pathways

Lewisite is an oily, colorless liquid that can appear amber to black in its impure form. It has the odor of geraniums. It is more volatile than the mustard agents. Lewisite in the air can cause damage to the eyes, skin, and airways by direct contact. Lewisite in water can lead to exposures from drinking water or from skin contact, and lewisite-contaminated food can be ingested. Lewisite remains as a liquid under a wide range of environmental conditions, from below freezing to very hot temperatures.

21.4.2 Toxicokinetics

Although the exact mechanism of biological activity is unknown, the trivalent arsenic in lewisite combines with the thiol groups in many enzymes.

21.4.3 Mechanism of Toxicity

Lewisite is readily absorbed from the skin, eyes, and respiratory tract, as well as after ingestion and through wounds. It causes blistering on the skin and mucous membranes on contact. After absorption, it causes an increase in capillary permeability, which produces hypovolemia, shock, and organ damage. Unlike the mustard agents, lewisite vapor or liquid causes immediate pain or irritation although lesions require up to 12 h to become full-blown cases.

21.4.4 Human Toxicity

Nasal irritation by lewisite begins at ~ 8 mg min m^{-3} and its odor is detected at ~20 mg min m^{-3}. Vesica-tion and death from lewisite inhalation is caused at the same concentration as mustard, which is 1500 mg min m^{-3}. The immediately dangerous to life health (IDLH) value of lewisite is 0.003 mg m^{-3}. Lewisite causes vesication at ~ 14 mg and the LD_{50} is 2.8 g on the skin.

Within 5 min after contact with liquid lewisite, a grayish area of dead epithelium is produced. Erythema and blister formation follows more rapidly that it does with mustard even though the full-blown lesion does not develop for 12-18 h. The lesion has more tissue necrosis and tissue sloughing than does a mustard agent lesion.

On the eyes, lewisite causes pain, tearing, and blepharospasm on contact. Edema of the conjunctiva and lids follows and the eyes may be swollen shut within an hour. Iritis and corneal damage may also occur. Within minutes, liquid lewisite causes severe eye damage on contact. Upon inhalation, the airway mucosa is the primary target and the damage progresses down the airways with pseudomembrane formation. Pulmonary edema may complicate exposure to lewisite. Runny nose, sneezing, hoarseness, bloody nose, sinus pain, shortness of breath, and cough also occur on inhalation. Lewisite causes an increase in permeability of systemic capillaries resulting in intravascular fluid loss, hypovolemia, shock, and organ congestion. This has been termed 'Lewisite shock' or hypotension. This also leads to hepatitis or renal necrosis with more prominent gastrointestinal effects of diarrhea, nausea, and vomiting.

The long-term effects of lewisite exposure do not include extensive skin burning as is seen with the mustard agents, but chronic respiratory disease may occur. Also, unlike the mustard agents, suppression of the immune systems does not occur, but extensive eye exposure may cause permanent blindness.

21.4.5 Clinical Management

To prevent or lessen lewisite damage, early decontamination within minutes after exposure must be instituted. Unlike mustard, lewisite does not cause damage to the hematopoietic organs, but fluid loss from increased capillary permeability necessitates careful attention to fluid balance.

British anti lewisite (BAL) or dimercaprol was developed as an antidote for lewisite. It is used in medicine as a chelating agent for heavy metals. Although BAL can cause toxicity itself, evidence suggests that BAL in oil administered intramuscularly will reduce the systemic effects of lewisite. BAL skin and ophthalmic ointment decrease the severity of skin and eye lesions when applied immediately after early decontamination, but neither of these ointments is currently manufactured.

21.5 Phosgene Oxime

Phosgene oxime was originally developed as a chem-ical warfare agent. It is sometimes grouped with the vesicant agents, but it is not a true vesicant in that it does not induce blisters. Phosgene oxime is an urticant or nettle agent that causes corrosive-type injuries. There is no evidence that this agent has ever been used as a chemical warfare agent.

- CHEMICAL ABSTRACTS SERVICE REGISTRY NUMBER:CAS 1794-86-1
- SYNONYMS: Dichloroformoxime; CX
- CHEi'vIlCAL/PHARMACEUTICAUOTHER CLASS: Urticant or nettle agent
- CHEMICAL FORMULA: CChNOH
- CHEMICAL STRUCTURE:

$$Cl-C=N-OH$$
$$|$$
$$Cl$$

21.5.1 Exposure Routes and Pathways

Phosgene oxime is a colorless solid or yellowish-brown liquid that can vaporize at room temperature. Due to its ability to rapidly change physical state, phosgene oxime can be absorbed through inhalation, dermal/ocular contact, or oral ingestion.

21.5.2 Toxicokinetics

Phosgene oxime appears to act directly and there have been no reported studies that have determined if any metabolism of phosgene *in vivo*.

21.5.3 Mechanism of Toxicity

The molecular mechanism of phosgene oxime toxic-ity is unknown.

21.5.4 Acute and Short-Term Toxicity

Most studies on the action of phosgene oxime have utilized animal studies. Human information has been obtained from accidental exposure to the chemical. Health effects following phosgene oxime exposure are dependent on the route of exposure.

Since phosgene oxime is an urticant/nettle agent, common physical effects include erythema, wheals, and urticaria. Phosgene oxime is a highly corrosive agent and the response resembles wounds caused by strong acids. Ocular contact results in severe pain, conjunctivitis, and keratitis. Direct dermal ex-posure to phosgene oxime causes immediate pain and blanching with an erythematous ring. In - 0.5 h a wheal will form followed by tissue necrosis. Extreme pain can persist for days. Absorption of phosgene oxime through the skin can result in pulmonary edema. Inhalation of phosgene oxime vapor will produce immediate irritation to the airways. Pulmonary edema, necrotizing bronchiolitis, and pulmonary thrombosis can also occur following inhalation or systemic absorption of phosgene oxime. There has been no human data on effects of phosgene oxime following ingestion, but animal studies suggest that hemorrhagic inflamma-tory lesions may occur throughout the gastrointestinal tract.

21.5.5 Chronic Toxicity

There have been no studies on chronic exposure to phosgene oxime. Thus, there is no data regarding the carcinogenicity or teratogenicity of phosgene oxime.

21.5.6 Clinical Management

Individuals who come in contact with phosgene oxime liquid or solid can contaminate those around them by release of vapor. Individuals who have been exposed to the vapor will not be able to contaminate others. Patients who come in contact with phosgene oxime will experience immediate pain and develop necrotic lesions. Since there is no antidote for phosgene oxime exposure, only supportive measures can be given. Patients arriving to the triage area must first be decontaminated to prevent cross-contamination. For inhalation exposures, the individual should be removed from the source of exposure. Oxygen must be administered to patients with significant respiratory symptoms. Artificial respiration should be given if necessary. For ocular treatment, eyes should be flushed with copious amounts of water. Topical antibiotics should be applied to reduce the risk of infections and adhesions. Topical anticholinergics should be applied to reduce the risk of future synechiae formation. Skin contact will require decontamination with large amounts of water. Treatment should be in the same fashion as with a chemical burn. If phosgene oxime has been ingested, emesis should not be induced. Parental analgesics such as morphine or meperidine may be administered to reduce pain.

21.5.7 Environmental Fate

Phosgene oxime does not accumulate in the soil. Small amounts that may be present can vaporize into the air or be degraded by soil bacteria. Once in vapor form, phosgene oxime remains in vapor form and will be inactivated by compounds in the atmosphere or broken down by bacteria. There is no evidence that phosgene oxime will accumulate in ground-water.

21.6 Cyanide

Cyanide compounds are widely used in industry. Sodium cyanide and potassium cyanide are used extensively in the extraction of gold and silver from low-grade ores. The cyanide ion can form a wide range of complex ions with metals. These complex metal cyanide ions are extensively used in electroplating. Cyanide compounds are also used in case-hardening of iron and steel, metal polishing, photography, and the fumigation of ships and ware-houses. Organic cyanide compounds are used in ynthetic rubber, plastics, and synthetic fibers; they are also used in chemical synthesis. Cyanides are used in rodenticide and fertilizer production.

In addition, cyanides can be found in the seeds of the apple, peach, plum, apricot, cherry, and almond in the form of amygdatin, a cyanogenic glycoside. Amygdatin (Laetrile) has been used as an antineoplastic drug, but such beneficial effects have not been scientifically proven.

- CHEMICAL ABSTRACTS SERVICE REGISTRY NUMBERS: CAS 57-12-5 (C); CAS 74-90-8 (Hydrogen cyanide); CAS 143-33-9 (Sodium cyanide); CAS 151-50-8 (Potassium cyanide)
- SYNO YMS: Carbon nitride ion; Cyanide anion; Cyanide ion; Cyanure (French); Hydrocyanic acid; Isocyanide; Hydrocyanic acid sodium salt; Hydrocyanic acid potassium salt
- CHEMICAL/PHARMACEUTICAL/OTHER CLASS: Cyanide is anyone of a group of compounds containing the monovalent combining group C. Inorganic cyanides are regarded as salts of hydrocyanic acid (hydrogen cyanide). Organic cyanides are usually called nitrites
- CHEMICAL FORMULAS: HCN (hydrogen cyanide); NaCN (sodium cyanide); KCN (potassium cyanide); CH_3C (acetonitrile)
- CHEMICAL STRUCTURE: $- C \equiv N$

21.6.1 Background Information

Cyanide poisoning causes a high incidence of severe symptomatology and fatality. Between 1926 and 1947, death rates from cyanide poisoning in America ranged between 79 and 416 per 10 million population and gradually declined thereafter. The availability of the antidote kit may have contributed to this decreasing death rate. There are numerous sources of potential cyanide exposure. With the increased use of plastic building materials, the potential hazards of cyanide poisoning as a component of smoke inhala-tion in closed space fires still exist.

21.6.2 Exposure Routes and Pathways

Humans may be exposed to cyanide in a number of different forms. These include solids, liquids, and gases. Sources include industrial chemicals, natural products, medications, and combustion products. Inhalation of toxic fumes and ingestion of cyanide salts, cyanide-containing fruit seeds, and cyanide waste-contaminated drinking water are the most common exposure pathways. The respiratory route represents a potentially rapidly fatal type of expo-sure. Exposure to cyanides may also occur via the dermal route in industrial workers.

21.6.3 Toxicokinetics

Cyanide is rapidly absorbed from the skin and all mucosal surfaces; it is most dangerous when inhaled because toxic amounts are absorbed with great rapidity through the bronchial mucosa and alveoli. Once absorbed, distribution of cyanide through the body is rapid. Within a few minutes, cyanide is dis-tributed through the body and its conversion to thiocyanate starts. The majority of cyanide in the body is protein-bound (60%). In sublethal doses, cyanide reacts with sulfane sulfur to form nontoxic thiocyanate through an enzymatic reaction involving rhodanase and mercaptopyruvate sulfur transferase. Within 3h, 90% of the dose of cyanide is converted to thiocyanate appering in blood. Cyanide is also trapped as cyano of vitamin B_{12}, oxidized to formate and carbon dioxide, and incorporated into cysteine. In nonfatal cases, metabolized cyanide (thiocyanate) is excreted in the urine. Although cyanide is volatile, excretion through the lungs is not a significant route of elimination of cyanide.

21.6.4 Mechanism of Toxicity

Cyanide is described as a cellular toxin because it inhibits aerobic metabolism. It reversibly binds with ferric (Fe^{3+}) iron-containing cytochrome oxidase and inhibits the last step of mitochondrial oxidative phosphorylation. This inhibition halts carbohydrate metabolism from citric acid cycle, and intracellular concentrations of adenosine triphosphate are rapidly depleted. When absorbed in high enough doses, respiratory arrest quickly ensues, which is probably caused by respiratory muscle failure. Cardiac arrest and death inevitably follow.

For this reason, cyanide action has been described as 'internal asphyxia'. Although some cyanide com-bines with hemoglobin to form a stable non oxygen-bearing compound, cyanhemoglobin, this substance is formed only slowly and in a small amount. Therefore, death is not due to cyanhemoglobin but to inhibition of tissue cell respiration.

Recent studies have shown that cyanide also inhibits the antioxidant defense enzymes (such as catalase, superoxide dismutase, and glutathione peroxidase) and stimulates neurotransmitter release. These effects of cyanide may also contribute to its acute toxicity. The prolonged energy deficit and the consequent loss of ionic homeostasis, which may result in activation of calcium signaling cascade and eventually cell injury, contribute to cyanide toxicity resulting from subacute exposure or in the postin-toxication sequela.

21.6.5 Acute and Short-Term Toxicity

21.6.6 Animal

Cyanide toxicity varies with the animal species, type of cyanide compound, route of uptake, metabolic state, and other factors. The LD_{so} for cyanide has been ported in various species. Potassium cyanide, if injected, has a 24 h LD_{so} of 6.7-7.9 mg kg I in mice. The lethal dose of potassium cyanide infused at a rate 0.1 mgkg I min I is 2.4mgkg- I in dogs breathing room air. When hydrogen cyanide is inhaled by mice. The LD_{50} is 177 ppm with a lethal time of 29 min. The time to death is greater than 17 min for exposure to less than 266 ppm, but falls to 40 s at 873 ppm. The LD_{50} for sodium cyanide is 4.6-15mgkg I in rats. Male gerbils are 50-fold more sensitive to methacrylonitrile, which is metabolized to cyanide in rodents, than Sprague-Dawley rats, and about fivefold more sensitive than albino Swiss mice. Single and repeated low-dose cyanide intoxication can result in demyelinating lesions of the cerebral white matter in monkeys, but high doses of cyanide are required to produce similar brain lesions in rat.

21.6.7 Human

Cyanide is a chemical asphyxiant, which renders the body incapable of utilizing an adequate supply of oxygen. Exposure to high dose of cyanide is often lethal. The lethal dose of cyanide in humans is 0.5-1.0 mg kg^{-1}. The lethal dose of hydrocyanic acid is ~ 50 mg for an adult and the lethal dose of the potassium or sodium salt is 200-500 mg. The threshold limit value (TLV) of HCN for inhalation is 4.7 ppm. This is defined as the maximum safe average exposure limit for a 15 min period by the Occupational Safety and Health Administration. Exposure to 20 ppm of HC in air causes slight warning symp-toms after several hours; 50 ppm causes disturbances within an hour; 100 ppm is dangerous for exposures of 30-60 min; and 300 ppm can be rapidly fatal unless prompt, effective first aid is administered. The median lethal dose for skin contamination is ~ 100 mg kg^{-1}.

Following the inhalation of toxic amounts of cya-nide, symptoms usually appear within a few seconds, whereas it may take a few minutes for symptoms to appear following oral ingestion or skin contamination by the salts. The symptoms include a flushed skin, tachypnea, and tachycardia. Stupor, coma, and sei-zure immediately precede respiratory arrest and cardiovascular collapse. Death shortly occurs. If large amounts have been absorbed, collapse is usually in-stantaneous-the patient falling unconscious and dying almost immediately. With smaller doses, weakness, giddiness, headache, nausea, vomiting, and palpitation usually occur. With the rise of the blood cyanide level, ataxia develops and is followed by lactic aci-dosis, convulsive seizures, coma, and death. At higher cyanide doses, cardiac irregularities are often noted, but heart activity always outlasts the respiration.

21.7 Chronic Toxicity

21.7.1 Animal

Ingestion of cyanogenic plants, such as cassava and sorghum, has been associated with development of goiter and tropical pancreatic diabetes in both human and animals. However, results from animal studies indicate this association in animals is controversial. Chronic cyanide exposure has been reported to re-duce memory along with reduction in the levels of dopamine and 5-hydroxytryptamine in the rat brain.

21.7.2 Human

Chronic low-level exposure to cyanide produces various signs and symptoms. Exposure to small amounts of cyanide compounds over long-term periods of time is reported to cause loss of appetite, headache, weakness, nausea, dizziness, and symptoms of irri-tation of the upper respiratory tract and eyes. The most widespread pathologic condition attributed to cyanide is tropical ataxic neuropathy associated with chronic cassava consumption. This is a diffuse degenerative neurological disease with peripheral and central signs. Cassava is the major staple food in various tropical areas; the plant has a high content of cyanogenic glycoside (linamarin). With continued ingestion over a period of time, tropical neuropathy gradually develops. The syndrome is characterized by optic atrophy, nerve deafness, and ataxia due to sensory spinal nerve involvement. Other signs include scrotal dermatitis, stomatitis, and glossitis. Chronic low-level exposure to cyanide may also lead to ultrastructural changes of heart muscle. In addi-tion, with chronic cyanide ingestion, the thyroid may be affected due to enhanced formation of thiocyan-ate. Thiocyanate can block uptake of iodide by the thyroid gland, and myxedema, thyroid goiter, and cretinism may occur. This chronic effect of cyanide may pass to the fetus through maternal exposure.

21.7.3 Clinical Management

To be of any value, treatment of cyanide poisoning must be rapid and efficient. The rapid and early recognition of cyanide poisoning is usually difficult because most of the clinical manifestations are non-specific. Potentially valuable cyanide blood levels are usually available for confirmation of diagnosis. Arterialization of venous blood has been used as a significant symptom of cyanide poisoning. If cyanide was ingested, removing the unabsorbed poison by ravaging the stomach with copious amounts of water through a gastric tube is necessary. This should be cominued until all odor of cyanide is gone from the lavage fluid. Artificial respiration with 100% oxygen is often used in the treatment of cyanide poisoning, although oxygen is not a specific antidote. It is theorized that oxygen therapy increases the rate of displacement of cyanide from cytochrome ox-idase, and the increased intracellular oxygen tension nonenzymatically converts the reduced cytochrome to the oxidized species, enabling the electron trans-port system to function again. The nitrite-thiosulfate antidotal combination is still one of the most ef-fective treatments of cyanide poisoning. If the victim is conscious and speaking, no treatment is necessary. If the victim is unconscious but breathing, an open ampoule of amyl nitrite can be placed under the victim's nose for 15 s and it can be repeated 5 to 6 times. A fresh ampoule should be used every 3 min until the victim regains consciousness. Amyl nitrite is a powerful cardiac stimulant and should not be used more than necessary. If the patient is not breathing, 0.3 g (10 ml of a 3% solution, adults) of sodium ni-trite should be administered intravenously at the rate of 2.5 ml min - I followed by 12.5 g (50 ml of a 25% solution) of sodium thiosulfate at the same rate. Inhalation of amyl nitrite should also be performed. Nitrite will convert hemoglobin to methemoglobin, which has higher affinity for cyanide than hemoglo-bin. A methemoglobin level of ~ 25% is desired for maintaining normal hemoglobin function and detoxification. Thiosulfate is a sulfur donor for converting cyanide to nontoxic thiocyanate. For chil-dren weighing less than 25 kg, sodium nitrite should be dosed on the basis of their hemoglobin level and weight. The patient should be observed for the next 24-48 h and if the signs of intoxication persist or reappear, injection of nitrite thiosulfate at one-half of the recommended dose should be repeated. Hydro-xocobalamine has been effectively used in France as an antidote for acute cyanide poisoning. Hydro-xocobalamine (vitamin B_{12a}) is currently approved by Food and Drug Administration, but is not pop-ularly used in the United States as an antidote for cyanide poisoning. Because of its extremely low adverse effect, hydroxocobalamine is ideal for out -of-hospital use in suspected cyanide intoxication. It is actively proposed to be used in the United States.

21.8 Ecotoxicology

The toxicity of cyanide in the aquatic environment or natural waters is a result of free cyanide, that is, as HCN and C -. Fish are extremely sensitive to cyanide. Most fish can tolerate a free cyanide stream concentration of 0.05 mg 1-, but some species are even more sensitive.

21.9 Exposure Standards and Guidelines

- Occupational Safety and Health Administration permissible exposure limit: time-weighted average (TWA) 5mg (CN) m^{-3}.
- American Conference of Governmental Industrial Hygienists TLV: CL 5 mg m^{-3} (skin)
- DFG MAK: $5mgm^{-3}$.
- National Institute of Occupational Safety and Health recommended exposure limit (cyanide) TWA CL $5mgm^{-3}$ per 10 min.

CASE STUDY (1)

1995 Tokyo sarin gas attacks and related incidents

On the morning of 20th March 1995, Aum members released sarin in a co-ordinated attack on five trains in the Tokyo subway system, killing 12 commuters, seriously harming 54 and affecting 980 more. At the cult's headquarters in Kamikuishiki on the foot of Mount Fuji, police found explosives, chemical weapons and biological warfare agents, such as anthrax and Ebola cultures, and a Russian MIL Mi-17 military helicopter. There were stockpiles of chemicals which could be used for producing enough sarin to kill four million people. Police also found laboratories to manufacture drugs such as LSD, methamphetamines, and a crude form of truth serum, a safe containing millions of dollars worth in cash and gold, and cells, many still containing prisoners.

On the evening of May 5th, 1995 a burning paper bag was discovered in a toilet in Shinjuku station in Tokyo, the busiest station in the world. Upon examination it was revealed that it was a hydrogen cyanide device which, had it not been extinguished in time, would have released enough gas into the ventilation system to potentially kill 20,000 commuters. Cyanide devices were found several more times in the Tokyo subway but none detonated.

CASE STUDY (2)

Bhopal disaster (December 3, 1984)

This is the worst environmental disaster in human history. A pesticide factory, Union Carbide Corporation, leaked large volumes of methyl isocyanate (raw material for production of the pesticide, Carbaryl into the atmosphere of Bhopal on December 3, 1984 at around midnight. Very soon the city was transformed into a gas chamber. Within a week 10,000 people died, 1000 people turned blind and lakhs of people continue to suffer from various diseases. This was mass murder in recent history and the victims/survivors are yet to receive their compensation from the Union Carbide Corporation or the government. It was the end result of negligence on the parts of the Central and State governments and factory manager on one hand and lack of awareness among the public and hospital doctors on the other hand.

SUGGESTED FURTHER READINGS

Eckstein M (2004) Cyanide as a chemical terrorism weapon. *JEMS* 29(8): Suppl 22-31.

Gracia R and Shepherd G (2004) Cyanide poisoning and its treatment. *Pharmacotherapy* 24(10): 1358-1365.

Hall AH and Rumack BH (1986) Clinical toxicology of cyanide. *Annals of Emergency Medicne* 15: 1067-1072.

Parnaik P (1999) *A Comprehensive Guide to the Hazard-ous Properties of Chemical Substances,* 2nd edn. New York: Wiley.

Sauer SW and Keirn ME (2001) Hydroxocobalamin: Improved public health readiness for cyanide disasters. *Annals of Emergency Medicine* 37: 635-641.

http://sis.nlm.nih.gov - US National Library of Medicine, Specialized Information Services, Chemical Warfare Agents.

http://www.bt.cdc.gov - US Centers for Disease Control and Prevention, Agency for Toxic Substances and Dis-ease Registry, Chemical Agents.

http://www.br.cdc.gov - (US) Centers for Disease Control and Prevention, Agency for Toxic Substances and Dis-ease Registry, Chemical Agents.

http://sis.nlm.nih.gov - (US) National Library of Medicine, Specialized Information Services, Chemical Warfare Agents.

http://www.bt.cdc.gov - US Centers for Disease Control and Prevention, Agency for Toxic Substances and Dis-ease Registry, Chemical Agents.

hrtp://sis.nlm.nih.gov - US National Library of Medicine, Spe-cialized Information Services, Chemical Warfare Agents.

http://www.atsdr.cdc.gov - Agency for Toxic Substances and Disease Registry. Toxicological Profile for Phosgene Oxime. http://www.biochemhazard.com- Biochemhazard.com.

Dominion Research Center (2000) Chemical Weapon Agent: Phosgene Oxime (CX) - Blister Agent. http://www.emedicine.com - eMedicine, CBRNE: Urti-cants, Phosgene Oxime (2003).

STUDY QUESTIONS

1. Define a chemical warfare agent? Classify war gases.
2. Write briefly on the toxicity of following:
   a. Mustard gas
   b. Nerve agents
   c. Lewisite
   d. Cyanide
3. Explain the effects of war gases with following examples:
   a. Bhopal gas tragedy
   b. Tokyo sarin gas incident

Chapter 22

BIOLOGICAL WARFARE THREATS

Biological warfare (BW), also known as **germ warfare**, is the use of any pathogen (bacterium, virus or other disease-causing organism) or toxin found in nature, as a weapon of war. BW may be intended to kill, incapacitate or seriously impede an adversary. It may also be defined as the material or defense against such employment.

The creation and stockpiling of biological weapons (offensive BW) was outlawed by the 1972 Biological Weapons Convention (BWC), signed by over 100 countries. The rationale behind the agreement is to avoid the devastating impact of a successful biological attack which could conceivably result in thousands, possibly even millions, of deaths and cause severe disruptions to societies and economies. Oddly enough, the convention prohibits only creation and storage, but not usage, of these weapons. However, the consensus among military analysts is that, except in the context of bioterrorism, BW is of little military use. Many countries pursue "defensive BW" research (defensive or protective applications) which are not prohibited by the BWC. Biological weapons are the most horrifying tools today. The Biological Weapons Convention (BWC) of 1975 instructs all signatories to forego any biological weapon program. They are relatively cheap, inconspicorous and anonymous. A small private airplane with 100 kilograms of anthrox can kill a million people on a day with moderate wind. A short description of biological warfare agents is given below:

22.1 Anthrax

Anthrax is a zoonotic disease caused by *Bacillus anthracis*. There are three types of this disease: cutaneous anthrax, inhalation anthrax, and gastrointestinal anthrax. About 95% of the human anthrax cases in the United States have been related to the former category. Cutaneous anthrax develops when a bacterial organism from infected animal tissues becomes deposited under the skin. When a patient contracts cutaneous anthrax, he develops a small elevated lesion on his skin which becomes a skin ulcer, frequently surrounded by swelling or edema.

Inhalation anthrax develops when the bacterial organism is inhaled into the lungs. A progressive infection follows. Since inhalation anthrax is usually not diagnosed in time for treatment, the mortality rate in the United States is 90-100%. A biological attack with anthrax spores delivered by aerosol would cause inhalation anthrax, an extraordinarily rare form of the naturally occurring disease.

276

A lethal dose of anthrax is considered to be 10,000 spores; 80 percent of a population that inhaled such a dose would die. Less than one millionth of a gram is invariably fatal within five days to a week after exposure. According to an estimate by the U.S. Congress's Office of Technology Assessment, 100 kilograms of anthrax, released from a low-flying aircraft over a large city on a clear, calm night, could kill one to three million people.

Vaccines are available against some forms of anthrax, but their efficacy against abnormally high concentrations of the bacteria is uncertain. A licensed, alum-precipitated preparation of purified *B.anthracis* protective antigen (PA) has been shown to be effective in preventing or significantly reducing the incidence of inhalation anthrax.

22.2 Botulism

Botulism is caused by intoxication with any of the seven distinct neurotoxins produced by the bacillus, *Clostridium botulinum*. In humans, disease results from only four (A, B, E, and F) of those seven types of neurotoxins. The toxins are proteins with molecular weights of approximately 150,000, which bind to the presynaptic membrane of neurons at peripheral cholinergic synapses to prevent release of acetylcholine and block neurotransmission. The blockade is most evident clinically in the cholinergic autonomic nervous system and at the neuromuscular junction. A biological attack with botulinum toxin delivered by aerosol would be expected to cause symptoms similar in most respects to those observed with food-borne botulism.

In pure form, the toxin is a white crystalline substance that is readily soluble in water, but decays rapidly in the open air. Symptoms of inhalation botulism may begin as early as 12-36 hours following exposure or as late as several days. Initial signs and symptoms include ptosis, generalized weakness, lassitude, and dizziness. Diminished salivation with extreme dryness of the mouth and throat may cause complaints of a sore throat. Urinary retention or ileus may also occur. Motor symptoms usually are present early in the disease; cranial nerves are affected first with blurred vision, diplopia, ptosis, and photophobia. Development of respiratory failure may be abrupt. Mucous membranes of the mouth may be dry and crusted. Neurological examination shows flaccid muscle weakness of the palate, tongue, larynx, respiratory muscles, and extremities. Deep tendon reflexes vary from intact to absent.

A pentavalent toxoid of *Clostridium botulinum* types A, B, C, D, and E is available under IND status. This product has been administered to several thousand volunteers and occupationally at-risk workers and induces serum antitoxin levels that correspond to protective levels in experimental animal systems. The currently recommended schedule (0, 2, and 12 weeks, then a 1 year booster) induces solidly protective antitoxin levels in greater than 90 percent of those vaccinated after 1 year.

22.3 Brucellosis

Brucellosis is a systemic zoonotic disease that, in humans, is caused by one of four species of bacteria: *Brucella melitensis*, *B. abortus*, *B. suis*, and *B. canis*; virulence for humans decreases somewhat in the order given. These bacteria are small gram-negative, aerobic, non-motile coccobacilli that grow within monocytes and macrophages. They reside quiescently in tissue and bone-marrow, and are extremely difficult to eradicate even with antibiotic therapy. Their natural reservoir is domestic animals, such as goats, sheep, and camels (*B. melitensis*); cattle (*B. abortus*); and pigs (*B. suis*). *Brucella canis* is primarily a pathogen of dogs, and only occasionally causes disease in humans upon contact with an infected dog's blood, semen, or placenta. Humans are infected when they ingest raw (unpasteurized), infected milk or meat, inhale contaminated aerosols, or have abraded skin or conjunctival surfaces that come in contact with the bacteria. Laboratory infections are quite common, but human-to-human transmission is rare and occurred through breastfeeding. Therefore, isolation of infected patients is not required. *Brucella* species long have been considered potential candidates for use in biological warfare. The organisms are readily lyophilized, perhaps enhancing their infectivity. Under selected environmental conditions (for example, darkness, cool temperatures, high CO_2), persistence for up to 2 years has been documented. When used as a biological warfare agent, *Brucellae* would most likely be delivered by the aerosol route; the resulting infection would be expected to mimic natural disease.

The initial symptoms of brucellosis are usually nonspecific, and the differential diagnosis is therefore very broad and includes bacterial, viral, and mycoplasmal infections. The systemic symptoms of viral and mycoplasmal illnesses, however, are usually present for only a few days, while they persist for prolonged periods in brucellosis. Brucellosis may be indistinguishable clinically from the typhoidal form of tularemia or from typhoid fever itself. The disease in humans is characterized by a multitude of somatic complaints, including fever, sweats, anorexia, fatigue, malaise, weight loss, and depression. Localized complications may involve the cardiovascular, gastrointestinal, genitourinary, hepatobiliary, osteoarticular, pulmonary and nervous systems. Without adequate and prompt antibiotic treatment, some patients develop a 'chronic' brucellosis syndrome with many features of the 'chronic fatigue' syndrome.

The recommended treatment is doxycycline (200 mg/day) plus rifampin (900 mg/day) for 6 weeks. Alternative effective treatment consists of doxycycline (200 mg/day) for 6 weeks plus streptomycin (1 gm/day) for 3 weeks. Trimethoprimsulfamethoxazole given for 4-6 weeks is less effective. In 5-10% of cases, there may be relapse or treatment failure. Laboratory infections with brucellosis are quite common, but there is no human-to-human transmission and isolation is not required.

22.4 Cholera

Cholera is a diarrheal disease caused by *Vibrio cholera*, a short, curved, gram-negative bacillus. Humans acquire the disease by consuming water or food contaminated with the organism. The organism multiplies in the small intestine and secretes an enterotoxin that causes a secretory diarrhea. When employed as a BW agent, cholera will most likely be used to contaminate water supplies. It is unlikely to be used in aerosol form. Without treatment, death may result from severe dehydration, hypovolemia and shock. Vomiting is often present early in the illness and may complicate oral replacement of fluid losses. There is little or no fever or abdominal pain.

Watery diarrhea can also be caused by enterotoxigenic *E. coli*, rotavirus or other viruses, noncholera *Vibrios*, or food poisoning due to ingestion of preformed toxins such as those of *Clostridium perfringens, Bacillus cereus*, or *Staphylococcus aureus*.

Treatment of cholera depends primarily on replacement of fluid and electrolyte losses. This is best accomplished using oral dehydration therapy with the World Health Organization solution (3.5 g NaCl, 2.5 g $NaHCO_3$, 1.5 g KC1 and 20 g glucose per liter). Intravenous fluid replacement is occasionally needed when vomiting is severe, when the volume of stool output exceeds 7 liters/day, or when severe dehydration with shock has developed. Antibiotics will shorten the duration of diarrhea and thereby reduce fluid losses.

Improved oral cholera vaccines are presently being tested. One vaccine, which has been discontinued in the U.S., is a killed suspension of *V. cholera* provided about 50% protection that lasts for no more than 6 months. The initial dose is two injections given at least 1 week apart with booster doses every 6 months.

22.5 Gas gangarene

Clostridium perfringens is a common anaerobic bacterium associated with three distinct disease syndromes; gas gangrene or clostridial myonecrosis; enteritis necroticans (pig-bel); and perfringens food poisoning. Each of these syndromes has very specific requirements for delivering inocula of *C. perfringens* to specific sites to induce disease, and it is difficult to imagine a general scenario in which the spores or vegetative organisms could be used as a biological threat agent. There are, however, at least 12 protein toxins elaborated, and one or more of these could be produced, concentrated, and used as a weapon. Waterborne disease is conceivable, but unlikely. The alpha toxin would be lethal by aerosol. This is a well characterized, highly toxic phospholipase C. Other toxins from the organism might be co-weaponized and enhance effectiveness. For example, the epsilon toxin is neurotoxic in laboratory animals.

Gas gangrene is a well-recognized, life-threatening emergency. Symptoms of the disease may be subtle before fulminant toxemia develops, and the diagnosis is often made at postmortem examination. The bacteria produce toxins that create the high mortality from clostridial myonecrosis, and which produce the characteristic intense pain out of proportion to the wound. Within hours, signs of systemic toxicity appear, including confusion, tachycardia, and sweating. Most *Clostridia* species produce large amounts of CO_2 and hydrogen that cause intense swelling, hence the term "gas" gangrene, resulting in gas in the soft tissues and the emission of foul-smelling gas from the wound. Clinical features include necrosis, dark red serous fluid, and numerous gas filled vesicles. The infection may progress up to 10 cm per hour, and early diagnosis and therapy are essential to prevent rapid progression to toxemia and death. Pulmonary findings might lead to confusion with staphylococcal enterotoxin B (SEB) initially. Liver damage, hemolytic anemia, and thrombocytopenia are not associated with SEB and the pulmonary findings should be reversible in SEB.

No specific treatment is available for *C. pefringens* intoxication. Early antibiotic treatment is effective, if undertaken before significant amounts of toxins have accumulated in the body. If not treated the bacteria enter the bloodstream causing fatal systemic illness. The organism itself is sensitive to penicillin, and consequently, this is the current drug of choice. Recent data indicate that clindamycin or rifampin may suppress toxin production and provide superior results in animal models. Prompt surgical debridement and broad spectrum, intravenous antibiotics are the mainstay of therapy. Hyperbaric oxygen has been seen as effective in prolonging survival in animal studies when coupled with other treatments.

22.6 Crimean-Congo Hemorrhagic Fever

Crimean-Congo hemorrhagic fever (CCHF) is a viral disease caused by the *Nairovirus*. The virus, first characterized in the Crimea in 1944, is transmitted by ticks, principally of the genus *Hyalomma*, with intermediate vertebrate hosts varying with tick species. In 1969 it was recognized that the pathogen causing Crimean hemorrhagic fever was the same as that responsible for an illness identified in 1956 in the Congo, and linkage of the two place-names resulted in the current name for the disease and the virus. The disease, next found in the Congo, occurs also in the Middle East, the Balkans, the former USSR, and eastern China. Little is known about variations in the virus properties over the huge geographic area involved. Humans become infected through tick bites, crushing an infected tick, or at the slaughter of viremic livestock. Even in epidemics, cases do not show narrow clustering and person-to-person spread is rare, though possible through contact with infectious blood or bodily fluids. CCHF would probably be delivered by aerosol if used as a BW agent.

Diagnosis of suspected CCHF is performed in specially-equipped, high biosafety level laboratories. Other viral hemorrhagic fevers, meningococcemia, rickettsial diseases, and similar conditions may resemble full-blown CCHF. Most fatal cases and half the others will have detectable antigen by rapid enzyme-linked immunosorbant assay (ELISA) testing of acute serum samples. IgM ELISA antibodies occur early in recovery. Polymerase chain reaction has recently been used in diagnosing CCHF.

Supportive therapy with replacement of clotting factors is indicated. Crimean-Congo hemorrhagic fever virus is sensitive to ribavirin *in vitro* and clinicians have been favorably impressed in uncontrolled trials. Immune sera have also been used for therapeutic purposes several times, but its value has not been demonstrated.

22.7 Ebola Hemorrhagic Fever

Ebola hemorrhagic fever is one of the most virulent viral diseases known to humankind, causing death in 50-90% of all clinically-ill cases. Consequently, it has figured prominently in popular discussions of biological weapons, although its practical applications as a biological threat agent remain speculative. The disease has its origins in the jungles of Africa and Asia and several different forms of Ebola virus have been identified and may be associated with other clinical expressions, on which further research is required.

The Ebola virus is transmitted by direct contact with the blood, secretions, organs or semen of infected persons. Transmission through semen may occur up to 7 weeks after clinical recovery, as with Marburg hemorrhagic fever. Health care workers have frequently been infected while attending patients. In the 1976 epidemic in Zaire, every Ebola case caused by contaminated syringes and needles died.

After an incubation period of 2 to 21 days, Ebola is often characterized by the sudden onset of fever, weakness, muscle pain, headache and sore throat. This is followed by vomiting, diarrhea, rash, limited kidney and liver functions, and both internal and external bleeding. Specialized laboratory tests on blood specimens (which are not commercially available) detect specific antigens or antibodies and/or isolate the virus. These tests present an extreme biohazard and are only conducted under maximum containment conditions.

No specific treatment or vaccine exists for Ebola hemorrhagic fever. Severe cases require intensive supportive care, as patients are frequently dehydrated and in need of intravenous fluids. Experimental studies involving the use of hyperimmune sera on animals demonstrated no long-term protection against the disease after interruption of therapy.

Suspected cases should be isolated from other patients and strict barrier nursing techniques practiced. All hospital personnel should be briefed on the nature of the disease and its routes of transmission. Particular emphasis should be placed on ensuring that high-risk procedures such as the placing of intravenous lines and the handling of blood, secretions, catheters, and suction devices are done under barrier nursing conditions. Hospital staff should have individual gowns, gloves, and masks. Gloves and masks must not be reused unless disinfected. Patients who die from the disease should be promptly buried or cremated.

Different hypotheses have been developed to try to uncover the cycle of Ebola. Initially, rodents were suspected, as is the case with Lassa Fever whose reservoir is a wild rodent (*Mastomys*). Another hypothesis is that a plant virus may have caused the infection of vertebrates. Laboratory observation has shown that bats experimentally infected with Ebola do not die and this has raised speculation that these mammals may play a role in maintaining the virus in the tropical forest.

22.8 Melioidosis

Melioidosis is an infectious disease of humans and animals caused by *Burkholderia pseudomallei* (formerly *Pseudomonas pseudomallei*), a gram-negative bacillus. It is especially prevalent in Southeast Asia but has been described in many countries around the world. The disease has a variable and inconstant clinical spectrum. A biological warfare attack with this organism would most likely be by the aerosol route.

Infection by inoculation results in a subcutaneous nodule with acute lymphangitis and regional lymphadenitis, generally with fever. Pneumonia may occur after inhalation or hematogenous dissemination of infection. It may vary in intensity from mild to fulminant, usually involves the upper lobes, and often results in cavitation. Pleural effusions are uncommon. An acute fulminant septicemia may occur characterized by rapid appearance of hypotension and shock. A chronic suppurative form may involve virtually any organ in the body.

Antibiotic regimens that have been used successfully include tetracycline, 2-3 g/day; chloramphenicol, 3 g/day; and trimethoprim-sulfamethoxazole, 4 and 20 mg/kg per day. Ceftazidine and piperacillin have enjoyed success in severely ill patients as well. In patients who are toxic, a combination of two antibiotics, given parenterally, is advised.

There are no means of immunization. Vigorous cleansing of abrasions and lacerations may reduce the risk of disease after inoculation of organisms into the skin. There is no information available on the utility of antibiotic prophylaxis after potential exposure, but before the onset of clinical symptoms.

22.9 Plague

Plague is a zoonotic disease caused by *Yersinia pestis*. Under natural conditions, humans become infected as a result of contact with rodents, and their fleas. The transmission of the gram-negative coccobacillus is by the bite of the infected flea, *Xenopsylla cheopis*, the oriental rat flea, or *Pulex irritans*, the human flea. Under natural conditions, three syndromes are recognized: bubonic, primary septicemia, or pneumonic. In a biological warfare scenario, the plague bacillus could be delivered via contaminated vectors (fleas) causing the bubonic type or, more likely, via aerosol causing the pneumonic type.

In bubonic plague, the incubation period ranges from 2 to 10 days. The onset is acute and often fulminant with malaise, high fever, and one or more tender lymph nodes. Inguinal lymphadenitis (bubo) predominates, but cervical and axillary lymph nodes can also be involved. The involved nodes are tender, fluctuant, and necrotic. Bubonic plague may progress spontaneously to the septicemia form with organisms spread to the central nervous system, lungs (producing pneumonic disease), and elsewhere. The mortality is 50 percent in untreated patients with the terminal event being circulatory collapse, hemorrhage, and peripheral thrombosis.

In primary pneumonic plague, the incubation period is 2 to 3 days. The onset is acute and fulminant with malaise, high fever, chills, headache, myalgia, cough with production of a bloody sputum, and toxemia. The pneumonia progresses rapidly, resulting in dyspnea, strider, and cyanosis. In untreated patients, the mortality is 100 percent with the terminal event being respiratory failure, circulatory collapse, and a bleeding diathesis.

In cases where bubonic type is suspected, tularemia adenitis, staphylococcal or streptococcal adenitis, meningococcemia, enteric gram negative sepsis, and rickettsiosis need to be ruled out. In pneumonic plague, tularemia, anthrax, and staphylococcal enterotoxin B (SEB) agents need to be considered. Continued deterioration without stabilization effectively rules out SEB.

Plague may be spread from person to person by droplets. Strict isolation procedures for all cases are indicated. Streptomycin, tetracycline, and chloramphenicol are highly effective if begun early. Significant reduction in morbidity and mortality is possible if antibiotics are given within the first 24 hours after symptoms of pneumonic plague develop.

A formalin-killed *Y. pestis* vaccine that was produced in the United States and extensively used is no longer manufactured. Efficacy against flea-borne plague was inferred from population studies, but the utility of this vaccine against aerosol challenge was unknown. Immunity was maintained through boosters every 1-2 years. Live-attenuated vaccines are available elsewhere but are highly reactogenic and without proven efficacy against aerosol challenge.

22.10 Q Fever

Q fever is a zoonotic disease caused by a rickettsia, *Coxiella burnetii*. The most common animal reservoirs are sheep, cattle and goats. Humans acquire the disease by inhalation of particles contaminated with the organisms. A biological attack would cause disease similar to that occurring naturally.

Following an incubation period of 10-20 days, Q fever generally occurs as a self-limiting febrile illness that includes headache, fatigue, and myalgias and lasts from 2 days to 2 weeks . Pneumonia occurs frequently, usually manifested only by an abnormal chest x-ray. A nonproductive cough and pleuritic chest pain occur in about one-fourth of patients with Q fever pneumonia. Patients usually recover uneventfully.

Q fever usually presents as an undifferentiated febrile illness, or a primary atypical pneumonia, which must be differentiated from pneumonia caused by mycoplasma, Legionnaire's disease, psittacosis or *Chlamydia pneumonia*. More rapidly progressive forms of pneumonia may look like bacterial pneumonias including tularemia or plague.

Tetracycline (250 mg every 6 hr) or doxycycline (100 mg every 12 hr) for 5-7 days is the treatment of choice. A combination of erythromycin (500 mg every 6 hr) plus rifampin (600 mg per day) is also effective.

Vaccination with a single dose of a killed suspension of *C. burnetii* provides complete protection against naturally occurring Q fever and >90% protection against experimental aerosol exposure in human volunteers. Protection lasts for at least 5 years. However, neither this vaccine nor any other is commercially available in the U.S. Administration of this vaccine in immune individuals may cause severe cutaneous reactions including necrosis at the inoculation site. Newer vaccines are under development. Treatment with tetracycline during the incubation period will delay but not prevent the onset of illness.

22.11 Ricin

Ricin is a glycoprotein toxin (66,000 daltons) from the seed of the castor plant. It blocks protein synthesis by altering the rRNA, thus killing the cell. Ricin's significance as a potential biological threat agent relates to its availability worldwide, ease of production, and extreme pulmonary toxicity when inhaled.

Overall, the clinical picture seen depends on the route of exposure. All reported serious or fatal cases of castor bean ingestion have taken approximately the same course: rapid onset of nausea, vomiting, abdominal cramps, and severe diarrhea with vascular collapse; death has occurred on the third day or later. Following inhalation, one might expect nonspecific symptoms of weakness, fever, cough, nausea, and hypothermia followed by hypotension and cardiovascular collapse. High doses by inhalation appear to produce severe enough pulmonary damage to cause death.

In oral intoxication, fever, gastrointestinal involvement, and vascular collapse are prominent, the latter differentiating it from infection with enteric pathogens. With regard to inhalation exposure, nonspecific findings of weakness, fever, vomiting, cough, hypothermia, and hypotension in large numbers of patients might suggest several respiratory pathogens.

Therapy is supportive and should include maintenance of intravascular volume. Standard management for poison ingestion should be employed if intoxication is by the oral route. There is presently no antitoxin available for treatment.

There is currently no prophylaxis approved for human use. Active immunization and passive antibody prophylaxis are under study, as both are effective in protecting animals from death following exposure by intravenous or respiratory routes.

22.12 Rift Valley Fever

Rift Valley Fever (RVF) is a viral disease caused by RVF virus. The virus circulates in sub-Saharan Africa as a mosquito-borne agent. Epizootics occur when susceptible domestic animals are infected, and because of the large amount of virus in their serum, amplify infection to biting arthropods. Deaths and abortions among susceptible species such as cattle and sheep constitute a major economic consequence of these epizootics, as well as providing a diagnostic clue and a method of surveillance. Humans become infected through the bite of mosquitoes, contact with infected blood, bodily fluids, or organs, or exposure to virus-laden aerosols or droplets. The human disease appears to be similar whether acquired by aerosol or by mosquito bite. A biological attack, most likely delivered by aerosol, would be expected to elicit the rather specific spectrum of human clinical manifestations and to cause disease in sheep and cattle in the exposed area. If disease occurred in the absence of heavy vector populations or without domestic animals as amplifiers of mosquito infection, a BW attack would also be a likely cause.

The occurrence of an epidemic with febrile disease, hemorrhagic fever, eye lesions, and encephalitis in different patients would be characteristic of RVF. Demonstration of viral antigen in blood by ELISA is rapid and successful in a high proportion of acute cases of uncomplicated disease or hemorrhagic fever. Other methods of diagnosis include polymerase chain reaction and virus propagation.

In hemorrhagic fever, supportive therapy may be indicated for hepatic and renal failure, as well as replacement of coagulation factors. The virus is sensitive to ribavirin *in vitro* and in rodent models. No studies have been performed in human or the more realistic monkey model to ascertain whether administration to an acutely ill patient would be of benefit.

Avoidance of mosquitoes and contact with fresh blood from dead domestic animals and respiratory protection from small particle aerosols are the mainstays of prevention. An effective inactivated vaccine for humans is not licensed for public use, but is available in limited quantities, particularly to laboratory and veterinary personnel.

22.13 Saxitoxin

Saxitoxin is the parent compound of a family of chemically related neurotoxins. In nature they are predominantly produced by marine dinoflagellates, although they have also been identified in association with such diverse organisms as blue-green algae, crabs, and the blue-ringed octopus. Human intoxications are principally due to ingestion of bivalve mollusks which have accumulated dinoflagellates during filter feeding. The resulting intoxication, known as paralytic shellfish poisoning (PSP), is known throughout the world as a severe, life-threatening illness requiring immediate medical intervention. In a BW scenario, the most likely route of delivery is by inhalation or toxic projectile. In addition, saxitoxin could be used in a confined area to contaminate water supplies.

After oral exposure, absorption of toxins from the gastrointestinal tract is rapid. Onset of symptoms typically begins 10-60 minutes after exposure, but may be delayed several hours depending upon the dose and individual idiosyncrasy. Initial symptoms are numbness or tingling of the lips, tongue and fingertips, followed by numbness of the neck and extremities and general muscular incoordination. Nausea and vomiting may be present, but typically occur in a minority of cases. Respiratory distress and flaccid muscular paralysis are the terminal stages and can occur 2-12 hours after intoxication. Death results from respiratory paralysis. Clearance of the toxin is rapid and survivors for 12-24 hours will usually recover. There are no known cases of inhalation exposure to saxitoxin in the medical literature, but data from animal experiments suggest the entire syndrome is compressed and death may occur in minutes.

Management is supportive and standard management of poison ingestion should be employed if intoxication is by the oral route. Toxins are rapidly cleared and excreted in the urine, so diuresis may increase elimination. Incubation and mechanical respiratory support may be required in severe intoxication. Timely resuscitation would be imperative, albeit very difficult, after inhalation exposure on the battlefield.

No vaccine against saxitoxin exposure has been developed for human use.

22.14 Smallpox

Smallpox virus, an orthopoxvirus with a narrow host range confined to humans, was an important cause of morbidity and mortality in the developing world until recent times. Although the World Health Organization declared the virus eradicated in 1980, eradication of the natural disease was completed in 1977 and the last human cases, associated with laboratory infections, occurred in 1978. The virus exists today in only 2 laboratory repositories, one in the U.S. and the other in Russia. Appearance of human cases outside the laboratory would signal use of the virus as a biological weapon. Under natural conditions, the virus is transmitted by direct (face-to face) contact with an infected case, by fomites, and occasionally by aerosols. Smallpox virus is highly stable and retains infectivity for long periods outside of the host. A related virus, monkeypox, clinically resembles smallpox and causes sporadic human disease in West and Central Africa.

The incubation period is typically 12 days (range, 10-17 days). The illness begins with a prodrome lasting 2-3 days, with generalized malaise, fever, rigors, headache, and backache. This is followed by defervescence and the appearance of a typical skin eruption characterized by progression over 7-10 days of lesions through successive stages, from macules to papules to vesicles to pustules. The latter finally form crusts and, upon healing, leave depressed depigmented scars. The case fatality rate is approximately 30% in unvaccinated individuals. Permanent joint deformities and blindness may follow recovery. Vaccine immunity may prevent or modify illness.

The eruption of chickenpox (*varicella*) is typically centripetal in distribution (worse on trunk than face and extremities) and characterized by crops of lesions in different stages of development. Chickenpox papules are soft and superticial, compared to the firm, shotty, and deep papules of smallpox. Chickenpox crusts fall off rapidly and usually leave no scar. Monkeypox cannot be easily distinguished from smallpox clinically. Monkeypox occurs only in forested areas of West and Central Africa as a sporadic, zoonotic infection transmitted to humans from wild squirrels. Person-to-person spread is rare and ceases after 1-2 generations. Mortality is 15%. Other diseases that are sometimes confused with smallpox include typhus, secondary syphilis, and malignant measles. Skin samples (scrapings from papules, vesicular fluid, pus, or scabs) may provide a rapid identification of smallpox by direct electron microscopy, agar gel immunoprecipitation, or immunofluorescence.

There is no specific treatment available although some evidence suggests that vaccinia-immune globulin is of some value in treatment if given early in the course of the illness.

Patients with smallpox should be treated by vaccinated personnel using universal precautions. Objects in contact with the patient, including bed linens, clothing, ambulance, etc.; require disinfection by fire, steam, or sodium hypochlorite solution.

22.15 Staphylococcal Enterotoxin B

Staphylococcal Enterotoxin B (SEB) is one of several exotoxins produced by *Staphylococcus aureus*, causing food poisoning when ingested. A BW attack with aerosol delivery of SEB to the respiratory tract produces a distinct syndrome causing significant morbidity and potential mortality.

The disease begins 1-6 hours after aerosol exposure with the sudden onset of fever, chills, headache, myalgia, and nonproductive cough. In more severe cases, dyspnea and retrosternal chest pain may also be present. Fever, which may reach 103-106° F, can last 2-5 days, but cough may persist 1-4 weeks. In many patients nausea, vomiting, and diarrhea will also occur. In moderately severe laboratory exposures, lost duty time has been <2 weeks, but, based upon animal data, it is anticipated that severe exposures will result in fatalities.

In foodborne SEB intoxication, fever and respiratory involvement are not seen, and gastrointestinal symptoms are prominent. The nonspecific symptoms of fever, nonproductive cough, myalgia, and headache occurring in large numbers of patients in an epidemic setting would suggest any of several infectious respiratory pathogens, particularly influenza, adenovirus, or mycoplasma. In a BW attack with SEB, cases would likely have their onset within a single day, while naturally occurring outbreaks would present over a more prolonged interval.

Treatment is limited to supportive care; humidified oxygen and steroids for pain control. No specific antitoxin for human use is available. There currently is no prophylaxis for SEB intoxication. Experimental immunization has protected monkeys, but no vaccine is presently available for human use.

22.16 Mycotoxicoses

The trichothecene mycotoxins are a diverse group of more than 40 compounds produced by fungi. They strongly inhibit protein synthesis, impair DNA synthesis, alter cell membrane structure and function, and inhibit mitochondrial respiration. Secondary metabolites of fungi, such as T-2 toxin and others, produce toxic reactions called mycotoxicoses upon inhalation or consumption of contaminated food products by humans or animals. Naturally occurring trichothecenes have been identified in agricultural products and have been implicated in the animal disease moldy corn toxicosis, or poisoning.

There are no well-documented cases of clinical exposure of humans to trichothecenes. However, strong circumstantial evidence has associated these toxins with alimentary toxic aleukia (ATA), the fatal epidemic seen in Russia during World War II, and with alleged BW incidents ("yellow rain") in Cambodia, Laos, and Afghanistan.

Consumption of these mycotoxins results in weight loss, vomiting, skin inflammation, bloody diarrhea, diffuse hemorrhage, and possibly death. The onset of illness following acute exposure to T-2 (intravenously or through inhalation) occurs in hours, resulting in the rapid onset of circulatory shock characterized by reduced cardiac output, arterial hypotension, lactic acidosis and death within 12 hours.

Clinical signs and symptoms of ATA were hemorrhage, leukopenia, ulcerative pharyngitis, and depletion of bone marrow. The purported use of T-2 as a BW agent resulted in acute exposure via inhalation and/or dermal routes, as well as oral exposure upon consumption of contaminated food products and water. Alleged victims reported painful skin lesions, lightheadedness, dyspnea, and a rapid onset of hemomhage, incapacitation and death. Survivors developed a radiation-like sickness including fever, nausea, vomiting, diarrhea, leukopenia, bleeding, and sepsis.

Specific diagnostic modalities are limited to reference laboratories. Because of their long "half-life" the toxic metabolites can be detected as late as 28 days after exposure.

Ascorbic acid (400-1200 mg/kg, inter-peritoneal (ip)) works to decrease lethality in animal studies, but has not been tested in humans. While not yet available for humans, administration of large doses of monoclinal antibodies directed against T-2 and other metabolites have shown prophylactic and therapeutic efficacy in animal models.

22.17 Tularemia

Tularemia is a zoonotic disease caused by *Francisella tularensis*, a gram-negative bacillus. Humans acquire the disease under natural conditions through inoculation of skin or mucous membranes with blood or tissue fluids of infected animals, or bites of infected deerflies, mosquitoes, or ticks. A BW attack with *F. tularensis* delivered by aerosol would primarily cause typhoidal tularemia, a syndrome expected to have a case fatality rate which may be higher than the 5-10% seen when disease is acquired naturally.

A variety of clinical forms of tularemia are seen, depending upon the route of inoculation and virulence of the strain. In humans, as few as 10-50 organisms will cause disease if inhaled or injected intradermally, whereas 10^8 organisms are required with oral challenge. Under natural conditions, ulceroglandular tularemia generally occurs about 3 days after intradermal inoculation (range 2-10 days), and manifests as regional lymphadenopathy, fever, chills, headache, and malaise, with or without a cutaneous ulcer. Gastrointestinal tularemia occurs after drinking contaminated ground water, and is characterized by abdominal pain, nausea, vomiting, and diarrhea. Bacteremia may be common after primary intradermal, respiratory, or gastrointestinal infection with *F. tularensis* and could result in septicemia or "typhoidal" tularemia. The typhoidal form also may occur as a primary condition in 5-15% of naturally-occurring cases; clinical features include fever, prostration, and weight loss, but without adenopathy. Diagnosis of primary typhoidal tularemia is difficult, as signs and symptoms are nonspecific and there frequently is no suggestive exposure history. Pneumonic tularemia is a severe atypical pneumonia that may be fulminant, and can be primary or secondary. Primary pneumonia may follow direct inhalation of infectious aerosols, or may result from aspiration of organisms in cases of pharyngeal tularemia. Pneumonic tularemia causes fever, headache, malaise, substernal discomfort, and a non-productive cough; radiologic evidence of pneumonia or mediastinal lymphadenopathy may or may not be present. A biological warfare attack with *F. tularensis* would most likely be delivered by aerosol, causing primarily typhoidal tularemia. Many exposed individuals would develop pneumonic tularemia (primary or secondary), but clinical pneumonia may be absent or non-evident. Case fatality rates may be higher than the 5-10% seen when the disease is acquired naturally.

The clinical presentation of tularemia may be severe, yet nonspecific. Differential diagnoses include typhoidal syndromes (e.g., salmonella, rickettsia, malaria) or pneumonic processes (e.g., plague, mycoplasma, SEB). A clue to the diagnosis of tularemia delivered as a BW agent might be a large number of temporally clustered patients presenting with similar systemic illnesses, a proportion of whom will have a nonproductive pneumonia. Identification of organisms by staining ulcer fluids or sputum is generally not helpful. Routine culture is difficult, due to unusual growth requirements and/or overgrowth of commensal bacteria.

Streptomycin (1 gm every 12 hours intramuscularly (IM) for 10-14 days) is the treatment of choice. Gentamicin also is effective (3-5 mg/kg/day parenterally for 10-14 days). Tetracycline and chloramphenicol treatment are effective as well, but are associated with a significant relapse rate. Although laboratory-related infections with this organism are very common, human-to-human spread is unusual and isolation is not required.

A live, attenuated tularemia vaccine is available as an investigational new drug (IND). This vaccine has been administered to more than 5,000 persons without significant adverse reactions and is of proven effectiveness in preventing laboratory-acquired typhoidal tularemia. Its effectiveness against the concentrated bacterial challenge expected in a BW attack is unproven. The use of antibiotics for prophylaxis against tularemia is controversial.

CASE STUDY I

1979 Sverdlovsk Incident

In late April, 1979, the city of Sverdlovsk experienced a loud explosion that was identified as originating from Military Compound 19. Several days later, residents downwind from this compound developed high fever and difficult breathing. Over the next several days, more cases were reported and fatalities rose sharply to around 40. Autopsies revealed severe pulmonary edema in addition to symptoms of serious toxemia. Local doctors announced an outbreak of pulmonary anthrax. On the other hand, government officials reported that the outbreak was caused by the illegal sale of contaminated meat from a cow suffering from the disease. Case fatalities did not display the symptoms of the usual gastric or skin anthrax, which would be more likely if contaminated beef had been handled or eaten. A nine story hospital was taken over by the military to handle exclusively the victims of the explosion. Vaccination and antibiotics were provided to patients and residents alike. The final death toll was estimated between 200 to 1,000. The victims were buried with special sanitary recautions, and relatives were not allowed to attend the funerals. Some western scientists, including those that doubted the evidence in Southeast Asia, accepted the explanation provided by the Soviet Ministry of Health. The Sverdlovsk incident remained unproven; yet all evidence available to the U.S. Government indicated that a massive accident had occurred at a BW production facility. Believers and non-believers of the Soviet explanation remained in a status quo situation until President Boris Yeltsin acknowledged in a press conference, prior to meeting with President Bush in the summer of 1992, Washington, D.C., that the Sverdlovsk incident was in fact a massive BW accident involving an aerosol of anthrax spores.

CASE STUDY II

Ebola fever

The Ebola virus was first identified in a western equatorial province of Sudan and in a nearby region of Zaire in 1976 after significant epidemics in Yamkubu, northern Zaire, and Nzara, southern Sudan. Between June and November 1976 the Ebola virus infected 284 people in Sudan, with 117 deaths. In Zaire there were 318 cases and 280 deaths in September and October. An isolated case occurred in Zaire in 1977 and a second outbreak in Sudan in 1979. In 1989 and 1990, a filovirus, named Ebola-Reston, was isolated in monkeys being held in quarantine in laboratories in Reston, VA, Alice, TX, and Pennsylvania. In the Philippines, Ebola-Reston infections occurred in a quarantine area near Manila for monkeys intended for exportation. A large epidemic occurred in Kikwit, Zaire in 1995 with 315 cases, 244 with fatal outcomes. One human case of Ebola hemorrhagic fever and several cases in chimpanzees were confirmed in Côte d'Ivoire in 1994-95. In Gabon, Ebola hemorrhagic fever was first documented in 1994 and recent outbreaks occurred in February 1996 and July 1996. In all, nearly 1,100 cases with 793 deaths have been documented since the virus was discovered. The natural reservoir of the Ebola virus seems to reside in the rain forests of Africa and Asia but has not yet been identified.

SUGGESTED FURTHER READINGS

NATO Handbook on the Medical Aspects of NBC Defensive Operations, Part II - Biological, (Army Field Manual 8-9) - U.S. Department of Defense, Departments of the Army, the Navy, and the Air Force, February 1996

Guidelines for the Surveillance and Control of Anthrax in Humans and Animals, (WHO/EMC/ZDI/98.6) - World Health Organization, 1998

WHO Recommended Guidelines for Epidemic Preparedness and Response: Ebola Haemorrhagic Fever (EHF), (WHO/EMC/DIS/97.7) - World Health Organization, 1997

U.S. Food & Drug Administration, Center for Food Safety & Applied Nutrition - The "Bad Bug Book": Foodborne Pathogenic Microorganisms and Natural Toxins Handbook

STUDY QUESTIONS

1. Write briefly on zoonotic diseases.
2. Describe in details the following biological warfare agent.
   i.   Antrax
   ii. Botulinum toxin
   iii. Brucellosie
3. Write briefly on the following diseases:
   i.   Gas gangarene
   ii. Congo-Crimean hemorhagic fever
   iii. Ebola fever
   iv. Melioidosis
4. Describe the following episodes as biological warfare case studies:
   i.   Sverdlovsk Incident
   ii. Ebola fever

Chapter 23

NATURAL TOXINS

Numerous species of microbes, plants and animals can produce toxins. In animals actively delivered poisons are called as venoms whereas those delivered passively are called as poisons. Venoms have been assigned a digestive function whereas poisons have a predatory function. Plant toxins have a worldwide distribution. The purpose of plant toxins is not well understood. Protection and survival are the obvious possibilities. Microbial toxins which include mycotoxins such as aflatoxin are also diverse in distribution and effects. This chapter describes briefly the microbial, animal toxins and plant toxins.

23.1 Microbial (Infectious) Diseases

Bacteria are the major source of food poisoning. They may cause following diseases:

23.1.1 Acute gastroenteritis

Common symptoms of acute gastroenteritis are abdominal pain, vomiting, diarrhea and dysentery with or without fever. Diarrhea kills about four million people in developing countries each year and remains a problem in developed countries as well. It might be a causative agent for gastroenteritis for 60% to 80% cases. Fried rice, cooked rice, noodles, pasta, pastries, meat, vegetables, sauces, puddings, milk and milk products are the sources of *Bacillus cereus* (Fig. 1). Contaminated water, unpasteurized milk, poultry are the sources of *Campylobacter* (Fig. 2). Home canned vegetables, fish, meat, honey, corn syrup lead to infection by *Clostridium botulinum* (Fig. 3). *Escherichia coli* occurs in contaminated water (Fig. 4). Cattle/cattle related foods, poultry, un/ undercooked eggs, sprouts, and peto (chicks, ducklings, all reptiles) are the sources of *Salmonella* sp. (Fig. 5). *Staphylococcus* sp. generally occur in proteinaceous foods, prepared foods, and salads (Fig. 6). *Vibrio cholera* comes from water (Fig. 7) and *Yersinia* sp (Fig. 8) infection occurs due to the consumption of contaminated pork.

Parasitic gastroenteritis is caused by protozoans like *Giardia lamblia* (Fig. 9), *Entamoeba histolytica* (Fig. 10) and *Balantidium coli* (Fig. 11).

291

23.1.2 Traveler's diarrhea

Although a heat labile enterotoxin derived from *E. coli* is the most common cause of traveler's diarrhea, other bacteria may also cause this disease. These infections are most commonly acquired through ingestion of contaminated food and water but may be transmitted by person to person contact. These agents are *E. coli.*, *Shigella* sp., *Salmonella* sp., *Aeromonas* sp., *Campylobacter jejuni, Viral agents* and protozoans like *Giardia lamblia* and *Entamoeba histolytica*.

23.1.3 Neurologic effects

Certain neurologic symptoms viz. blurred vision, diplopia, dysarthria, dysphagia, and descending paralysis are typical of botulinum toxin. Seizures are typical of *Shigella*. Marine toxins also produce neurologic symptoms.

23.1.4 Nephrotoxicity

Nephrotoxicity from ingested foods is uncommon. *E. coli* 0157:H7 may produce acute gastroenteritis hemolysis and anemia, thrombocytopenia and azotemia, the hemolytic anemic syndrome.

23.1.5 Other food related illness

Mushrooms are capable of producing acute GI illness with or without additional toxicity. Spicy foods, such as capsaicin-containing peppers or horseradish may produce severe oesopharyngeal or abdominal pain and *syncope tyramine* is found in wines and aged cheese.

23.2 Marine Venoms and Poisons

A remarkable number of marine organisms, venomous animals produce venom in a specialized gland. Venom is a mixture of mainly protein and peptide toxins. In contrast, poisonous animals may have special glands that produce toxins but more often accumulate toxic compounds from the environment in their bodies. These substances are known as poisons and have to be ingested to be effective because the animal has no specialized organ to deliver them. They are further classified as *neurotoxic, hematoxic, cytotoxic* or *myotoxic*.

23.2.1 Invertebrates

Human poisoning may occur from several invertebrates viz. Cnidaria (jelly fish), Porifera (sponges), Echinodermata, Mollusca and Annelida.

The phylum Cnidaria contains approximately 10,000 species. All four classes of Cnidaria contain venomous species. Almost 100 species are medically important. The class hydrozoa has worldwide distribution and includes *Physalia*, (Portuguese and

Pacific man of war or blue bottle), the hydroids (Millipora, stinging corals), limnomedusae and Gonionemus (Fig. 12). Physalia causes thousands of envenomations in Florida, South America, parts of Asia and Africa, Australia and Portugal. Physalia has large gas filled floats which suspend multiple tentacles. These tentacles bear nematocysts. Each nematocyst contains a small amount of venom. A physical or chemical stimulus triggers the release of hollow, sharply pointed thread like tube from the contained nematocyst. This process is very rapid (1/100 second) and results into the penetration of the tube into the skin and delivering venom simultaneously.

Hydroid stings are minor but can occur even after superficial contact. It causes discomfort in humans *Lytocarpus* (Fig. 13) and *Aglaophenia* are locally called as fireweeds. They have featherlike branches lined with nematocyst bearing polyps. A number of other medically important hydrozoans include *Gonionemus* (Fig. 14) and *Olindias*. *Gonionemus* causes envenomations in Japan. *Olindias* stinging is known to occur in Brazil, Uruguay and Argentina. Stings cause erythematous and edematous reactions and mild pain.

The class cubozoa contains box jelly fish. They possess a cube shaped body with tentades attached to each of the four corners. Chirodropid box jelly fish (*Chironex fleckeri*) is the world's most venomous animal (Fig. 15). It is found along the northern coast of Australia, another group of box jelly fish belong to the order Carybdidae. The group is identified by *Carukia barnesi* from Australia which causes IRUKANDJI syndrome.

23.2.2 Irukandji syndrome

It was first described in 1940s and is characterized by generalized pain hypertension, nausea, vomiting and anxiety. Initially the sting causes mild local pain and a patch of erythema. After 20 to 30 minutes, severe generalized pain in the abdomen, back and muscles occurs. The pain is associated with systemic features like tachycardia, sweating, piloerection, agitation and in severe case pulmonary edema. At least in Australia, several jelly fish may be responsible for lrukandji like effects.

23.2.3 Scyphozoa (True jelly fish)

Several species of true jelly fish are known to be poisonous. They include sea nettle, hair jelly fish, blubber jelly fish, mauve stingers and moon jelly fish.

23.2.4 Sea anemones

Sea anemones are often found in tidal pools. They have numerous nematocysts on their tentacles. Stings are characterized by local pains. The pain may last for few hours and local tenderness may remain for days.

23.2.5 Echinoderms

The phylum Echinodermata contains organisms like sea urchins, star fish and sea cucumbers. Echinoderms contain a variety of toxins including steroid glycosides and terpenes but only a few of these animals are known to cause envenomation. In the small group of sea-urchins (Fig. 16), the venom apparatus varies from short sharp spines with venom glands on their tips. Systemic features have been reported including nausea, vomiting, parasthesiae, muscular paralysis, hypotension and respiratory distress.

Sea-stars only cause a minor traumatic injury. The animals are covered with sharp rigid spines that can passively deliver a variety of substances when they penetrate skin. The region may become dusky or discolored particularly with multiple spine injuries.

Sea-cucumbers do not have a venom apparatus but direct contact may induce a contact dermatitis. Injury to the cornea and conjunctiva may cause intense inflammation.

23.2.6 Sponges

Sponges are the simplest of the multicellular organisms. They do not have a specialized venom gland but many species produce **crinotoxins** (slimes or surface liquids) and a few species can cause skin irritation and dermatitis. Application of sponge extract to the rabbit cornea results in opacity and blindness demonstrating potential hazard to eyes. *Neofibularia mordens* (Fig. 17) causes the most severe stings in Australasian region. Stings from several sponges have been reported from United States, Hawaii and the Caribbean islands.

23.2.7 Annelida

Phylum Annelida includes the bristle worms (Fig. 18). Their body is segmented and covered with bristle like setae that become erect on contact. These setae or spines detach from the worm and are able to penetrate human skin. Initially there is a burning sensation followed by intense skin inflammation. Localized safe tissue edema and itchiness may occur. In some cases the effect may last for many weeks.

23.2.8 Molluska

The phylum Molluska contains unsegmented soft bodied invertebrates that include the snails and slugs, chitons, bivalves, octopi, squids, and related species. Two classes of mollusks have a venom apparatus the gastropods (cone snails and nudibranchs) and the cephalopods (Octopi). Cone snail shells mainly occur in the Pacific Ocean. Cone snails are often classified as (a) piscivorous (b) molluskivorous (c) vermivorous. The venom they produce consists of a mixture of neurotoxic components called as **conotoxins**. They have a broad range of pharmacologic effects, including blockade of Na^+, Ca^+, and K^+ channels and antagonism at nicotinic and serotonin receptor sites.

The only cephalopod that have been reported to cause significant envenomation belongs to the genus *Hapalochlaena* (Fig. 19), the blue ring Australian octopus. The octopus can inject its neurotoxin which is stored in modified venom glands. The symptoms are nausea, dizziness, malaise, jerky muscular movements and finally respiratory failure.

23.2.9 Fish

Several species of fish are known to be poisonous. They include catfish, stonefish, lion fish, bull trout, weever fish, stingrays, puffer fish, shell fish and sea hares.

Approximately 1000 species of catfish inhabit fresh and software although many do not contain venomous spines. Spines at dorsal and pectoral fins can inflict wounds in fishermen and less commonly in water sports participants. The venom glands are attached in the leading edges of the spines and are contained within integumentary sheaths. Penetration of the spine into the skin and subdermal tissues simultaneously causes the rupture of the spine integument and venom is then injected into the wound Many spines have a series of rajor sharp teeth that act as barbs making extraction of these structures difficult. Serious envenomations have been reported from Arabian Gulf Catfish (*Aurius thalasinus*) (Fig. 20) and the oriental or striped cat fish (*Plotosus lineatus*) (Fig. 21). The strike causes muscular spasm, respiratory distress, neurotoxicity, hemolysis and death in laboratory animals but these effects are rarely reported in humans.

Stone fish (*Synanceia* spp) (Fig. 22) occur through out tropical warm temperate oceans. The venom apparatus of stonefish is highly developed and consists of paired venom glands that are associated with 13 dorsal spines of the fish. Each spine produces 6 mg of venom. It can cause cardiotoxicity, myotoxicity and neurotoxicity.

Lionfish (Fig. 23) are probably the most commonly reported fish to cause stings in Untied States. They are the most colorful venomous fish with large fan like fins. Therefore, they are kept in aquariums. Most of the injures have been reported from aquarium keepers.

Scorpion fish (Fig. 24) occur worldwide but most commonly in tropical and temperate oceans and seas. At least 80 members of the fish have been reported to be poisonous. They exhibit camouflage and may be stationary or slow moving. Their venom apparatus varies between species with most having 10 to 15 dorsal spines, 2 pelvic and 3 anal spines associated with venom glands.

Soldier-fish (*Gymmapistes marmoratus*) occur in southern Australia. The venom apparatus consists of 13 dorsal, 3 anal, 2 pectoral, and 4 opercular spines. The venom has the same effects as described for stone fish.

The bull trout (*Notesthes robusta*) (Fig. 25) lives in tidal estuaries of eastern Australia and belongs to the Scorpaenidae family. They occur in rocks and weeds. It has 15 spines that erect when fish is disturbed. Their venom is called nocitoxin that contain a 170 kd protein.

Weever fish are marine fish found in the Mediterranean and European coast of Chile. They are bottom dwellers and sting when stepped on. Weever fish have five obviously visible dorsal spines and a few less visible ones that are responsible for stings. These spines can pierce a leather boot. The venom is proteinaceous and has hemolytic properties. Sting may cause pain and local inflammation for upto a week.

Stingrays (Fig. 26) have a characteristic dorsoventrially flattened appearance with pectoral flaps that they use for propulsion. Gills are confined to the ventral surface. The venom apparatus of stingrays is a whip like striking organ (spine) attached to the dorsal surface of the fish. The sting can produce a significant laceration and less commonly a punctured wound. The traumatic wound itself is dangerous and increases the risk of infection. Venom is simultaneously injected. It may cause cardiovascular collapse as well as neurologic effects when injected intravenously or intraperitoneally to rats. Divers sustain injuries to the chest or abdomen.

There are other several venomous fish. Scats occur in the Indo Pacific Ocean. Stings cause immediate pain. They are kept in aquariums. Zebra fish has also been implicated in spine injuries. Australian shark has venomous spines at front of each of the dorsal fin. Puffer fish poisoning is mainly confined to South East Asia and is more common in Japan where fugu is a delicacy and mainly consumed as a food.

Shellfish poisoning is a medical and economic problem that affects many fisheries specially in Japan, Southeast Asia, Central and North America, North Africa and Europe. Shellfish are the vectors for a number of infections including viral and bacterial infections, allergies and toxin poisoning. Four major toxic syndromes that result from shell fish poisoning are – paralytic shellfish poisoning, diarrhetic shellfish poisoning and encepalopathic shellfish poisoning.

23.2.10 Amphibians

Amongst Amphibians, toads are the only poisonous animals. They are found all over the world except Madagaskar, New Guinea, New Zealand and Polynesia. Their skin secretes the venom. It may be stored in the parotid glands behind the eyes. Toxicity can occur if toad's skin or secretions are handled or ingested.

Dried and powdered toad skin have been used as cardiac medicines, hallucinogens, expectorants and diuretics and also for the treatment of toothaches, sinusitis, and bleeding gums. Eyes, nose and throat pain and irritation may follow on contact exposure. Cardiac arrest, atrial fibrillation bradycardia, ventricular fibrillation and hypotension have been described. Seizures have been reported, as have salivation and vomiting.

Another amphibian known to be poisonous is poison dart frog. They are found in tropical rain forests of central America and south America. The most poisonous amongst them is *Phyllobates terribilis* (Fig. 27). Poison dart frogs secrete **batracotoxin** that depolarizes electrical membranes through increasing the permeability to sodium ions.

Salamanders (Fig. 28) secrete a poison known as **samandrin**. It is a potent neurotoxin. Newts secrete a potent toxin TTX in their skin. Poisoning is caused by the ingestion of flesh, viscera or skin. Ascending paralysis develops which leads to respiratory paralysis.

23.2.11 Reptiles (lizards)

Amongst lizards, helodermatods are known to be poisonous. They are large, slow moving and primarily nocturnal. Adult lizards can reach to the length of 55 cm. The skin has bead like scales, which can be black or brown in colour. The lizard has a set of venom glands which can be found in the anterior lower jaw. *Heloderma suspectum* (Fig. 29) and *H. cinctum* inhabit the southwestern parts of the United States. *H. horridum* is native to Mexico and Guatemala and is generally larger than the sub species found in U.S. The lizard has a powerful bite, and the teeth may cause several small puncture wounds in the skin of the victim. Envenomation by adult helodermatids

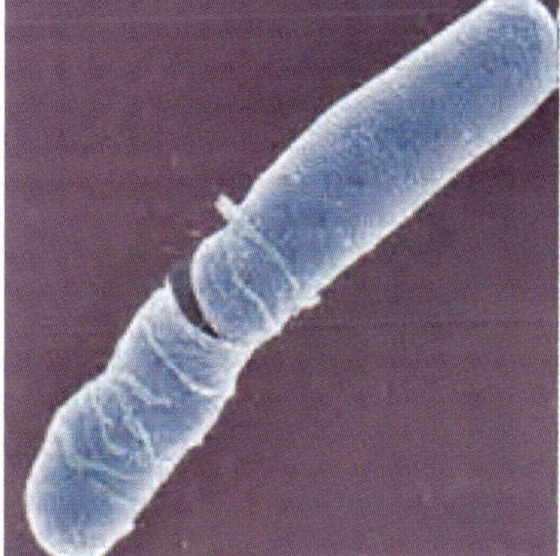

Fig. 1: *Bacillus cereus*

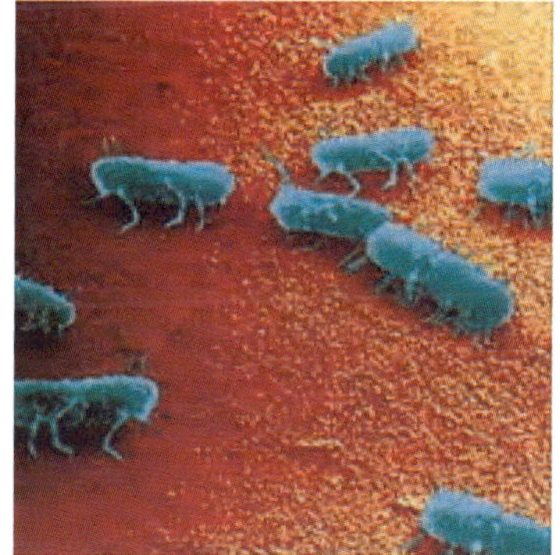

Fig. 5: *Salmonella*

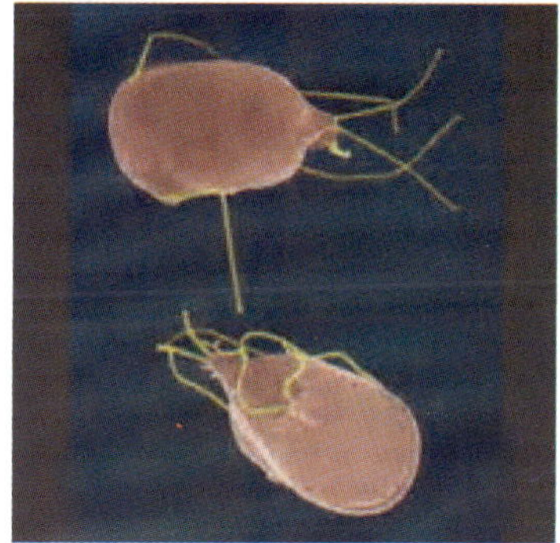

Fig. 9: *Giardia lamblia*

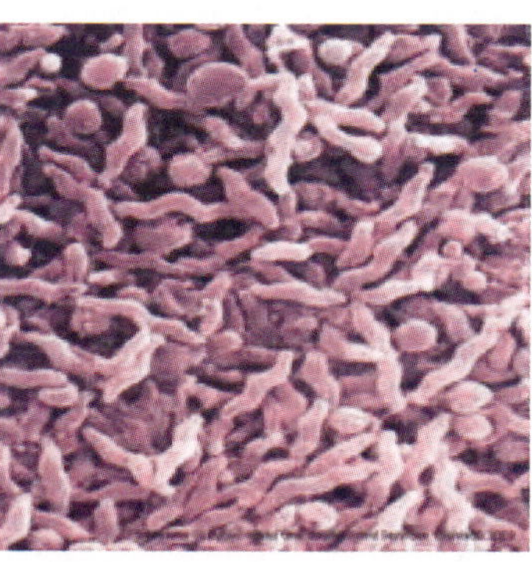

Fig. 2: *Campylobacter sp.*

Fig. 6: *Staphylococcus*

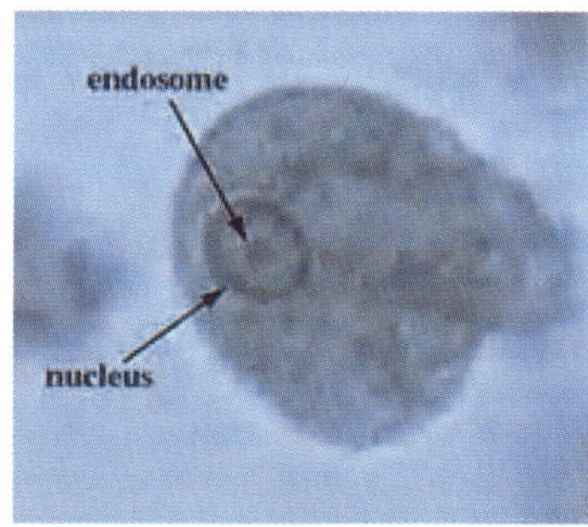

Fig. 10: *Entamoeba histolytica trophozoite*

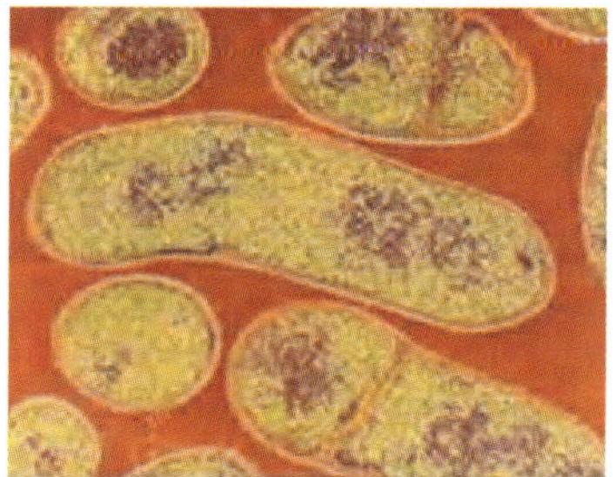

Fig. 3: *Clostridium botulinum*

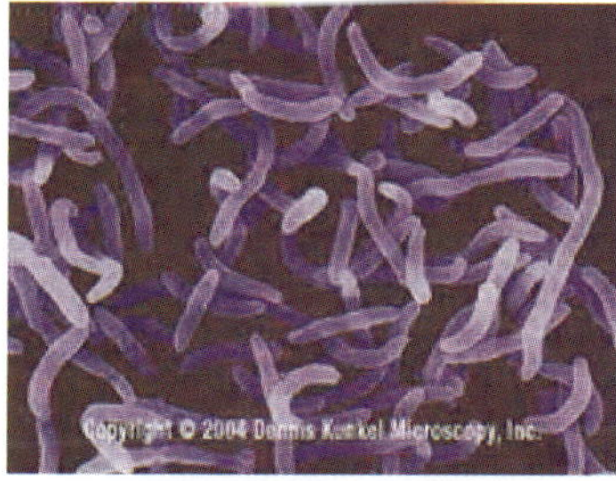

Fig. 7: *Vibrio cholerae*

Fig. 11: *Balantidium coli*

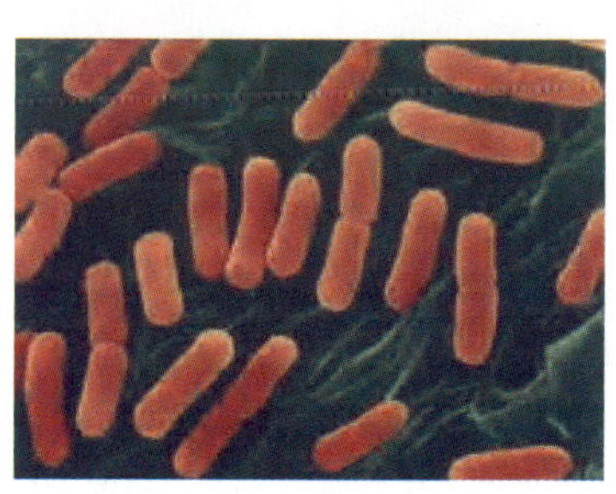

Fig. 4: *E. Coli Escherichia*

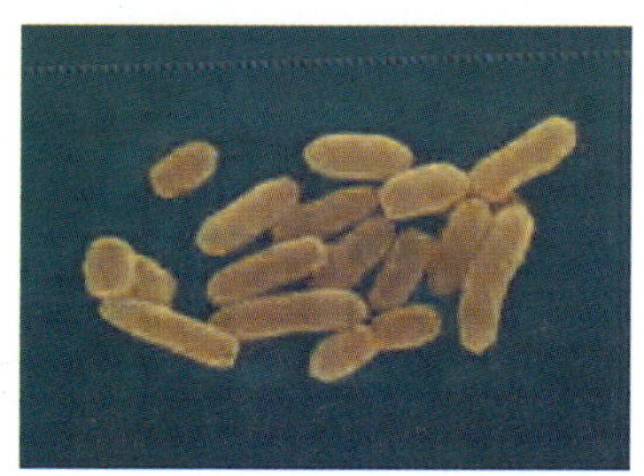

Fig. 8: *Yersinia-pestis*

Fig. 12: *Physalia physalis*

Fig. 13: *Lytocarpus philippinus*

Fig. 14: *Gonionemus side view*

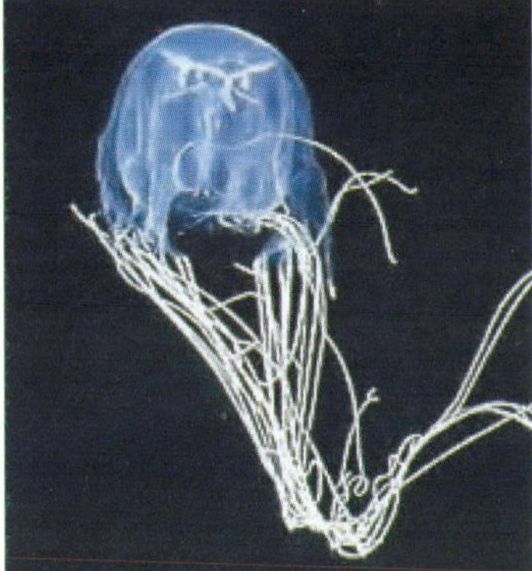

Fig. 15: *Chironex fleckeri.*

Fig. 16: *Sea urchin*

Fig. 17: *Neofibularia mordens*

Fig. 18: *Bristle worms*

Fig. 19: *Ringed Octopus*
(Hapalochlaena maculosa)

Fig. 20: *Arius thalassinus*

Fig. 23: *Lion fish*

Fig. 21: *Plotosus lineatus*

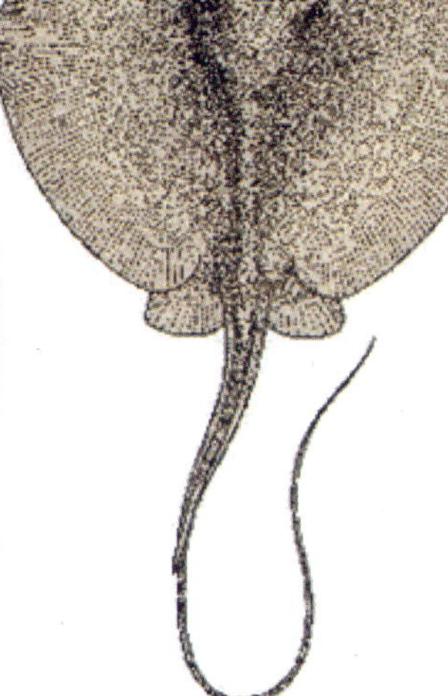

Fig. 26: *Stingray*

Fig. 24: *Scorpion fish*

Fig. 22: *Synanceia verrucosa*

Fig. 25: *Bull trout*

Fig. 27: *Phyllobates terribilis*

Fig. 28: *Salamander*

Fig. 29: *Heloderma*

Fig. 30: *Thelotornis*

Fig. 33: *Boomslang Dispholidus types*

Fig. 36: *Coral snake*

Fig. 31: *Rhabdophis cheysarga*

Fig. 34: *Naja naja (Indian Cobra)*

Fig. 37: *Atractaspis microlepidota*

Fig. 32: *Malpolon monspessulanus*

Fig. 35: *Bungarus*

Fig. 38: *Rose Throated Viper*

Fig. 39: *Scorpion*

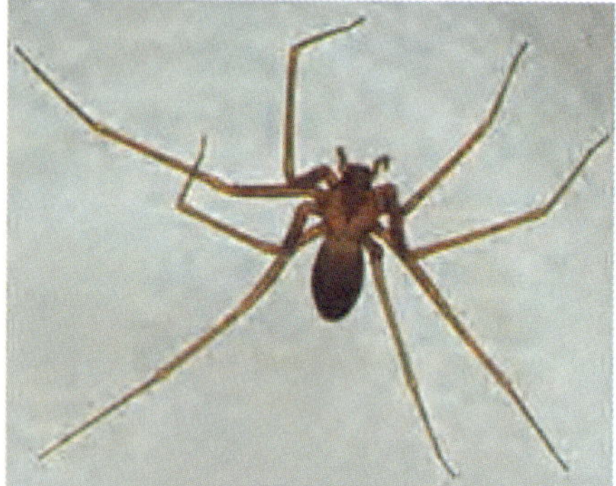

Fig. 40: *Loxosceles rufescens*

Fig. 41: *Asian bee (left), European honeybee*

Fig. 42: *Hornet*

Fig. 43: *Solenopsis*

Fig. 44: *Scolopendra*

Fig. 45: *Giant millipede*

Fig. 46: *Ricinus Communis*

Fig. 49: *Atropa belladonna*

Fig. 53: Apricot

Fig. 47: *Digitalis*

Fig. 50: *Conium maculatum*

Fig. 51: *Selsnepe Cicuta virosa*

Fig. 54: *Manihot esculenta*

Fig. 48: *Aconitum sp.*

Fig. 52: *Hydrangea paniculata*

Fig. 55: *Solanum tuberosum*

Fig. 56: *Cantharis vesicatoria*

Fig. 57: *Ephedra viridis*

produce pain and local swelling at the site of the bite. Lymphangitis, hypotension, and weakness may occur. The venom contains serotonin, amine oxidase, phospholipase, hyaluronidase, protease and salivary kallikrein. Unique compounds include gilatoxin and helothermine. No antivenom is available for helodermatid envenomation.

23.2.12 Snakes

Of the approximately 3000 species of snakes in the world, about 600 or 20% are venomous. Nearly all of them possess specialized teeth called as fangs. Venomous snakes are found on all continents except Anarctica. Venomous snakes belong to four families: Colubridae, Elapidae, Atractaspididae and Viperidae.

Family Colubridae is the largest family of snakes, with 1864 species in a global distribution. Most of them are nonvenomous but a minority is venomous. They include vine or bird snakes (*Thelotornis sp.*) (Fig. 30), red necked keelback (*Rhabdophis sp.*) (Fig. 31). Montpelier snake (*Malpolon monspessulanus*) (Fig. 32), Argentine black headed snake (*Elapomorphus bilineatus*) and Boomslang (*Dispholidus typus*) (Fig. 33).

Next family of elapid snakes encompass a wide variety of species with 297 species on a global distribution, including African and Asian cobras, mambas, kraits and coral snakes, Australian snakes and sea snakes. In most species fangs have an enclosed groove, acting like a hypodermic needle. The snake has the ability to control venom release. Indian cobra (*Naja naja*) (Fig. 34), king cobra (*Ophiophagus hannah*), kraits (*Bungarus sp.*) (Fig. 35), water cobra (*Naja melanoleuea*), tree cobras (*Pseudohaje spp*), burrowing cobra (*Paranaja multifasciata*) American coral snakes (*Micrucoides euryxanthus*) (Fig. 36) Asian coral snake (*Maticora spp.*) are a few major species of medically important elapid snakes.

Family Atracaspididae have a limited distribution in Africa and the Middle East. They are fossorial, usually hunt the prey beneath the surface (Fig. 37). Their fangs come out of the side of the mouth to allow a sideways strike. Most are probably harmless to humans but a few in the genus Atractaspis have toxic venom with unusual components notably the **sarafotoxins**.

Vipers (*Viperidae*) are the second family of venomous snakes. All have mobile fangs in front of the mouth on hinged maxillae, allowing the fang to fold away against the root of the mouth when not in use. Their fangs are longer than those of elapid snakes. These long fangs are often coupled with large venom glands, allowing large quantities of venom to be injected but **"dry bites"** can also occur. There are 3 typical vipers (Fig. 38) and pit vipers. Pit organ is located in front of their head. It allows infrared detection of the warm blooded prey in total darkness. Rattlesnakes are best known amongst pit vipers.

23.3 Snake Venoms

Venoms are produced in specialized venom glands, usually linked to the base of fangs by a duct. Venom glands have evolved from digestive glands and many venom components viz. phospholipases are the examples of this ancestral relationship. Venoms fulfill three main functions.

 i. Immobilization of prey

   ii.   Digestion of prey

   iii. Deterrence of predators

Nevertheless, venoms are complex mixtures of toxins. Each toxin may have several distinct actions on different systems of the body. They can suitably be classified as – neurotoxins, myotoxins, cardiotoxins coagulopathic toxins, nephrotoxins and necrotoxins.

23.3.1 Neurotoxins

Classic snake venoms cause flaccid and respiratory paralysis. They damage the human presynaptic terminal axon. They stop the release of neurotransmitters. A few snakes secrete dendrotoxins. They are further categorized the potassium channel neurotoxins.

23.3.2 Myotoxins

A few snakes produce systemic **myolysins** which in humans may result in potential lethal hydrolysis and intracellular damage to individual muscle cells within 60 minutes of reaching the target organ and by 24 hours, cell destruction is complete. The snakes that cause systemic myolysis included Australian snakes (sea snakes, tiger snakes, mulga snakes), South American pit vipers and Sri Lankan Russel's pit viper.

23.3.3 Cardiotoxins

Cardiotoxins of phospholipase A2 (PLA2) type are known to be present in snake venoms, scorpions, jelly fish and cone shells. The mechanism of these toxins is diverse and needs confirmation.

23.3.4 Coagulopathic toxins

Disturbances in hemostasis are the major effects of snake venoms. Certain snake venoms induce precoagulation of blood. These procoagulants are usually multicomponent molecules. Their structure mimics normal components of human hemostasis, particularly a part or prothrombokinase complex (factor Xa, Va, phospholipid & Ca).

Some venoms contain anticoagulants that directly inhibit the portions of clotting cascade resulting in prolonged clotting time.

A number of viperoid snake venoms contain fibrinolytic agents including proteinases. Fibrinolytic enzymes splitt off either the Aá or Bâ or both sets of fibrinogen fibrinopeptides. Other fibrinolytic agents are the plasminogen activators. They often consume fibrinogen thus effectively causing anticoagulation.

Venoms do contain platelet active agents. They promote or inhibit platelet aggregation through a variety of direct or indirect mechanisms.

Hemorrhagic activity is a prominent feature of many viper venoms. The viperid zinc metalloproteinases are amongst the best characterized enzymes. They cause capillary leakage by degrading blood vessel basement membranes, resulting in haemorrhagic necrosis.

23.3.5 Nephrotoxins

A few snakes appear to possess primary nephrotoxins in their venom. Snakebite may cause permanent renal injury viz. cortical necrosis. Several factors may contribute in renal injury.

23.3.6 Necrotoxins

A variety of venomous animals can cause local tissue necrosis through diverse mechanisms. Vipers, pit vipers and some cobras commonly cause local tissue injury.

23.4 Arthropoda

23.4.1 Scorpions

More than 1000 scorpion species are known worldwide. Stings of at least 20 of them cause serious envenomations. In humans, in 1998, 12845 scorpion stings were reported by poison centres in the United States. Scorpions, like Arachnids are predators. Venom is stored in a "telson" or bulb at the tip of the tail to which is attached a stinger (Fig. 39). Venom is injected into the prey by a tail strike to paralyse it and enable subsequent feeding. Venom is also injected during defensive stings, including those provoked by humans. Because scorpions are predominantly nocturnal, most envenomations occur at night. Several species of scorpions commonly enter homes, putting inhabitants at risk of accidental stings, particularly at night.

Scorpion venom contains a set of polypeptides directed differently against insect, crustacean, and vertebrate nervous system. They involve ion channels including those for transport of sodium, potassium and calcium. Scorpion venoms may comprise of 65 to 68 amino acid chains, all of which contain eight highly conserved cysteine residues that are involved in the formation of four disulfide bridges. Voltage-dependent ion channel toxicity appears to affect human axonal function leading to simultaneous sensory (local pain) nuscarinic (hyper secretion), nicotinic (tachycardia, hypertension) and neuromotor (cranial and peripheral motor hyperactivity) effects.

23.4.2 Spiders

More than a dozen families of spiders throughout the world have been reported to be poisonous to humans. Six of these families are distributed in United States. Most of them have four pairs of eyes. The genus *Loxosceles* (Fig. 40) is the most venomous and by far the best characterized of **dermonecrotic** have been described, the effects of envenomation are indistinguishable from species to species. Amino acid sequence of the venom has revealed only minor sequence differences in species. The spider uses its venom to paralyze the inset prey. Venom components include hyaluronidase which maximizes dermal inflammation. Other components include S-ribonucleotide, phosphohydrolase, alkaline phosphalase and sphingomyelinase D. After envenomation, the dermal endothelial epithelial cells produce secrete and secrete the chemokines-interleukin-8, growth related oncogene-á and monocyte chemo attaractant protein-1. The major systemic manifestations induced by multiple venom effects include endothelial cell amage, red cell hemolysis, and coagulopathy. The clinical signs of enevomation vary from a mild local reacton to systemic symptoms and (on rare occasion) death depending upon the quantity of venom inoculation and number of host variables.

23.4.3 Hymenoptera (Bees, wasps and ants)

Many arthropods in addition to spiders and scorpions cause injuries to humans. Order hymenoptera contains the bees, wasps and ants. Deaths from bees, wasps and ant stings continue to occur in most countries including western countries.

Although many thousand of species of bees exist, only *Apis* is known to be dangerous. Mass envenomations occur due to African honey bee. In Asia, the giant Asian rock bee (*Apis dorsata*) and the Asian honey bee (*Apis cerana*) have caused mass envenomations (Fig. 41).

The principal component of the venom is MELLITIN and a phospholipase A2(PLA_2). Other components include hyaluronidase, apamin, mast cell degranulating peptide and small amounts of histamine and catecholamines. Amongst then PLA_2 is the most lethal.

Most stings are characterized by initial local pain that may be severe associated with the development of localized swelling and erythema. Local reactions may occur that appear like cellulitis, with erythema and edema peaking at 48 hours and persisting for seven days. Deaths from mass envenomations from *A. mellifera* have been reported in Europe and Hawaii.

23.4.4 Wasps

The family vespidae includes three main types of venomous wasps i.e. hornets, yellow jackets and paper wasps. Only these three groups exist in large colonies and can cause systemic envenomation. The vespa wasp includes approximately 20 species and are the most dangerous because of their greater venom toxicity and ability to deliver venom (Fig. 42). The majority of them occur in eastern Asia. Only female vasps can sting. They have unbarbed hollow stings that also serve as an ovipositor in the queen. Vespid venoms contain three types of substances (a) high molecular weight protein enzymes such as PLA_2 and hyaluronidases, (b) biologic active amines (histamine, serotonin, tyramine, and catecholamines and (c) small peptides (kinins, mastoparans and chemotactic peptides. Some wasps have neurotoxins such as mandaratoxin from *V. mandarina*. The patient shows nausea, vomiting, fatigue, fever, and malaise. Severe envenomation is characterized by intravascular hemolysis, rhabdomyolysis, common effects include acute respiratory distress syndrome, hypertension, myocardial damage, shock and coma. Death may occur with a few as 20 to 50 stings.

23.4.5 Ants (Formicidae)

More than 10,000 species of ants have been described. Only a few are medically important. In Australia, the most medically significant are the jack jumper and bull ants (*Myrmecia* sp). Other important species include the fire ants (*Solenopsis* sp.) of

North America (Fig. 43). *Solenopsis* occur in many parts of the world. Most species of the ants grab the skin with their jaws and either sting or spray venom which causes local pain, etching, erythema and swelling.

23.4.6 Chilopoda (Centipedes)

They are widely distributed animals. They are nocturval and prefer to live in dark and moist places. The venom apparatus of most centepedes consist of a pair of poison claws, which are the first pair of legs modified into biting appendages. The most medically important amongst centepedes is *Scolopendra* (Fig. 44). It achieves the length upto 20 cm. The poisoning by Seolopendra has been poorly defined. It was found to cause severe local and radiating pain and significant swelling of the bitten area and the hand.

23.4.7 Diplopoda (Millipedes)

The millipedes are a small group that rarely cause injuries to humans. Giant millipedes of tropical regions produce irritant secretions. These secretions contain quinonoids and p-benzoquinones. Quinonoids cause minor effects on contact with the skin. They include pain, itching and local skin eruptions. *Giant millipedes* (Fig. 45) occur in Africa, Papua New Guinea, Irian Joya and the Caribbean inducing Mexico and Panama.

23.4.8 Ticks

Ticks are medically important. They are vectors for some infectious diseases, including lyme disease. Tick salina is known to contain the neurotoxins that are responsible for paralysis Allergic reactions to Australian ticks are well reported. Ticks can cause local reactions to severe systemic allergy.

23.5 Plant Poisons

Poisoning from plants is a common fear but an extremely rare event. It is true that a few plant species in certain localities can produce serious toxicity like dander (*Nerium oleander*), foxglove (*Digitalis purpurea*), jequirity pea (*Abrus precatorius*), castor bean (*Ricinus communis*), water hemlock (*Cicuta maculata*), Jerusalem cherry (*Solanaum pseudocapsicum*), free tobacco (*Nicotina glauca*), jimsonweed (*Datura stramonium*), autumn crocus (*Colchicum autumnale*) and hepatotoxic mushrooms (*Amantia phalloides* and *A. virosa*).

In most cases, exceptional circumstances are required to produce severe poisoning. Even then, each year, a few deaths from plant poisoning do occur all over the world.

23.5.1 Yew (*Taxus canadensis*)

This is a very common ornamental hedge. All parts of *Taxus* except the red fruit contain poisons i.e. taxine A and taxine B within 1 hour of consumption it causes severe gastroenteritis which may be followed by convulsions, shock, coma and death.

It has also been classified as cardiotoxic. It has been recognized as an abortifacient by many women.

23.5.2 Castor bean (*Ricinus communis*)

This plant contains ricin, a potent cellular protein toxin. Its seeds containing the toxin can be swallowed whole without injury, if, however, the seeds are crushed and the toxin is released, then severe gastroenteritis results. It may lead to CNS depression, cardiac dysrhythimias, coma and death (Fig. 46).

23.5.3 Rosary pea (*Abrus precatorius*)

It produces a toxin called abrin which is similar to ricin. It inhibits protein synthesis. All parts of the plant are toxic with seeds containing the higher amount of toxin.

23.5.4 Lily of the valley (*Convallaria majalis*)

Convallarin and convallotoxin are found in this common garden plant. Convallotoxin and other glycosides act by inhibiting the enzyme Na^+/K^+ ATPase, *Convallaria* causes bradycardia and uncoordinated heartbeat.

23.5.5 Toxglove (*Digitalis purpurea*)

It produces a toxin called as "digitoxin" Digitoxin has been very valuable in medicine, but while it has saved many lives, it has also produced significant toxicity. It damages kidney. Many murders have been committed with digitalis as vehicle of death (Fig. 47).

23.5.6 Oleander (*Nerium oleander*)

Its poisoning is based on cardiac glycoside. It predominantly causes GI and cardiac symptoms. Even meat roasted on the twigs of this plant becomes poisonous. Bees sometime use oleander pollen for their honey. The honey too, prepared in this manner has been found to be poisonous.

23.5.7 Monkshood (*Aconitum napellus*)

This plant is also known as wolfsbane (Fig. 48). It is used by some as an herbal medicine under the name of aconite. It contains two alkaloid toxins i.e. aconine and aconitine. Upon ingestion, cardiac and neurologic symptoms have been reported. Several French recruits died from eating monkshood while on a training exercise during World War – I.

23.5.8 Black hellebore (*Helleborus niger*)

This plant is also known as henbane. The entire plant is poisonous and contains hellebrin, helleborin, and saponins. It is a GI irritant, but its major effect is death from cardiac arrest.

23.5.9 Death camas (*Zygadenus venenosus*)

They are called as grayanotoxins and include veratrine and zygadenine. They cause bradycardia and hypotension as well as cholinergic symptoms namely salivation,lacrimation, rhinorrhea and emesis. They have been mistaken by campers for onion.

23.5.10 Azalea (*Rhododendron*)

These plants are omnipresent. They share many properties. They produce a specific poison called andromedotoxin. This toxin causes GI stress, respiratory difficulty and bradycardia. Some birds eat these bushes which renders their flash poisonous.

23.5.11 Mushrooms

Five thousand mushroom species are known to occur in United States alone. Approximately 2% of these species are toxic. Poisonous mushrooms contain toxins that are as diverse as mushrooms, themselves. The mushroom portion of the fungus is the reproductive structure that grows from underground mycelium as a densely packed cap and stipe of interwoven hyphal strands. This structure contains the spores that germinate and form new mycelia. Variation in size, shape, color, spore and other microscopic structures aids in the identification of mushroom species.

There are about eight groups of compounds secreted by different species of mushrooms.

i. Cyclopeptide group – *Amantia phalloides., A verna, A. virosa, A. bisoporegera and Galerina sp.*
ii. Monomethyl hydrazine group – *Amantia muscaria and A. patherina*
iii. Coprine group - *Gyromitra sp.*
iv. Muscarine group – *Clitocybe and Inocybe sp.*
v. Ibotenic acid and muscimol group – *Coprimus sp.*
vi. Hallucinogen group – *Psilocybe, Panaeolus, Gymnopilus*
vii. Gastrointestinal group – *Chlorophyllum molybditis*
viii.Renal failure group – *Cortinarius sp.*

The cyclopeptide amatoxins of *Amantia* and *Galerina* can cause severe hepato-renal dysfunction. *A phalloides* is the prominent European poisonous mushroom. Group II mushrooms cause gastritis, less often hemolysis, hepato-renal dysfunction, convulsions and death follows ingestion. Coprine mushrooms of group III cause hyperacefaldehydemia. Muscarine mushrooms are reported to secrete muscarine. TI causes cholinergic excesss syndrome within 30 minutes. Ibotenic acid species (*A. muscaria*) that contain psychoactive isoxazole derivatives, produce hallucinations. They cause muscle spasm, confusion, intoxication drowsiness and sleep. The main toxin of hallucinogenic mushrooms is psilocybin. This has lysergic acid diethylamide like properties and produce alterations in autonomic function, motor reflexes, behavior and perception gastrointestinal weakness, nausea, vomiting and diarrhea.

23.5.12 Anticholinergic poisoning by plants

Anticholinergic poisons interfere with the progpagation of nerve impulse. They are found in the following plants:

    i. Deadly nightshade (*Atropa belladonna*)

All parts of Atropa are dangerous including roots, leaves and berries. Berries contain the highest content of toxic alkaloids. These are known as atropine, hyoscine and hyoseyamine and refers to the practice during Renaissance of placing an extract of Atropa inorder to achieve dilated pupils, regarded by many as the attaractive feature (Fig. 49).

    ii. Jimsonweed (*Datura stramonium*)

Jimsonweed is also called as thornapple, stink weed and Devil's trumpet. It contains all the three alkaloids as described above. Severe cases may lead to loss of sight, convulsions, coma and death. Many soldiers died after eating this plant when famine broke out in 1666 in the early American colony, Jamestown in Virginia.

    iii. Mandrake (*Mandragora officinarum*)

This plant, in addition to atropine and hyoscyamine, contains momdragorin which is considered to promote fertility and had aphrodisiac properties. It was also associated with witchcraft and women, who possessed mandrake were executed in 17th century in Germany.

23.5.13 Neurotoxic plants

23.5.13.1 Hemlock (*Conium maculatum*)

The toxins found in this plant are piperidine alkaloids, coniine and gamma coniceine. Their primary lethal consequences are respiratory failure. They also produce nicotinic effects viz. salivation, mydriasis, tachycardia followed by bradycardia. Pluto, the great Greek philosopher has described the death of Socrates due to hemlock (Fig. 50).

23.5.13.2 Water hemlock (*Cicuta virosa*)

Water hemlock is a weed commonly found along lakes and streams. Its toxin is known as cicutoxin that specifically works upon brain and spinal cord. It causes rapid onset of status epilepticus (Fig. 51).

23.5.13.3 Curare (*Chondrodendron tomentosum*)

The red Indians in Amazon and South America dip their arrows with this poison which paralyzes the skeletal muscles of their prey. Death results from respiratory failure.

23.5.14 Cyanogenic plants

All those plants that are able to form cyanide under certain conditions are called as cyanogenic plants. Most cyanogenic substances are glycosides-meaning thereby that a carbohydrate moiety is part of their structure. Fortunately the cyanogens are not found in fruits but present in leaves, stem and bark. Examples are apple, apricot, cherry, peach, and black berry.

23.5.14.1 *Hydrangea paniculata*

Hydrangea contains two cyanogenic glycosides known as hydrangin and amygdalin. Symptoms of poisoning are nausea and gastroenteritis (Fig. 52).

23.5.14.2 Apricot

This fruit contains amygdalin. Interestingly amygdalin has become celebrated as an alleged cancer cure and is most commonly called as laetrile. Laetrile is more likely to cause harm rather than benefit (Fig. 53).

23.5.14.3 Cassava (*Manihot esculenta*)

This is a common dietary component in many parts of the world. It contains a cyanogenic glycoside known as linamarin. Tropical ataxic neuropathy has been observed in Nigeria and epidemic spastic paraperesis has been observed in certain parts of equatorial Africa (Fig. 54).

23.5.15 Hepatotoxic plants

Akee when unripe, contains hypoglycin, a compound believed to be teratogenic and a cause of toxic hypoglycemic syndrome (also called *Jamaican vomiting sickness*). Akee is a staple food of Jamaica and British west Indies. Raw or spoiled fruit is to be avoided.

23.5.16 Solanaceous plants

About 1700 species of these plants are known till date. They all contain salanaceous alkaloids. Many solanaceous alkaloids contain the same basic aglycone but differ in the number and type of carbohydrate molecules. Solanine occurs in the common potato. *Solanum tuberosum.* Levels of solanine above 20 ppm are dangerous. Solanine is a cholinesterase inhibitor. It causes cholinergic symptoms such as salivation, trembling, progreisive weakness and paralysis (Fig. 55).

23.6 Herbal and Indigenous Remedies

It is not unusual for people who use traditional medicines and seek advice from traditional ethnic practioners before resorting to more conventional medical advice. As a matter of fact, indigmoous medicines have been used for millennia dating back to 1500-2000 BC. It should not be surprising to know that many medicines used today such as morphine, cocaine, colchicines, reserpine, vineristine and paclistaxel were initially derived from plants. The beneficial effects of many plants/plant produces are anecdotal and based on years of use in traditional medicine. However, many herbal products have been shown to cause adverse effects and death. Toxicity of a few selected herbs is described below.

23.6.1 Aconite

Aconites are the dried root stocks or tubers of Aconitum plants. They have been used to treat rheumatism, bruises, fractures, hemiplegia, diarrhea and abdominal pain. The principal toxic ingredients are C19-diterpenoid alkaloids. They bind to sodium channels in excitable tissues causing persistent activation and sodium influx during the plateu phase of depolarization.

23.6.2 Anticholinergic plants

Plants such as belladonna, henbane, jimsonweed and mandrake contain alkaloids such as scopolamine, hyoscyamine, or atropine. These are often used to treat GI and respiratory disorders or as hallucinogens. The toxicity occurs in the form of confusion, agitation, tachycardia, dry flushed skin, fever, urinary retention and decreased bowel sounds.

23.6.3 *Aristolochia*

It is also known as birthwort, snakeroot, guangfangi and pelican flower. Consumption of this plant in Belgium in a weight loss preparation resulted into nephropathy. A high prevalence of urothelial cancer was also observed in patients with end stage nephropathy.

23.6.4 Ayurvedic medicines (Bhasmas)

Ayruvedic system of medicine in India uses preparations such as vegetable products, animal products, minerals, precious stones and metals. The betel nut is chewed by more then 200 million people throughout the world. It is used as a mild stimulant similar to caffeine or tobacco-and as a digestive acid. Betel juice is red and chronic use creates dark stains on the teeth, gingival and oral mucosa. Chewing is associated with oral leukoplakia, submucous fibrosis, and squamous cell carcinoma. Betal use also appears to be addictive.

23.6.5 Polister beetle (Spanish fly)

There are more than 2000 species of beetles including Spanish fly (Cantharis vesicatoria), that contain the toxic ingredient canthasidin. Crystals of cantharides are colorless, odorless, glistening and water insoluble. They are soluble in oils. Canthandin is extremely toxic and causes damage to nearly every organ system. Burning of lips, mouth and pharynx may occur minutes after ingestion. Tongues becomes swollen. Genitourinary involvement is prominent. Acute tubular necrosis and renal failure may ensure (Fig. 56).

23.6.6 Chinese or Asian patent medicines (CPMs)

CPMs are a mixture of multiple products used to treat a variety of conditions. They may contain herbal extracts, minerals, animal parts and medciations. Extracts of Chinese cucumber root (*Trichosanthes kirilowii*) have been used as on abortifacient. *Ephedra viridis* has been used as a respiratory stimulant (Fig. 57). Ginkgo, the world's oldest living *tree sp.* have been used in traditional Chinese medicine for more than a thousand years. Flavenoids and terpenoids found in the leaf are associated with various pharmacologic actions. It is believed that its constituents scavenge free radicals and prevent oxidative damage and decrease inflammation. Gingko is generally considered safe. Adverse effects include GI upset, nausea, vomiting and headache.

23.7 Professional Poisoners

Poisoners of the Middle ages are better known in history than their colleagues of antiquity. Murder by administration of a poison was considered an art by the fifteenth century. Schools taught the art of poisoning. These practioners in Venice were called the council of ten.

The Borgia family of Florence was well known to possess the knowledge of this art. A mixture called la cantarella, whose true nature is not known, was usually used to poison. La cantarella might have contained arsenic and phosphorus. Pope Alexander VI, Rodrigo Borgia and his son Cesare, are alleged to have killed many members of the nobility and the church with la cantarella.

Catherine de Medici, queen of France (1547-1559) brought poisoning skills from Italy to France. She used to make poisoning experiments with sick, poor and criminals. She succeeded in making several political poisonings. Other celebrated practioners of this art during the late Middle Ages included Madama Giulia Toffana. She killed more than 600 victims, probably using arsnenic. Another French women known at that time for this art were Marchioness de Brinvilliers and Catherine Deshayes. De Brinvilliers used several poisons viz. mercuric chloride, arsenic, lead, copper sulphate and antimony.

Deshayes specialized in family planning and executed 2000 infants and an uncounted number of husbands. She is alleged to have used a mixture of aconite, arsenic, belladonna and opium. Because she practiced sorcery she was burnt to death as a witch.

23.7.1 Poison control centres

Poison control centres with their mission of providing helpful information and tracking the national incidences of poisoning, have the most extensive databank on overdose and poisoning. By 1996, there were 67 Poison Control Centres (PCC) in the United States. It is estimated that these centres serve 232 million people.

PCC try as far as possible, to evaluate the outcome for each toxic exposure. Outcomes are classified as no effect, minor effect, moderate effect, major effect, and death. Data collected by PCC are of great value to healthcare planners and others interested in disease trends and public health policy in the most effective manner.

SUGGESTED FURTHER READINGS

Richard C. Dart (1997). Medical Toxicology (IIIrd Edition). Lippincott Williams & Wilkins.

John Joseph Fenton (2002). Toxicology- A case oriented approach. CRC Press.

STUDY QUESTIONS

1. Give a detailed account of poisons found in marine animals.
2. Discuss the toxicity of following animal groups:
   i. Snakes
   ii. Arthropods
3. Write in details on plant poisons.
4. Discuss the following diseases:
   i. Infectious diseases
   ii. Irukandji syndrome
   iii. Diseases caused by indigenous plants
5. Write briefly on the following:
   i. Mushrooms
   ii. Professional poisoners
   iii. Poison control centres
   iv. Chinese medicines

GLOSSARY

A

Absorbance (A; optical density). A measure of the amount of radiation, at a particular wave-length, absorbed by a sample.

$$A = \mathrm{Log}\left(\frac{I_0}{I}\right)$$

where, I_0 is the radiation intensity incident upon the sample and I is the radiation intensity passing through the sample.

Absorbed dose. The amount of a chemical that enters the body of an exposed organism.

Absorption. The uptake of water or dissolved chemicals by cells or organisms.

ACGIH. American Conference of governmental and Industrial hygienists

Acid deposition. The wet and dry air pollutants that lower the pH of deposition and subsequently of the environment.

Acid-rain. A condition in which natural precipitation becomes acidic after reacting chemically with pollutants in the air.

Acquired immune deficiency syndrome (AIDS). A condition in humans in which the immune system suffers a progressive failure, leaving the victim susceptible to opportunistic infections. It is caused by the human immunodeficiency virus (HIV), a slow-acting retrovirus that invades and kills T_4 helper cells that are integral to the immune system.

Acute exposure. Exposure, Acute.

Acute toxicity. Toxicity manifested within a relatively short time interval after toxicant exposure (i.e., as short as a few minutes to as long as several days). Such toxicity is usually caused by a single exposure to the toxicant.

Acute toxicity testing. In the past, such tests were usually concerned with lethality estimated by the LD_{50} or LC_{50} tests. At present acute tests include those for eye and skin irritation and sensitization and changes in autonomic and cardiovascular function.

Additive effect. That situation in which the combined effect of two or more chemicals is equal to the sum of the individual effects.

Aflatoxins. A family of mycotoxins produced by the mold *Aspergillus flavus* and related fungi; included among them are carcinogens and hepatotoxicants.

Agency for Toxic Substance and Disease Registry (ATSDR). A division of the US Public Health Service charged, under the Superfund Amendments and Reauthorization Act (SARA).

Air pollution. Gaseous or aerosol material in the air not considered to be a normal constituent or excess of a normal minor constituent such as sulphur dioxide, carbon dioxide, nitrogen dioxide, dust etc.

Ames test. An *in vitro* test for mutagenicity and, by implication, carcinogenicity, using mutant strains of the bacterium *Salmonella typhimurium,* that can be used as a preliminary screen of chemicals for assessing potential carcinogenicity.

Antagonism. In toxicology, that situation in which the toxicity of two or more compounds administered together, or sequentially, is less than that expected from consideration of their toxicities when administered alone.

Antioxidants. Chemicals that hinder oxidation, frequently serving as free radical scavengers. Important examples include vitamins C and E, butylated hydroxyanisole (BHA) and butylated hydroxytoluene (BHT). These are frequently used as food preservatives and dietary supplements.

Apoptosis. The process of programmed cell death. This is a normal part of such processes as embryonic development, metamorphosis, and immune cell differentiation. It is characterized by high levels of energy consumption, condensation of nuclear chromatin, internucleosomal cleavage of DNA, and activation of highly conserved signal transduction pathways Molecular biological techniques now permit rapid assessment of toxicant- induced apoptosis.

Aquatic toxicology. A branch of toxicology that deals with adverse effects of toxicants on aquatic organisms (marine, estuarine or freshwater) and on aquatic ecosystems; largely a study of water pollution and its ecological effects.

Atmosphere. The gaseous envelope surrounding a planet. The earth's atmosphere consists of nitrogen (79.1%), oxygen (20.9%) by volume with about 0.03% carbon dioxide and traces of noble gases (argon, krypton, xenon, neon and helium) plus water vapour, traces of ammonia, organic matter, ozone, various salts and suspended solid particles.

B

Bioaccumulation. The accumulation of a chemical by organisms from water directly or through consumption of food containing the chemical. Efficient transfer of chemical from food to consumer, through two or more trophic levels, results in a systematic increase in tissue residue concentrations from one trophic level to another.

Biochemical Oxygen Demand (BOD). (a) It is the dissolved oxygen required by organisms for the aerobic decomposition of organic matter present in water. (b) The amount of oxygen used for biochemical oxidation by a unit volume of water at a given temperature for a given time. On an average basis, the demand for oxygen is proportional to the amount of organic waste to be degraded aerobically. BOD value can be used as a measure of waste strength and degree of water pollution, the more organic matter the sample contains, the more oxygen is used by its microorganisms.

Bioconcentration. Accumulation of a chemical in an organism to levels greater than in the surrounding medium. Most often used for accumulation in aquatic organisms of chemicals found it the water in which they exist.

Biodegradation. The environmental destruction of toxicants as a result of microbial and/or fungal action. This is an extremely important mechanism for the detoxication of environmental pollutants in soil and water, such as pesticides, organometallics, plasticizers and petroleum products, as well as other industrial and municipal wastes.

Biological amplification (bio-magnification). The concentration of a persistent substance (e.g. organochlorine insecticide) by the organisms of a food chain so that at each successive trophic level the amount of the substance relative to the biomass is increased.

Biological half-life. The time required for a 50% reduction in the concentration of a particular chemical in the body.

Biological monitoring. Monitoring current exposure or internal load by measurement of a biological parameter in the exposed individuals. This may be by measuring either the chemical and/or its metabolites in blood or urine or by determining some related enzyme activity in the blood.

Biomarker. A parameter (pharmacokinetic, physiological or pharmacological) that can be used to predict a toxic event in an individual animal and also can be used to extrapolate to a similar toxic endpoint across species.

Biomass. Total weight of living matter in a defined system such as a population, ecosystem, etc. Biomass is usually expressed as dry weight per unit volume or area.

$C.T = K$

C

Cancer. Malignant cellular growth consisting of genetically and functionally modified cells capable of invading other tissues.

Carbamate (a group of pesticides). A group of organic compounds with a wide range of biological activity, used as herbicides, fungicides, or insecticides. They have a low toxicity to mammals and persists in soil for only a short time. They include chlorpropham, carbaryl, barban, asulam, diallate, triallate, zineb, maneb, methamsodium, etc.

Carbon dioxide (CO_2). A minor constituent of air comprising about 0.32% of the atmosphere. Carbon dioxide is essential to living systems, released by respiration and to a much lesser extent by volcanic activity and removed from the atmosphere by photosynthesis.

Carbon monoxide (CO). Colourless gas found in trace quantities in the natural atmosphere and produced by incomplete combustion, notably in motor cars and cigarettes. Able to form a stable compound with blood haemoglobin, the carboxyhaemoglobin. It is harmless in small doses.

Carboxyhaemoglobin (COHb). The product of combination of CO with blood haemoglobin.

Carcinogens. Cancer inducing substances, e.g. some hydrocarbons such as 3,4 benzpyrene, asbestos, vinyl chloride and radioactive substances.

Carcinogen, chemical. Chemicals that can induce tumors without prior or subsequent expo-sure to other chemical or physical agents.

Carcinogen, epigenetic (non-genotoxic carcinogen). An agent causing an increase in malig-nant tumors that does not directly interact with DNA, but may cause changes in methylation pat-terns or the tertiary structure of DNA.

Carcinogen, genotoxic. A chemical that acts through genetic mechanisms by interacting with DNA, causing gene mutation or duplication, or a change in the chromosome structure or number.

Cell injury. Any process that harms or wounds a cell. Toxicants may cause cell injury directly or secondarily via disruption of normal physiological mechanisms.

Chemical Oxygen Demand (COD). The weight of oxygen taken up by the organic matter in a sample of water, expressed as parts per million of oxygen taken up from a solution of boiling potassium dichromate in two hours. The test is used to assess the strength of sewage and trade wastes.

Chlorofluorocarbons (CFCs). These are the compounds of carbon and halogens. There are several CFCS, being used mainly as propellants in aerosol cans and as refrigerants in refrigerators, deep freezers and air conditioners, and also used in plastic foams.

Chronic toxicity. The adverse effects manifested after a long time period of uptake of small quanti-ties of a toxicant. The dose is small enough that no acute effects are manifested, and the time period is frequently a significant part of the expected normal lifetime of the organism. The most serious manifestation of chronic toxicity is carcinogenesis, but other types of chronic toxicity are also known (e.g., reproductive or neural effects).

D

Dichloro-diphenyl-trichlorothane (DDT). Persistent organochlorine insecticide introduced in the 1940s and used widely because of its low toxicity to mammals and cheapness of manufacture. It is very stable, relatively insoluble in water but very soluble in fats.

Dissolved oxygen. The amount of oxygen available for biochemical activity within a given volume of water expressed as mg/l, g/m^3, or ppm by weight.

Disulfide bond (S-S bond). A covalent bond between two sulfur heteroatoms. Disulfide bonds are important in establishing crosslinks between polypeptide chains or between loops of a single poly-peptide chain and serve to stabilize the three-dimensional structure of proteins. They are formed by the oxidation of the sulfhydryl groups of two cysteine moieties, thus forming a cystine resi-due.

Diuresis. Increased urine production by the kidneys or urine passage by the bladder. Diuresis may be a consequence of chemical toxicity.

Dysplasia. A disordered growth or development. The term is used most commonly with reference to the growth patterns of cells. It does not necessarily imply precancerous changes (e.g., some forms of dwarfism are due to bone dysplasias), Dysplasia is often used interchangeably with atypia when referring to cells with morphological characteristics that fall short of outright anaplasia,

E

Ecosystem. Ecological system formed by the interaction of coacting organisms and their environment (a community of interdependent organisms together with the environment which they inhabit and with which they interact).

Effluent. General term for a fluid emitted by a source.

Electrophilic. Describing chemicals that are attracted to, and react with, electron-rich centers in other molecules. Many activation reactions pro-duce electrophilic intermediates such as epoxides, that exert their toxic action by forming covalent bonds with nucleophilic substituents in cellular macromolecules such as DNA or proteins.

Endemic. Confined to a given region and having originated there (or a species which occurs continuously in a given area).

Environment. Physical, chemical, biotic and cultural conditions, and their ramifications, collectively comprise "the environment".

Exposure, cumulative. A computation or assessment of all of the exposure to a toxicant which has been experienced by the organism or population up until a specific point in time.

Exposure, subchronic. Repeated doses spread over an intermediate time range (Le., one to three months). Doses may be repeated single doses or continuous low-level doses in food, drinking water or air.

F

Fatty liver. A liver containing more than 5% by weight of lipid with visible lipid accretions in the cells visible under light microscopy. It may be caused by chemical toxicity (e.g., ethionine, cyclo. hex amide) or nutritional deficiency (e.g., lowcho. line).

Fetotoxicity. The deleterious effects exhibited by a fetus (i.e., an animal from the completion of organogenesis to birth, the time equivalent to the second and third trimester of human development) as a result of exposure to a toxic agent. Fetotoxicity manifestations include: (1) lethality; (2) growth impairment (e.g., reduced birth weight); (3) physiological dysfunctions.

Fibrosis (scarring). The deposition of collagen within an organ or tissue. The collagen deposition is generally preceded by proliferation of fibroblasts. Fibrosis is a common tissue response to acute or chronic injury and is the hallmark of the repair phase of the inflammatory response.

Fluorides. Fluorine containing compounds occur widely in nature or emitted from many industrial operations.

Fog. Visible moisture in the atmosphere that reduces horizontal visibility to below 1000 m. Fog is caused by the cooling of relatively warm, moist air when it encounters a land or sea surface that is colder, so reducing the temperature of the air in immediate contact with the surface to below the dew point.

Haber's rule is applied to gaseous toxicants and is approximated in many, but not all, cases.

H

Haber's rule. Where C is the concentration and T is the time, then the product K (toxic effect) is con-stant.

Hazardous waste. Any waste that is potentially dangerous to environmental health because of chemical reactivity, flammability, explosiveness, and so on.

Herbicide. Synthetic organic chemical or other compound used to control plant growth or to kill weeds.

Hexachlorobenzene (HCB). A fungicide for seed grain and vegetable seeds.

I

Indoor air pollution. The concept that the air of modem buildings, which are closed and whose air is recirculated, may be polluted with chemicals which are not allowed to escape. Examples of these pollutants would be tobacco smoke, pest control chemicals and chemicals vaporized from carpets, paints, insulation, furniture, varnishes, etc. Also known as "sick building syndrome". Similar con-cern also exists for the quality of air in aircraft.

Industrial waste. Liquid, solid or gaseous waste originating from the manufacture of specific products.

International Union of Toxicology (IUTOX). A group with both member societies and individ-ual members that fosters international cooperation in toxicology primarily through the organization of International Congresses of Toxicology.

L

LC50 (median lethal concentration). The con-centration of a chemical that, when in the environ-ment of a test organism, is estimated to be fatal to 50% of those organisms under the stated condi-tions of the test. The LC50 is usually used for esti-mating acute lethality of chemicals to aquatic organisms or of airborne chemicals to terrestrial animals. As with LD50 determinations, it is important that both biological and physical condi-tions be narrowly defined in order to reduce vari-ability. LD50 and LC50 values are standards for comparison of acute toxicity between toxicants and between species.

LD50 (median lethal dose). The quantity of a chemical compound that, when applied directly to test organisms, is estimated to be fatal to 50% of those organisms under the stated conditions of the test.

Leukaemia. A progressive, malignant disease of the blood-forming organs, marked by distorted proliferation and development of leucocytes and their precursors in the blood and bone marrow.

London smog. A contraction of fog and smoke that characterized air pollution episodes in London. Fog, with chemical and physical pollutants, remained there for four days in 1952 and 5 days in 1962. In 1952, 4000 people died while in 1962 about 700 died.

M

Malignancy (cancer). A state of tumor charac-terized by its non-encapsulation, invasiveness, poor differentiation, possessing common cell divi-sion and rapid growth characteristics that are anaplastic to various degrees giving rise to metas-tases.

Maximum allowable concentration (MAC). The concentration of a pollutant considered (in regulations) harmless to healthy adults during their working hours, assuming that they are not in contact with the pollutant the rest of the time.

Maximum permissible concentration. The concentration of a radioisotope in air, water, milk, etc. which will deliver not more than the maximum permissible dose to a critical organ when breathed or consumed at a normal rate.

Mercury. Element (Hg). Liquid metals, which damage the nervous system on inhalation or ingestion, especially of its organic compounds. Used extensively in chemical and plastic industries. Acute mercury poisoning is marked by severe abdominal pain, vomiting, bloody diarrhea with watery stools, and corrosion and ulceration of the digestive tract.

Metallothionein (MT). A low-molecular-weight (6500-7000 daltons) cytosolic protein found in many eukaryotic species. It has a high metal con-tent, usually containing bound zinc, cadmium or copper. Of the total amino acid composition, 30% consists of cysteinyl residues. The metal ions are bound to the sulfhydryl groups of the protein by mercaptide bonds. *De novo* synthesis of MT is induced by dietary or parenteral administration of Cd, Zn, Cu, Hg or Au. While little is known about the mechanisms of MT induction by nonmetallic inducers (e.g., growth factors), MT is highly expressed during liver regeneration. The possibil-ity that the MT could participate in a DNA synsthesis-related process through donation or abstraction of Zn to and from transcription factors has been inferred from *in vitro* studies. Overex-pression of MT is often accompanied by increased resistance towards a variety of alkylating agents and chemotherapeutic drugs, involving several mechanisms that are dependent on the metal com-position of MT.

Meuse valley incident. In air pollution incident, which occurred in Meuse Valley, France, in December 1930 when cold weather and catabatic winds, together with a temperature inversion, caused fog, contaminated with industrial pollutants, to persist for several days. Several hundred people became ill, about 60 died, and many cattle had to be slaughtered.

Microenvironment. The immediate environ-ment in contact with, or close to an organism.

Microfilament. Intracellular fiber of polymerized.

Minamata. The disease of the central nervous system caused by mercury poisoning has been named 'Minamata' after a bay and town in Japan where it occurred in 1959 as a result of consuming the fish and shellfish caught in the bay. The pollution was due to the discharge of effluents from the Chisso factory, manufacturing acetaldehyde and vinyl chloride. The pollutant accumulated as dimethyl mercury, which was ingested by marine organisms. During 1953 to 1960, 43 persons died and many others rendered unfit by disease.

Mist. Liquid (water) droplets smaller than 10 micron in diameter suspended in the atmosphere reducing visibility to between 1 and 2 km.

N

Nausea. An unpleasant sensation with a tendency to vomit.

Neoplasm. A heritably altered, relatively autono-mous growth of tissue. A neoplasm is composed of abnormal cells, the growth of which is more rapid than that of other tissues and is not coordinated with the growth of other tissue.

Nucleophilic. Describing electron-rich substitu-ents on organic molecules which react rapidly with electrophiles to form covalent bond.

Nutrition Toxicology. Important aspect toxicology that deals with the effect of diet on the expression of toxicity and the mechanism of such effects. Areas such as the effects of dietary defi-ciencies on the activity and induction of xeno-biotic-metabolizing enzymes have been studied. Currently there is much interest on the effect of diet on the expression of carcinogenesis and the application of this information to cancer preven-tion programs.

O

Occupational Exposure Limit (OEL). Generally the time-weighted average concentrations of an airborne toxicant to which a worker may be exposed during a defined work period.

Organochlorine. An organic compound containing chlorine. Many are used as the active gradients for pesticides with a high persistence, due to their chemical stability and low solubility in water. Pesticide applications include DDT, aldrin and lindane.

Organophosphorus pesticides. It is a group of non-pesticide compounds. The compounds appear to inhibit the action of cholinesterase enzyme which cancels chemical message within the nervous system. Since break down rapidly, they harm wildlife for a short duration.

P

Parkinsonis disease. A group of neurological disorders marked by hypokinesis, tremor, and muscular rigidity.

Pesticide. Chemical agent that kills plant or animal 'pests'. This is a general term embracing insecticides, herbicides, fungicides, etc.

Phagocytosis. A process by which the cell mem-branes of certain cells flow around particles and engulf them into the interior of the cell.

Pollution. Addition of some exogenous substances in the environment which are harmful for organisms including human beings.

Polychlorinated biphenyl (PBC). A group of closely related chlorinated hydrocarbons whose principal use has been as liquid insulators in high voltage transformers. Their use being reduced owing to evidence of their persistence and toxicity in the environment.

Q

Quantitative structure-activity relationships (QSAR). The relationship between the physical and/or chemical properties of chemicals and their ability to cause a particular effect, enter into particular reactions, etc. The goal of QSAR stud-ies in toxicology is to develop procedures whereby the toxicity of a compound can be predicted from its chemical structure by analogy with the proper-ties of other toxicants of known structure and toxic properties.

R

Risk assessment, environmental (ecological). The process that evaluates the probability that adverse ecological effects may occur (or are occur-ring) as a result of one or more stressors. The stres-sors may be one or more chemicals and/or physical or biological changes.

S

Sewage. The contents of sewers carrying the water borne wastes of a community.

Smog. The world was coined in 1905 by H.A. Des Voeux, founder-president of the British National Smoke Abatement Society, to denote combination of smoke and natural fog which, in urban areas, may have unpleasant and even disastrous consequences. The Great London smog of 1952 caused 4000 excess deaths and led to the enactment of Clean Air Acts of 1956 and 1968, which provided for the banning of smoky fuels within specified areas. The word smog has since been applied to other air pollution effect not necessarily connected with smoke, such as 'Los-Angeles Smog' which arises from nitrogen oxides and hydrocarbons emitted by motor vehicles and the photochemical action of sunlight.

Sulphur dioxide (SO_2). A constituent of products of combustion of a wide variety of fuels, but particularly heavy fuel oil and coal. It is also released in volcanic eruptions.

Sulphuric acid (H_2SO_4). In the atmosphere, it is formed by the combination of sulphur trioxide with water.

T

Three Mile Island. Island is the unquehanna River, near Harrisburg, Pennsylvania, that furnishes the since of a nuclear reactor which failed on March 28, 1979 in the most serious incident experienced by the nuclear industry up to that time.

Torrey Canyon disaster. An incident which occurred on March 18, 1967 when the Liberian oil tanker Torrey Canyon went around on the Seven Stones reef, liberating her cargo of Kuwait crude oil. On the afternoon of 24 March, there was a slick of oil some 64 km long of Lands End with an average width of perhaps 16 km.

Toxicology. Study of the harmful effects of toxic substances on living organisms in any ecosystem.

W

Waterborne diseases. Diseases such as cholera, typhoid fever, dysentery, gastroenteritis and hepatitis which are commonly transmitted through contaminated water supplies.